Staphylococcus aureus Toxins

Staphylococcus aureus Toxins

Special Issue Editor

William R. Schwan

MDPI • Basel • Beijing • Wuhan • Barcelona • Belgrade

Special Issue Editor
William R. Schwan
University of Wisconsin-La Crosse,
Department of Microbiology
USA

Editorial Office
MDPI
St. Alban-Anlage 66
4052 Basel, Switzerland

This is a reprint of articles from the Special Issue published online in the open access journal *Toxins* (ISSN 2072-6651) from 2018 to 2019 (available at: https://www.mdpi.com/journal/toxins/special_issues/Staphylococcus_toxins)

For citation purposes, cite each article independently as indicated on the article page online and as indicated below:

LastName, A.A.; LastName, B.B.; LastName, C.C. Article Title. *Journal Name* **Year**, *Article Number*, Page Range.

ISBN 978-3-03921-425-9 (Pbk)
ISBN 978-3-03921-426-6 (PDF)

Contents

About the Special Issue Editor

William R. Schwan graduated from Quincy College with a degree in Biology. He received his Master's degree in Microbiology from Iowa State University and Ph.D. in Microbiology and Immunology from Northwestern University. After obtaining his Ph.D., Dr. Schwan conducted his first Postdoctoral Fellowship with Werner Goebel at the Universitaet Wuerzburg in Wuerzburg, Germany. He conducted another Postdoctoral Fellowship at the Food and Drug Administration (FDA) in Bethesda, MD, under the guidance of Dennis Kopecko. Following his Postdoctoral Fellowship at the FDA, he worked as Scientist for Pathogenesis Corporation attempting to develop new drugs against S. aureus and Pseudomonas aeruginosa infections. Over the last 21 years, Dr. Schwan has worked at the University of Wisconsin-La Crosse, rising to the rank of full Professor. He has been part of several National Institutes of Health research grants and has published over 50 papers.

Editorial

Staphylococcus aureus Toxins: Armaments for a Significant Pathogen

William R. Schwan

Department of Microbiology, University of Wisconsin-La Crosse, 1725 State St., La Crosse, WI 54601, USA; wschwan@uwlax.edu

Received: 30 July 2019; Accepted: 1 August 2019; Published: 3 August 2019

Staphylococcus species are common inhabitants of humans and other animals. *Staphylococcus aureus* is responsible for hundreds of thousands of skin and soft tissue infections [1] and is a significant cause of bloodstream infections in humans [2]. Several different exotoxins are produced by *S. aureus*, including enterotoxins, hemolysins, and leukocidins that kill white blood cells [3,4]. These exotoxins lead to direct damage to the host, modulate immune defenses, and may have an indirect effect on the host by mounting a cytokine storm. Early studies on staphylococcal exotoxins include work by Bayliss who identified that *S. aureus* enterotoxins caused vomiting [5], Marks investigation of hemolysins [6], and Blobel et al. demonstrating the lethal properties of leukocidins on human white blood cells [7]. When community-associated methicillin-resistant (CA-MRSA) *S. aureus* emerged, sequencing of the first CA-MRSA strain showed the acquisition of new pathogenicity islands harboring new exotoxin genes that contribute to human disease [8]. Thus, *S. aureus* continues to evolve in part by acquiring new exotoxin genes.

This special issue is focused on staphylococcal toxins and the impact they have on mammalian health. Several studies examined the role toxins play in *S. aureus* pathogenesis, whereas other papers explored new therapeutics aimed at limiting the action of the toxins. A couple of review papers give good overviews as to why toxins are crucial for specific aspects of *S. aureus* infections.

A paper by Brctl ct al. investigated the in vivo regulation of staphylococcal superantigen-like protein 1 [9]. They showed transcription of the *ssl1* gene increased when *S. aureus* was grown under nutrient deprived conditions that included early growth in murine abscesses. This represented the first time transcription of a staphylococcal superantigen-like gene was studied in vivo within *S. aureus* infecting animals.

A couple of papers dealt with *S. aureus* alpha-toxin. The Keogh et al. paper examined the role peptidyl-prolyl cis/trans isomerases (PPIases) have in regulating alpha-toxin and their contribution to *S. aureus* virulence in murine abscess and systemic models of infection [10]. A *ppiB* mutant that no longer encodes one of the PPIases produced less alpha-toxin and phenol-soluble modulins (PSMs) compared to the unmutated parent strain. A change in the host cell membrane that affected alpha-toxin activity was explored in the work done by Ziesemer et al. [11]. When sphingomyelinase was used to pre-treat airway epithelial cells, alpha-toxin heptamer formation was blocked, which led to a loss of transmembrane pore formation.

Two papers within this issue are centered on *S. aureus* enterotoxins. Grispoldi et al. studied enterotoxin production in *S. aureus* within canned meat [12]. The time between seaming and sterilization was examined. Enterotoxin was detected within one to two days depending on the incubation temperature of the canned meat. Heat treatment killed the *S. aureus*, but active enterotoxins could be present that would not be detected by serology. In the other enterotoxin paper, Fang et al. demonstrated that the application of purified staphylococcal enterotoxin C to murine mammary glands caused a significant increase in proinflammatory cytokines within those mammary glands compared to a phosphate buffered saline control [13]. Application of an anti-staphylococcal enterotoxin C antibody reduced both the inflammation and tissue damage within the mammary gland, strengthening the argument that staphylococcal enterotoxin C produced by *S. aureus* has an important role to play in mastitis development.

S. aureus is one of the leading causes of eye infection in humans. The role of toxins in *S. aureus* eye disease is well known. Astley et al.'s review paper describes the historic role of alpha-toxin in cornea damage [14]. Other hemolysins (i.e., delta-toxin and gamma toxin) and leukocidins have significantly lesser roles in regard to eye damage compared to the effect of alpha-toxin. The review notes that not much is known about the role of PSMs in corneal damage, a topic that needs further investigation.

Two papers within this special issue are focused on atopic dermatitis. Traisaeng et al. describes how butyric acid produced by one species of *Staphylococcus* was able to inhibit the growth of a *S. aureus* isolate derived from a patient with atopic dermatitis [15]. A derivative of the butyric acid was synthesized and led to a reduction in *S. aureus* mediated inflammation. The review by Seiti Yamada Yoshikawa et al. outlines the role of toxins in atopic dermatitis, a chronic skin disease that involves a significant inflammatory response [16]. They discuss how alpha toxin compromises E-cadherin integrity and interacts with sphinomyelin that in turn leads to lysis of keratinocytes. Newer studies they present suggest that staphylococcal toxins not only promote inflammation, but they may also serve as a counterbalance to some of the host regulatory mechanisms.

The paper by Habib et al. used a bioinformatic approach to find toxin–antitoxin systems in *S. aureus* [17]. They found 39% of the toxin–antitoxin systems are within the seven *S. aureus* pathogenicity islands. Furthermore, a new *S. aureus* toxin–antitoxin system was identified where the antitoxin is a transcriptional autoregulator and the toxin inhibits the autoregulation.

Papers by Kailasan et al. [18] and Ouyang et al. [19] explored treatment options for *S. aureus* infections. Kailasan et al. generated a library of leukotoxin gene mutations that targeted functional domains of the leukocidin protein that were in turn used to make polyclonal antibodies to the lead toxoid candidate. A combination of antibodies to various toxins that included the new anti-leukocidin antibody completely neutralized the cytotoxic properties of the *S. aureus*. On the other hand, Ouyang et al. tested a bibenzyl compound for its anti-virulence capabilities against *S. aureus*. Mice treated with the drug had a significantly higher survival rate than untreated mice.

Lastly, the paper by Tuchscherr et al. examined several strains of *S. aureus* for their genotype and virulence capabilities [20]. These strains were split into low-cytotoxicity and high-cytotoxicity arms for use in two murine infection models. Both groups persisted within the mice and were able to cause infections. The low-cytotoxicity strains grew to higher numbers and were not cleared to the same extent as the high-cytotoxicity strains, suggesting this adaptation could be important in the development of chronic infections.

Funding: This research received no external funding.

Acknowledgments: The editor is grateful to all of the authors who contributed their work to this Special Issue. Special thanks goes to the rigorous evaluations of all of the submitted manuscripts by the expert peer reviewers who contributed to this Special Issue. Lastly, the valuable contributions, organization, and editorial support of the MDPI management team and staff are greatly appreciated.

Conflicts of Interest: The author declares no conflict of interest.

References

1. Suaya, J.A.; Mera, R.M.; Cassidy, A.; O'Hara, P.; Amrine-Madsen, H.; Burstin, S.; Miller, L.G. Incidence and cost of hospitalizations associated with Staphylococcus aureus skin and soft tissue infections in the United States from 2001 to 2009. *BMC Infect. Dis.* **2014**, *14*, 296. [CrossRef] [PubMed]
2. Cahill, T.J.; Baddour, L.M.; Habib, G.; Hoen, B.; Salaun, E.; Pettersson, G.B.; Schafers, H.J.; Prendergast, B.D. Challenges in infective endocarditis. *J. Am. Coll. Cardiol.* **2017**, *69*, 325–344. [CrossRef] [PubMed]
3. Archer, G.L. Staphylococcus aureus: A well-armed pathogen. *Clin. Infect. Dis.* **1998**, *26*, 1179–1180. [CrossRef] [PubMed]
4. Dinges, M.M.; Orwin, P.M.; Schlievert, P.M. Exotoxins of Staphylococcus aureus. *Clin. Microbiol. Rev.* **2000**, *13*, 16–34. [CrossRef] [PubMed]

5. Bayliss, M. Studies on the mechanism of vomiting produced by Staphylococcus. *J. Exp. Med.* **1940**, *72*, 669–684. [CrossRef] [PubMed]

6. Marks, J. The standardization of staphylococcal alpha anti-toxin, with special reference to anomalous haemolysins including delta-lysin. *J. Hyg.* **1951**, *49*, 52–66. [CrossRef] [PubMed]

7. Blobel, H.; Wenk, K.; Kanoe, M. Effects of Panton-Valentine leukocidin of Staphylococcus aureus on leukocytes from patients with leukemia. *Infect. Immun.* **1971**, *3*, 507–509. [PubMed]

8. Baba, T.; Takeuchi, F.; Kuroda, M.; Yuzawa, H.; Aoki, K.; Oguchi, A.; Nagai, Y.; Iwama, N.; Asano, K.; Naimi, T.; et al. Genome and virulence determinant of high virulence community-acquired MRSA. *Lancet* **2002**, *359*, 1819–1827. [CrossRef]

9. Bretl, D.J.; Watkins, H.; Schwan, W.R. Regulation of the staphylococcal superantigen-like protein 1 gene of community-associated methicillin-resistant Staphylococcus aureus in murine abscesses. *Toxins* **2019**, *11*, 391. [CrossRef] [PubMed]

10. Keogh, R.A.; Zapf, R.L.; Trzeciak, E.; Null, G.G.; Wiemels, R.E.; Carroll, R.K. Novel regulation of alpha-toxin and the phenol-soluble modulins by peptidyl-prolyl cis/trans isomerase enzymes in Staphylococcus aureus. *Toxins* **2019**, *11*, 343. [CrossRef] [PubMed]

11. Ziesemer, S.; Moller, N.; Nitsch, A.; Muller, C.; Beule, A.G.; Hildebrandt, J.P. Sphingomyelin depletion from plasma membranes of human airway epithelial cells completely abrogates the deleterious actions of *S. aureus* alpha-toxin. *Toxins* **2019**, *11*, 126. [CrossRef] [PubMed]

12. Grispoldi, L.; Popescu, P.A.; Karama, M.; Gullo, V.; Poerio, G.; Borgogni, E.; Torlai, P.; Chianese, G.; Fermani, A.G.; Sechi, P.; et al. Study on the growth and enterotoxin production by Staphylococcus aureus in canned meat before retorting. *Toxins* **2019**, *11*, 291. [CrossRef] [PubMed]

13. Fang, R.; Cui, J.; Cui, T.; Guo, H.; Ono, H.K.; Park, C.H.; Okamura, M.; Nakane, A.; Hu, D.L. Staphylococcal enterotoxin C is an important virulence factor for mastitis. *Toxins* **2019**, *11*, 141. [CrossRef] [PubMed]

14. Astley, R.; Miller, F.C.; Mursalin, M.H.; Coburn, P.S.; Callegan, M.C. An eye on Staphylococcus aureus toxins: Role in ocular damage and inflammation. *Toxins* **2019**, *11*, 356. [CrossRef] [PubMed]

15. Traisaeng, S.; Herr, D.R.; Kao, H.J.; Chuang, T.H.; Huang, C.M. A derivative of butyric acid, the fermentation metabolite of Staphylococcus epidermidis, inhibits the growth of a Staphylococcus aureus strain isolated from atopic dermatitis patients. *Toxins* **2019**, *11*, 311. [CrossRef] [PubMed]

16. Seiti Yamada Yoshikawa, F.; Feitosa de Lima, J.; Notami Sato, M.; Alefe Leuzzi Ramos, Y.; Aoki, V.; Leao Orfali, R. Exploring the role of Staphylococcus aureus toxins in atopic dermatitis. *Toxins* **2019**, *11*, 321. [CrossRef] [PubMed]

17. Habib, G.; Zhu, Q.; Sun, B. Bioinformatic and functional assessment of toxin-antitoxin systems in Staphylococcus aureus. *Toxins* **2018**, *10*, 473. [CrossRef] [PubMed]

18. Kailasan, S.; Kort, T.; Mukherjee, I.; Liao, G.C.; Kanipakala, T.; Williston, N.; Ganjbarsh, N.; Venkatasubramamiam, A.; Holtsberg, F.W.; Karauzam, H.; et al. Rational design of toxoid vaccine candidates for Staphylococcus aureus leukocidin AB (LukAB). *Toxins* **2019**, *11*, 339. [CrossRef] [PubMed]

19. Ouyang, P.; He, X.; Yuan, Z.-W.; Yin, Z.-Q.; Fu, H.; Lin, J.; He, C.; Liang, X.; Lv, C.; Shu, G.; et al. Erianin against Staphylococcus aureus infection via inhibiting sortase A. *Toxins* **2018**, *10*, 385. [CrossRef] [PubMed]

20. Tuchscherr, L.; Pllath, C.; Siegmund, A.; Deinhardt-Emmer, S.; Hoerr, V.; Svensson, C.M.; Figge, M.T.; Monecke, S.; Loffler, B. Clinical *S. aureus* isolates vary in their virulence to promote adaptation to the host. *Toxins* **2019**, *11*, 135. [CrossRef] [PubMed]

Article

Regulation of the Staphylococcal Superantigen-Like Protein 1 Gene of Community-Associated Methicillin-Resistant *Staphylococcus aureus* in Murine Abscesses

Daniel J. Bretl [1], Abdulaziz Elfessi [2], Hannah Watkins [1] and William R. Schwan [1,*]

[1] Department of Microbiology, University of Wisconsin-La Crosse, La Crosse, WI 54601, USA
[2] Department of Mathematics and Statistics, University of Wisconsin-La Crosse, La Crosse, WI 54601, USA
* Correspondence: wschwan@uwlax.edu

Received: 12 April 2019; Accepted: 2 July 2019; Published: 4 July 2019

Abstract: Community-associated methicillin-resistant *Staphylococcus aureus* (CA-MRSA) causes substantial skin and soft tissue infections annually in the United States and expresses numerous virulence factors, including a family of toxins known as the staphylococcal superantigen-like (SSL) proteins. Many of the SSL protein structures have been determined and implicated in immune system avoidance, but the full scope that these proteins play in different infection contexts remains unknown and continues to warrant investigation. Analysis of *ssl* gene regulation may provide valuable information related to the function of these proteins. To determine the transcriptional regulation of the *ssl1* gene of CA-MRSA strain MW2, an *ssl1* promoter::lux fusion was constructed and transformed into *S. aureus* strains RN6390 and Newman. Resulting strains were grown in a defined minimal medium (DSM) broth and nutrient-rich brain-heart infusion (BHI) broth and expression was determined by luminescence. Transcription of *ssl1* was up-regulated and occurred earlier during growth in DSM broth compared to BHI broth suggesting expression is regulated by nutrient availability. RN6390 and Newman strains containing the *ssl1::lux* fusion were also used to analyze regulation in vivo using a mouse abscess model of infection. A marked increase in *ssl1* transcription occurred early during infection, suggesting *SSL1* is important during early stages of infection, perhaps to avoid the immune system.

Keywords: mouse abscess; methicillin-resistant *Staphylococcus aureus*; lux fusion; superantigen-like protein; gene regulation; defined minimal medium

Key Contribution: Transcription of the *S. aureus ssl1* gene significantly increases during the early stage of abscess formation in mice.

---

1. Introduction

Staphylococcus aureus is one of the most clinically significant human pathogens and is responsible for a wide variety of infections, ranging from skin and soft tissue infections (e.g., boils, furuncles, etc.) to more invasive diseases (e.g., bacteremia, endocarditis, pneumonia, toxic shock syndrome, and necrotizing fasciitis) [1]. Presently, *S. aureus* causes close to 700,000 infections annually in the United States and is the leading cause of nosocomial infections in the United States [1]. Many healthcare facilities in the United States have endemic problems with methicillin-resistant *S. aureus* (MRSA), with ~60% *S. aureus* infections caused by MRSA strains [2]. In 1997, community-associated methicillin-resistant *S. aureus* (CA-MRSA) strains emerged in the United States, causing infections in people with no known risk factors [3] and have become the major cause of skin and soft tissue infection [2].

The disease-causing ability of CA-MRSA can be attributed to an impressive list of virulence factors. Many of these factors are shared with all *S. aureus* strains and include lipases, nucleases, proteases, hyaluronidase, collagenase, exfoliative toxins, leukocidins, and four hemolysins [4]. Along with these virulence factors, CA-MRSA strains have been shown to carry various novel putative toxin genes, including staphylococcal enterotoxin homologues (*seg*2, *sel*2, *sec*4, and *sek*2), *bsa* (a bacteriocin), *cna*, *ear*, and *lpl*10 [5,6]. Most of these novel toxins have yet to be fully characterized but may represent a repertoire of factors necessary for the disease-causing capability of MRSA in the community.

Another example of less understood *S. aureus* virulence factors is a family of toxin genes originally described as staphylococcal exotoxin-like (*set*) genes [7]. These genes have since been renamed staphylococcal superantigen-like (*ssl*) to reflect their structural similarity to the staphylococcal superantigens [8]. There is a total of fourteen *ssl* genes, of which at least a subset is found in nearly all *S. aureus* strains tested [5–7,9,10]. SSL proteins share the typical superantigen structure, including a characteristic oligonucleotide/oligosaccharide binding (OB-fold) domain linked to a beta-grasp domain [7,11,12]. However, while genuine superantigens are pyrogenic, increase the lethality of endotoxin up to 100,000-fold [4], and non-specifically stimulate T-cells by cross-linking major histocompatibility complex-II molecules with T-cell receptors [12], SSL proteins do not have any of these superantigen capabilities [13–15]. Rather, functional analysis of members of the SSL family has shown a link to immune avoidance. For example, SSL7 binds efficiently to IgA, which limits leukocyte activation, and also binds complement C5 protein preventing its cleavage into C5a and C5b, which reduces the proinflammatory activity [11,16,17]. Furthermore, other SSL proteins also play a role in immune avoidance by binding efficiently to phagocytic cells, preventing rolling adhesion of neutrophils and attachment of IgA [10,18]; by binding to tenascin, affecting cell motility of keratinocytes [19]; and by inhibiting toll-like receptor 2 [20,21].

SSL proteins are found exclusively in *S. aureus*, but not all *S. aureus* strains have the same array of SSL proteins [15]. Genome sequencing shows that the *ssl* genes share 36% to 67% homology [5,6,22]. Allelic differences of individual *ssl* genes among *S. aureus* strains show 81% to 95% homology at the nucleotide level and 77% to 94% homology at the amino acid level [15]. However, allelic differences have not been shown to affect the function of the SSLs. Transcription studies have shown that for most of the *ssl* genes there is up-regulated transcription during the stationary phase of growth in nutrient rich media [7,11,15]. Furthermore, SSL proteins are recognized by the human immune system as shown by seroconversion in human patients [14,15,23]. The binding of human antibodies to SSL proteins seems to be specific and not cross-reactive [23].

SSL proteins are likely to contribute to the virulence of MRSA in the community setting. Both dominating strains of CA-MRSA, MW2, a prototypical USA400 clone of CA-MRSA first described after an outbreak in North Dakota and Minnesota in 1998 [3], and USA300 clones, the other dominating CA-MRSA PFGE type [6,10,24] have all fourteen *ssl* genes. Furthermore, in a comparison of virulence genes found in commensal or clinical MSSA isolates versus CA-MRSA (MW2) isolates, it was found that 100% of the MW2 isolates tested were positive for the *ssl*1 gene, which was significantly greater than the number of *ssl*1 positive MSSA isolates [25]. The following study demonstrates that the expression of *ssl*1 is enhanced in nutrient poor broth medium compared to nutrient rich medium, suggesting regulation of this gene responds to nutrient limitation that may be encountered within the host. Furthermore, we demonstrate that *ssl*1 expression occurs early in a murine abscess model of infection, consistent with the hypothesis that the SSL proteins are important for immune avoidance.

2. Results

2.1. Expression of ssl1 is Up-Regulated in Nutrient-Poor Conditions

To examine the expression of *ssl*1, a vector containing an *ssl1::lux* promoter fusion was constructed in the shuttle plasmid pXen5 [26]. The *ssl*1 promoter was amplified from genomic DNA of the *S. aureus* MW2 strain. Previous research has shown that most *ssl* genes, including *ssl*1, showed increased expression during the stationary phase [7,15]. However, previous reports of *ssl* gene expression were done in nutrient rich media. To test whether expression would also occur in nutrient poor defined minimal medium (DSM), the reporter fusion construct was transformed into *S. aureus* strain RN4220 (resulting in strain WS4108). This strain was inoculated into both DSM and brain-heart infusion (BHI) broth and examined for growth and luminescence. Initially, there was measurable luminescence following growth in both media, demonstrating that *ssl1* gene expression occurs under both nutrient poor and nutrient rich conditions and provided the rationale to test expression in virulent *S. aureus* strains (data not shown).

The *ssl1::lux* promoter fusion was subsequently integrated into the chromosome of *S. aureus* strain RN6390, resulting in strain WS0501. RN6390 has a genetically similar background as RN4220, but is more virulent and conducive for in vivo infection models [27]. Expression of *ssl1* was then assessed in this strain, and in all other experiments discussed below, by harvesting 1 ml of culture at given time points and normalizing the raw luminescence to total colony forming units resulting in the reported relative luminescence units (RLU). Absolute optical density was different, but the overall growth curve pattern was similar, between growth conditions. Despite any subtle differences in growth, all RLU values reflect the per cell expression when normalized to colony forming units. Expression of *ssl1* in the virulent WS0501 strain was markedly different depending on the growth environment. In nutrient rich broth, expression of *ssl1* was relatively low throughout the entirety of the growth curve (Figure 1A), with minor peaks of expression as the culture reached late log phase growth (320 RLU, OD_{600} = 1.418) and again during stationary phase (280 RLU, OD_{600} = 2.106). In contrast, in nutrient poor medium (Figure 1B), expression of *ssl1* was markedly higher, ranging from two ($p < 0.05$) to eleven-fold ($p < 0.001$) higher RLU per growth curve time point compared to the same time points during growth in BHI. For example, after two hours of growth in DSM expression of *ssl1* (1200 RLU) was 9.2-fold greater than expression at the same time point following growth in nutrient rich BHI (130 RLU) ($p < 0.001$). Moreover, peak expression of *ssl1* in DSM occurred during log phase growth and there was not a second peak during stationary phase, though expression remained higher at these time points relative to expression in nutrient rich broth. To determine if the expression of *ssl1* was strain-dependent we integrated the same *ssl1::lux* reporter fusion into *S. aureus* Newman. Expression of *ssl1* in this Newman strain construct (strain NS3513) was consistent with the *ssl1* expression demonstrated in the RN6390 background (Figure 2). For example, after 4 h of growth in DSM, expression was 3.25-fold higher (2700 RLU) compared to expression in BHI (830 RLU). Expression remained higher after 6 h in DSM, before both cultures reduced expression during late stationary phase. Thus, both strains showed higher *ssl1* transcription when grown in DSM broth compared to BHI broth.

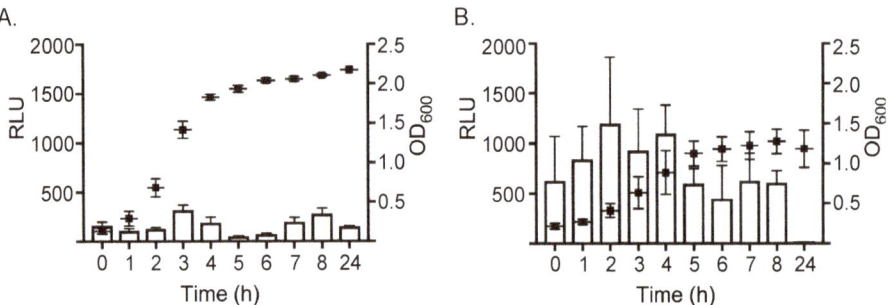

Figure 1. Expression of staphylococcal superantigen-like (*ssl1*) in strain WS0501 (RN6390 background) grown in (**A**) brain-heart infusion (BHI) broth or (**B**) defined minimal medium (DSM) broth. One milliliter aliquots were removed every hour for 8 h and at 24 h to measure luminescence (white column) and optical density (600 nm) (black diamond). Relative luminescence units (RLU) were divided by the viable colony forming units and multiplying the per cell luminescence by 10^6 to obtain the final RLU shown. The data represents the RLU means + standard deviation of at least three trials.

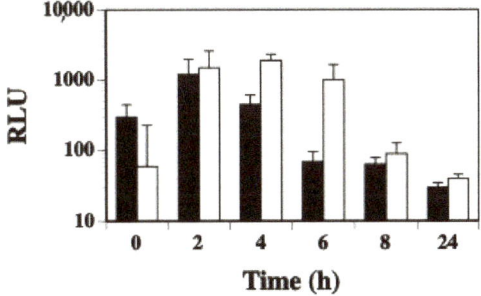

Figure 2. Expression of *ssl1* in strain NS3513 (Newman background) grown in BHI broth (black column) and DSM broth (white column). One milliliter aliquots were removed every 2 h for 8 h and at 24 h to measure luminescence. Luminescence (RLU) was determined as was presented before. The data represents the RLU means + standard deviation of at least three trials.

2.2. Transcription of Other ssl Genes Is Up-Regulated in Nutrient Poor Conditions

The luminescence results indicated that *ssl1* transcription was elevated in *S. aureus* grown in a nutrient poor growth environment (DSM broth) compared to growth in a nutrient rich environment (BHI broth). To determine if this elevated transcription in nutrient poor conditions occurred for other *ssl* genes, a quantitative real time-polymerase chain reaction (qRT-PCR) procedure was used that targeted the *ssl5* and *ssl8* genes. Total RNA was collected at early stationary growth, when we observed high levels of *ssl1* expression. Consistent with the expression observed for *ssl1*, both *ssl5* (3.2-fold increase, $p < 0.04$) and *ssl8* transcript abundance (4.4-fold increase, $p < 0.003$) significantly increased in strain RN6390 grown in DSM broth versus BHI broth (Figure 3). Thus, nutrient deprivation elicits an increase in the transcription of several *ssl* genes and the *ssl* genes may share common mechanisms of regulation.

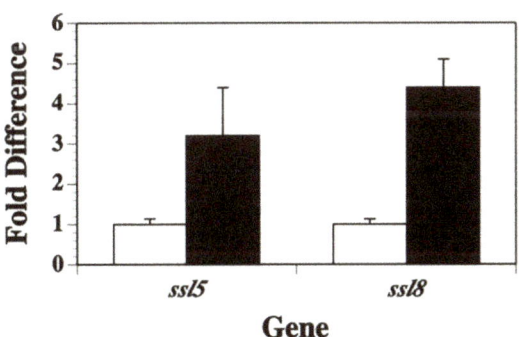

Figure 3. Quantitative reverse transcribed-polymerase chain reaction results of *S. aureus* strain RN6390 *ssl5* and *ssl8* transcription grown in BHI broth (white column) compared to DSM broth (black column). Transcription following growth in BHI broth was set as the baseline values. The data represents the mean + standard deviation from three separate runs.

2.3. Agr Has a Minor Contribution to the Modulation of ssl1 Expression in S. aureus

The accessory global regulator (*agr*) system of *S. aureus* controls the up-regulation of many toxin genes during stationary phase growth [28,29]. To test whether Agr was involved in regulating expression of *ssl1*, the *ssl1::lux* fusion was moved into the *agr* mutant strain RN6911 by transduction. Strains RN6911 and RN6390 are identical except for the *agr* mutation. Both strains were grown in BHI and DSM broth over a 24 h time period. In BHI broth, the *ssl1* expression was 2.3-fold lower in the *agr* mutant strain compared to the wild-type strain at the 2 h time point (Figure 4A, $p = 0.06$) and 2.7-fold lower in the *agr* mutant versus wild-type at 8 h ($p = 0.099$). After 24 h, the wild-type culture had a significantly higher transcription of *ssl1* compared to the *agr* mutant (12-fold, $p = 0.046$). When both strains were grown in DSM (Figure 4B), they had similar 0 h *ssl1* transcription levels (*agr* mutant 1700 RLU versus 1114 RLU for wild-type, $p = 0.141$). After 2 h of growth in DSM, the *agr* mutant showed 1.5-fold higher *ssl1* transcription and the wild-type strain 6.3-fold higher *ssl1* transcription versus their 0 h expression numbers, respectively. However, the *ssl1* expression increase was not significant when either comparing both strains at the 2 h reading or compared to their respective 0 h measurements. Both strains had the highest *ssl1* expression after 4 h in DSM broth (*agr* mutant 3759 and wild-type 8229). No significant difference was observed between the two strains at this time point ($p = 0.197$). The only significant difference in *ssl1* expression for the strains grown in DSM was also at the 24 h time point where the *ssl1* expression in the *agr* mutant was 1.2 RLU and in the wild-type it was 21.8 RLU ($p = 0.036$). Thus, *agr* may contribute to modulation of *ssl1* regulation, but not nearly to the extent that has been demonstrated for other toxins. Additionally, expression of *ssl1* still occurred in the absence of *agr* indicating additional unknown transcription factors regulate *ssl1*.

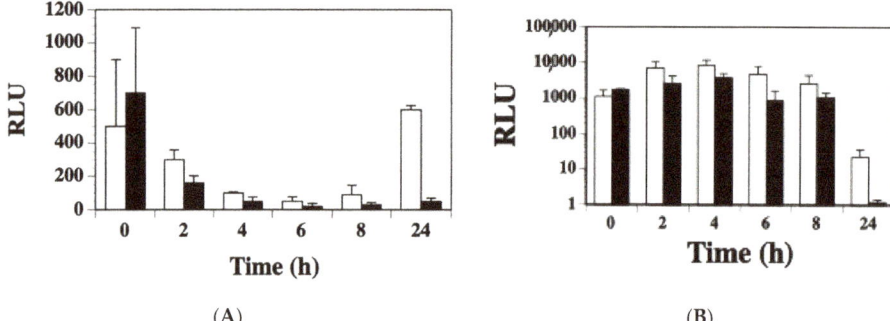

Figure 4. Expression of *ssl1* in strains WS0501 (RN6390 background, white column) and WS2601 (RN6390 agr mutant; black column) grown in (**A**) BHI broth or (**B**) DSM broth. One milliliter aliquots were removed every 2 h for 8 h and at 24 h to measure luminescence. RLU were determined as was presented before. The data represents the RLU means + standard deviation of at least three trials.

2.4. SaeS Positively Regulates ssl1 Expression in S. aureus

Due to the observation that Agr did not greatly affect the regulation of *ssl1*, we sought to determine if another known regulatory system played a role. Other investigators have shown that SaeS positively regulates transcription of other *ssl* genes [30–33]. To examine whether SaeS positively regulated transcription of the *ssl1* gene, the *ssl1::lux* fusion was transduced into strain WS0604 that has an *saeS* mutation. At all time points examined, strain WS0604 displayed significantly lower *ssl1* expression ($p < 0.05$) compared to the wild-type strain when grown in both BHI broth (Figure 5A) and DSM broth (Figure 5B). Therefore, consistent with observations elsewhere, *ssl1* expression is positively regulated by SaeS in nutrient replete conditions, but also under nutrient poor conditions.

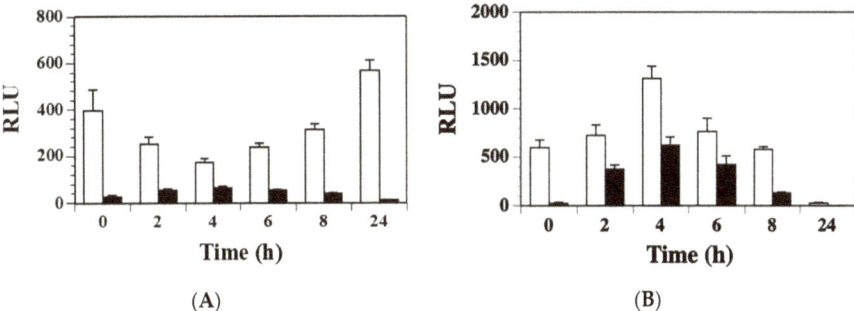

Figure 5. Expression of *ssl1* in strains WS0501 (RN6390 background, white column) and WS0604 (RN6390 *saeS* mutant; black column) grown in (**A**) BHI broth or (**B**) DSM broth. One milliliter aliquots were removed every 2 h for 8 h and at 24 h to measure luminescence. RLU were determined as was presented before. The data represents the RLU means + standard deviation of at least three trials.

2.5. Expression of ssl1 Occurs Early in S. aureus Found in Murine Abscesses

To date, there is no direct evidence of any *ssl* gene expression in an animal model, although these proteins are produced during human infections indicated by antibody binding of human sera in reaction to these proteins [14,15,23]. Since CA-MRSA causes skin and soft tissue infections, *ssl1* expression in *S. aureus* growing in mammalian tissue was tested. Mouse thighs were inoculated with virulent strain WS0501 (*S. aureus* RN6390 background) followed by sacrifice of mice and collection of tissue at 0 h, 8 h, 24 h, 48 h, and 72 h post-infection. As was done in vitro, RLUs were normalized to viable colony forming units and the median values determined (Figure 6A,B). A baseline value at 0 h collected soon after inoculation showed 16,941 RLU for strain WS0501, indicating that the per cell expression of *ssl1* increased quickly in the host. After 8 h post-inoculation, this had increased further to 298,375 RLU, representing a 17.6-fold increase compared to the 0 h time point ($p = 0.021$). At 24 h post-inoculation, there was a significant decline to 2477 RLU, representing a 120.4-fold drop from the 8 h time point ($p = 0.015$). After 48 h post-inoculation, the expression decreased slightly to 1,061 RLU. Finally, after 72 h post-inoculation, the expression rose again to 5404 RLU, representing a 5.1-fold increase over the 48 h time point. We again wanted to minimize strain-specific effects on *ssl1* expression, so we repeated this experiment in the Newman strain background, strain NS3513. As was seen in vitro, *ssl1* expression was consistent between both strains in vivo. For example, there was a significant 1.94-fold increase in the median RLU value for *ssl1* transcription from 0 h to 8 h in the murine thighs ($p = 0.016$). There was also a second peak of expression later during infection at 72 h where the median RLU was 18,229.

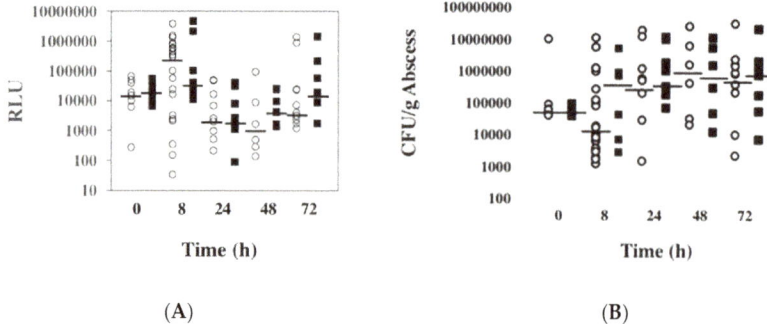

(A)   **(B)**

Figure 6. Expression of *ssl1* in a murine abscess model of infection. (**A**) Murine thighs were inoculated with either strain Newman (black column) or RN6390 (white column) and the thighs were collected at time points 0, 8, 24, 48, and 72 h post-inoculation. Each mouse is represented as the luminescence divided by the colony forming units (CFU) and then multiplied by 10^6 (RLU). The black lines represent the median values of at least six mice for each strain per time point. (**B**) The CFU/g of abscess for each mouse infected with WS0501 (RN6390 background, white circles) or NS3513 (Newman background, black squares) at 0, 8, 24, 48, and 72-h post-inoculation. Black lines represent the median values for each strain at each time point.

For mice infected with the RN6390 strain background, the CFU/g abscess dipped when comparing the 0 h time point to the 8 h time point (Figure 4A). However, the colony numbers increased markedly at 24 h and reached their peak after 48 h post-inoculation. On the other hand, mice infected with the Newman strain background exhibited a more than 10-fold increase in CFU/g going from 0 h to 8 h, reaching a peak also at 48 h post-inoculation.

Collectively, this data demonstrated that there is robust *ssl1* expression during infection and abscess formation, with peak expression early during the infection when access to nutrients within the host tissue may be limited.

3. Discussion

In recent years, there has been an increase in *S. aureus* skin and soft tissue infections [1]. This increase of infections in community settings is certainly multifactorial and linked to the multitude of virulence factors encoded by *S. aureus*. Amongst these other virulence factors, CA-MRSA strains encode more of the *ssl* genes compared to the repertoire of other *S. aureus* strains, so the virulence of these strains may be influenced by the SSL proteins [3,6,10,24,25]. While the structure and corresponding functions of several SLL proteins have been described [19,20,34–36], there is less known about the regulation of the *ssl* genes. Previous work has demonstrated that there is up-regulated transcription of the *ssl* genes during stationary phase growth in nutrient rich media, consistent with other known staphylococcal toxins [7,15]. In this study, we investigated the regulation of the *ssl1* gene in *S. aureus* grown in a nutrient-rich broth (BHI) and a nutrient-poor broth (DSM), and in murine abscesses.

To study *ssl1* gene regulation, an *ssl1::lux* promoter fusion was constructed and moved into the chromosome of multiple *S. aureus* strains. Importantly, the promoter was constructed from *S. aureus* MW2, and thus reflects the expression of this gene from a CA-MRSA strain. A modest up-regulation of *ssl1* transcription was observed during the stationary phase growth in BHI broth, consistent with previous research elsewhere [7,15]. However, *ssl1* transcription was different in the nutrient poor, DSM broth with peak expression occurring during log phase growth. Furthermore, overall expression of *ssl1* during growth in DSM broth was higher throughout the entirety of the growth curve as compared to growth in BHI broth. We observed similar results for expression of *ssl5* and *ssl8* in independent qRT-PCR experiments. These results suggest expression of the *ssl* genes may share similar regulatory mechanisms.

While the growth phase may be a factor in nutrient rich media, limited nutrient availability likely plays a more important role in the up-regulation. Emphasizing this point, Agr seems to have a limited modulatory effect on *ssl1* expression. The *agr* (accessory global regulator) gene codes for a component of a quorum-sensing system [28,29] that up-regulates many secreted proteins during post-exponential phase growth, while simultaneously down-regulating cell-surface-expressed factors [37]. It has also been shown that *agr* mutants are attenuated, solidifying its place as an important regulator of virulence factors [38]. Some investigation into the role of Agr in regulation of the *ssl* genes has been done. For example, a previous study performed by Laughton et al. [39] demonstrated that Agr did not control *ssl11* expression. Using similar methods and identical parental strains of *S. aureus*, our results are largely consistent with those results. However, we did observe a small reduction of *ssl1* transcription in an *agr* mutant strain, albeit not to statistically significant levels, which suggests a possible positive, modulatory role in regulation of *ssl1*. Agr-independent expression of staphylococcal toxins is not without precedent. For example, both the staphylococcal enterotoxin A and J expression are independent of Agr [40,41]. Overall, it is likely that Agr and stationary phase-dependent quorum sensing does not regulate *ssl1* expression, rather it is the reduction of nutrients that is the signal to increase expression. The nutrient poor conditions in DSM broth may better mimic nutrient conditions present at the onset of a *S. aureus* infection in or on the human body [42].

Growth in minimal medium may trigger a stress response that would involve SaeRS or Rot (repressor of toxins) [39,40]. Several studies have shown Sae regulator-binding sites in the promoter regions of *ssl* genes [43,44]. The SaeRS two-component system (TCS) has also been shown to be important for *S. aureus* virulence using animal models of infection [30,45,46]. Moreover, the SaeRS TCS is important for avoidance of human neutrophil-mediated killing [46]. Rot and SaeR have been shown in strain Newman and other *S. aureus* strains to work synergistically to activate *ssl* promoters, including the *ssl1* promoter [30–32,47]. Transcriptional activation of *ssl1* in vivo (discussed below) by Rot and SaeR in *S. aureus* growing in murine thigh abscesses at an early time point could help to explain the immune cell avoidance reported by Voyich et al. [46]. In this study, we have shown that SaeS does indeed appear to be a positive regulator of *ssl1* transcription. A *saeS* mutation significantly reduced *ssl1* transcription in the RN6390 strain at all time points tested during growth in both nutrient rich and nutrient poor conditions. The role of SaeRS and Rot in the regulation of *ssl1* and other *ssl* genes during nutrient poor, other environmental conditions, and in vivo is the topic of future study.

Beyond the role of nutrients in regulation in vitro, our study demonstrates evidence for the expression of *ssl1* in *S. aureus* infected animals. In vivo regulation of the other *ssl* genes in vivo is unknown and requires further study. It is clear that some of the SSL proteins must be produced during human infections, demonstrated by human patient sera binding to at least one of the SSL proteins [15]. However, not all SSL proteins were recognized by the sera, including SSL1. Whether this lack of antibody binding to the SSL proteins is due to low expression levels in vivo or a result of the complex interaction between the bacterial proteins and the host immune system is not known. In our study, we show that *ssl1* is indeed expressed during infection of mammalian tissue. The highest *ssl1* expression was 8 h post-inoculation, followed by a subsequent decrease in *ssl1* expression at 24 and 48 h. This peak expression early during infection corresponds to a time when the bacteria are establishing infection and therefore correlates with the evidence that the SSL proteins are important for immune avoidance, such as binding IgA and C5 proteins or inhibiting phagocytic cells from reaching the site of infection [10,11,16,17,19,48–50]. In fact, more recent studies have shown that SSL1 limits neutrophil chemotaxis and migration via matrix metalloprotease inhibition [35] and can cleave human recombinant cytokines [36]. Our data further suggests that once the infection is established, the SSL1 protein may no longer be central for survival of the *S. aureus* cells and is significantly repressed. However, as the abscess matures, *ssl1* expression may again increase as there is a secondary need for the SSL1 protein to sustain the infection at this late stage. Collectively, our *ssl1* expression data is consistent with the model that the SSL proteins are important for infection and that their expression may be regulated in part by the availability of nutrients within the host.

4. Conclusions

This study investigated transcriptional regulation of *S. aureus ssl1* following growth in vitro and in vivo in murine abscesses. The in vitro analysis demonstrated that *ssl1* transcription was higher in *S. aureus* growing in nutrient limited medium (DSM) compared to nutrient rich medium (BHI broth). Maximum transcription was during log-phase growth in nutrient poor conditions. Importantly, we have shown that the *ssl1* gene is also transcriptionally regulated in *S. aureus* growing within murine abscesses with peak expression early during infection (8 h), which correlates with the likely impairment of neutrophil function by *SSL1* early during an infection.

5. Materials and Methods

5.1. Bacterial Strains, Plasmids, and Growth Conditions

Various *Staphylococcus aureus* strains were used to assess *ssl1* expression under various conditions including RN4220, RN6390, RN6911, and Newman. Strain NE1296 (*saeS* mutation) was obtained from the Network on Antimicrobial Resistance in Staphylococcus aureus (NARSA) strain repository (Table 1), representing part of the Nebraska Transposon Mutant Library [51]. Luria broth or agar was used for growth of *E. coli*. All *S. aureus* strains were grown in either BHI media or in a DSM originally described by Rudin et al. [52], but modified later to better support the growth of *S. aureus* [53]. Antibiotics were used at the following concentrations: erythromycin, 300 µg/mL for *E. coli* and 5 µg/mL for *S. aureus*; kanamycin, 250 µg/mL. The pXen5 plasmid, containing an erythromycin resistance gene, an origin of replication for *E. coli*, a temperature-sensitive origin of replication for *S. aureus*, as well as a Tn*4100* transposon that contains a promotorless *lux* operon and a kanamycin resistance gene, was used for cloning [26].

Table 1. Bacterial strains and plasmids.

Strain or Plasmid	Relevant Characteristics	Reference
	Strains	
	E. coli	
DH5α	General cloning strain	Invitrogen
	S. aureus	
MW2	CA-MRSA USA400 virulent strain	[5]
RN4220	Transformation efficient strain	[27]
RN6390	Virulent strain	[27]
RN6911	Agr inactive, RN6390 with *agr::tetM*	[54]
Newman	Virulent strain	[55]
NE1296	JE2 strain with *saeS* mutation	[51]
WS4108	RN4220 *ssl1::lux*	This study
WS0501	RN6390 *ssl1::lux*	This study
WS0604	RN6390 *ssl1::lux, saeS::mariner*	This study
WS2601	RN6911 *ssl1::lux*	This study
NS3513	Newman *ssl1::lux*	This study
	Plasmids	
pXen5	TS [a] origin, Tn*4001*, promoterless *lux*	[26]
pXssl	*ssl1::lux* on pXen5	This study

[a] TS = Temperature-sensitive.

5.2. Construction of the ssl1::lux Fusion

Genomic DNA from *S. aureus* strain MW2 grown to stationary phase in BHI at 37 °C was extracted with a commercial kit (Edge Biosystems, Gaithersburg, MD, USA) with a lysostaphin (Sigma-Aldritch, St. Louis., MO, USA) addition at the first step. Primers to amplify the promoter region were designed based on the MW2 genome [5]. The promoter DNA amplification was done with a PE9700 Thermal Cycler (Perkin Elmer, Wellesley, MA, USA) and *Taq* polymerase under the following conditions: 35 cycles, 94 °C for 30 s, 55 °C for 30 s, and 72 °C for 1 min. The forward primer was 5′-CACTGAATTCCCACTTCTGGAATACGTTTG-3′ and the reverse primer was 5′-CATTGGTACCACCTGTTGCTAACATTCCCA-3′. The forward primer included an EcoRI restriction site at the 5′ end and a KpnI restriction site at the 5′ end of the reverse primer. The *ssl1* promoter was cloned upstream of a promoterless *lux* operon in plasmid pXen5 [26]. Resulting *E. coli* transformants were screened for luciferase activity using a Femtomaster FB12 luminometer (Zylux Corporation, Maryville, TN, USA). Several clones exhibited luminescence and plasmid DNAs were extracted from clones that were positive for luminescence. One plasmid, named pXssl, was electroporated into electrocompetent *S. aureus* strain RN4220 as previously described [56]. Transformed *S. aureus* cells were selected on BHI agar containing erythromycin and kanamycin. Resulting clones were grown in overnight cultures and checked for luminescence activity. A successful clone was incubated at 42 °C to silence the temperature-sensitive origin of replication and allow the Tn*4100* transposon with the *ssl1::lux* fusion to move into the chromosome. A final clone, WS4108, was obtained based on a loss of erythromycin resistance, due to the movement of the transposon into the chromosome, and positive luminescence.

5.3. Transduction of S. aureus

The *ssl1::lux* fusion in strain WS4108 was transduced into strains RN6390, RN6911, and Newman and the *saeS::mariner* mutation was transduced into strain WS0501 [57]. Briefly, a staphylococcal bacteriophage Φ80α lysate was prepared using WS4108 cells. The phage lysate was then used to transduce the various staphylococcal strains, plating the transduced cells onto BHI containing kanamycin. Colonies that arose were grown overnight and examined for luminescence.

5.4. In vitro Testing of the ssl1::lux Fusion

The *S. aureus* strains containing the *ssl1::lux* fusion were grown overnight in BHI and an aliquot of each overnight culture was used to inoculate fresh media the next day for testing of *ssl1* expression by luminescence analysis. All strains were grown in both BHI and DSM broth with kanamycin at 37 °C with shaking. At given time intervals spanning the growth curve, a 1 mL aliquot was removed from the culture and tested for luminescence and optical density (600 nm) as described above. Viable colony forming units were obtained by serial dilution with aliquots plated onto BHI agar plates and incubated at 37 °C overnight. RLU were determined by dividing the raw luminescence units by the viable colony forming units. Because the per cell luminescence was low, all values were multiplied by 10^6 to obtain the final RLU values.

5.5. RNA Extraction and Quantitative Reverse Transcribed-Polymerase Chain Reaction (qRT-PCR)

Total RNA was isolated from *S. aureus* strain RN6390 cells grown to early stationary phase in DSM or BHI broth using a High Pure RNA Isolation kit (Roche Diagnostics, Indianapolis, IN, USA) with an additional lysostaphin treatment step to help lyse the *S. aureus* cell walls and DNase I digestion. RNA samples were analyzed on a Nanodrop machine (Thermo Scientific, Waltham, MA, USA) to assess concentration and purity as well as run on 0.8% agarose gels to confirm concentration and integrities of the RNAs. The cDNAs were synthesized from 2 µg of total RNA using a First-Strand Synthesis kit (Life Technologies, Carlsbad, CA, USA) according to manufacturer's instructions. All of the RT-qPCRs were performed using the iTaq Universal SYBR Supermix kit according to manufacturer's instructions (BioRad, Hercules, CA, USA). Primers used in this study have been used in other studies: *ftsZ* [58], *ssl5* [33], and *ssl8* [33] and were synthesized by Integrated DNA Technologies (Coralville, IA, USA). A CFX96 machine (BioRad, Hercules, CA, USA) was used throughout the study. The *ftsZ* housekeeping gene was used as a standardization control. Each RT-qPCR run followed the minimum information for publication of quantitative real-time PCR experiments guidelines [59]. The RT-qPCRs were done at least three times under the following conditions: 94 °C, 20 s; 55 °C, 30 s; and 72 °C, 1 min for 35 cycles. The level of target gene transcripts in RN6390 cells was compared to the *ftsZ* gene. Crossover points for all genes were standardized to the crossover points for *ftsZ* in each sample using the $2^{-\Delta\Delta CT}$ formula [60].

5.6. Murine Abscess Model of Infection

The Institutional Animal Care and Use Committee at the University of Wisconsin-La Crosse approved the animal handling protocol of this study (Protocol 4-07, approved on 7 May 2007). A previously used murine abscess model of infection was used to test in vivo expression of the *ssl1::lux* fusion in strains WS0501 and NS3513 [61]. Briefly, cultures were grown to mid-logarithmic phase, diluted to 2×10^7 CFU/mL, and mixed 1:1 with Cytodex beads (Sigma, St. Louis., MO, USA). Female Swiss Webster mice (Harlan) that were 8 to 16 weeks old were injected intramuscularly with 50 µL of the mixture into the right posterior thigh in batches of four or five mice per time point repeated at least once. At 0, 8, 24, 48, and 72 h post-infection, mice were sacrificed, and the infected thigh muscle tissues collected. Each thigh tissue was homogenized in one mL of phosphate-buffered saline (PBS) and luminescence measured. Luminescence readings were normalized by subtracting the background RLU of an uninfected control and then dividing by the viable colony forming units. Any mice that did not demonstrate a successful infection indicated by bacterial counts below the detection limit (300 CFU) or below the background luminescence, were excluded from the analysis. The results represent the averages from at least six mice.

5.7. Statistical Analysis

The results of the *ssl1* expression in mice were analyzed by a one-way ANOVA. To further analyze the differences between the mean RLU of each time point, a least significant differences post hoc test was performed. For in vitro analysis in the growth media, Student's *t*-tests were performed. *p*-values of <0.05 were considered significant.

Author Contributions: D.J.B. carried out the experiments, D.J.B. and W.R.S. designed the experiments and wrote the paper, A.E. and D.J.B. ran the statistical analyses, and H.W. ran growth curve analyses.

Funding: This research was supported by a UWL Graduate Student Research, Service, and Education Leadership Grant to D.J.B., a UWL Michael and Kathi McGinley Endowed Undergraduate Student Research Grant to H.W., and a National Institutes of Health grant [grant number AI061T385].

Acknowledgments: We thank Kenneth Bayles, Jean Lee, Richard Novick, NARSA, and Xenogen Corp. for supplying strains and plasmids used in this study and the Statistical Consulting Center for a statistical analysis consultation.

Conflicts of Interest: The authors declare no conflict of interest.

References

1. Suaya, J.A.; Mera, R.M.; Cassidy, A.; O'Hara, P.; Amrine-Madsen, H.; Burstin, S.; Miller, L.G. Incidence and cost of hospitalizations associated with *Staphylococcus aureus* skin and soft tissue infections in the United States from 2001 to 2009. *BMC Infect. Dis.* **2014**, *14*, 296. [CrossRef] [PubMed]

2. Klein, E.Y.; Sun, L.; Smith, D.L.; Laxminarayan, R. The changing epidemiology of methicillin-resistant *Staphylococcus aureus* in the United States: A national observational study. *Am. J. Epidemiol.* **2013**, *1771*, 666–674. [CrossRef] [PubMed]

3. Centers for Disease Control and Prevention. Four pediatric deaths from community-acquired methicillin-resistant *Staphylococcus aureus*–Minnesota and North Dakota, 1997-1999. *Morbid. Mortal. Wkly. Rep.* **1999**, *48*, 707–710.

4. Dinges, M.M.; Orwin, P.M.; Schlievert, P.M. Exotoxins of *Staphylococcus aureus*. *Clin. Microbiol. Rev.* **2000**, *13*, 16–34. [CrossRef] [PubMed]

5. Baba, T.; Takeuchi, F.; Kuroda, M.; Yuzawa, H.; Aoki, K.; Oguchi, A.; Nagai, Y.; Iwama, N.; Asano, K.; Naimi, T.; et al. Genome and virulence determinant of high virulence community-acquired MRSA. *Lancet* **2002**, *359*, 1819–1827. [CrossRef]

6. Diep, B.A.; Gill, S.R.; Chang, R.G.; Phan, T.H.; Chen, J.H.; Davidson, M.G.; Lin, F.; Lin, J.; Carleton, H.; Mongodin, E.F.; et al. Complete genome sequence of USA300, an epidemic clone of community-acquired methicillin-resistant *Staphylococcus aureus*. *Lancet* **2006**, *36*, 731–739. [CrossRef]

7. Williams, R.J.; Ward, J.M.; Henderson, B.; Poole, S.; O'Hara, B.P.; Wilson, M.; Nair, S.P. Identification of a novel gene cluster encoding staphylococcal exotoxin-like proteins: Characterization of the prototypic gene and its gene product, SET1. *Infect. Immun.* **2000**, *68*, 4407–4415. [CrossRef] [PubMed]

8. Lina, G.; Bohach, G.A.; Nair, S.P.; Hiramatsu, K.; Jouvin-Marche, E.; Mariuzza, R. Standard nomenclature for the superantigens expressed by *Staphylococcus*. *J. Infect. Dis.* **2004**, *189*, 233–236. [CrossRef]

9. Holtfreter, S.; Bauer, K.; Thomas, D.; Feig, C.; Lorenz, V.; Roschack, K.; Friebe, K.; Selleng, S.; Lovenich, S.; Greve, T.; et al. *egc*-encoded superantigens from *Staphylococcus aureus* are neutralized by human sera much less efficiently than are classical staphylococcal exotoxins or toxic shock syndrome toxin. *Infect. Immun.* **2004**, *72*, 4061–4071. [CrossRef]

10. McCarthy, A.J.; Lindsay, J.A. *Staphylococcus aureus* innate immune evasion is lineage-specific: A bioinformatics study. *Infect. Genet. Evol.* **2013**, *19*, 7–14. [CrossRef]

11. Langley, R.; Wines, B.; Willoughby, N.; Basu, I.; Proft, T.; Fraser, J.D. The staphylococcal superantigen-like protein 7 binds IgA and complement C5 and inhibits IgA-Fc(alpha)RI binding and serum killing of bacteria. *J. Immunol.* **2005**, *174*, 2926–2933. [CrossRef] [PubMed]

12. Papageorgiou, A.C.; Achaya, K.R. Microbial superantigens: From structure to function. *Trends Microbiol.* **2000**, *8*, 369–375. [CrossRef]

13. Al-Shangiti, A.M.; Naylor, C.E.; Nair, S.P.; Briggs, D.C.; Henderson, B.; Chain, B.M. Structural relationships and cellular tropisms of staphylococcal superantigen-like proteins. *Infect. Immun.* **2004**, *72*, 4261–4270. [CrossRef] [PubMed]

14. Arcus, V.L.; Langley, R.; Proft, R.; Fraser, J.D.; Baker, E.N. The three-dimensional structure of a superantigen-like protein, SET3, from a pathogenicity island of the *Staphylococcus aureus* genome. *J. Biol. Chem.* **2002**, *277*, 32274–32281. [CrossRef] [PubMed]

15. Fitzgerald, J.R.; Reid, S.D.; Ruotsalainen, E.; Tripp, T.J.; Liu, M.Y.; Cole, R.; Kuusela, P.; Schlievert, P.M.; Jarvinen, A.; Musser, J.M. Genome diversification in *Staphylococcus aureus*: Molecular evolution of a highly variable chromosomal region encoding the staphylococcal exotoxin-like family of proteins. *Infect. Immun.* **2003**, *71*, 2827–2838. [CrossRef] [PubMed]

16. Lorenz, N.; Clow, F.; Radcliff, F.J.; Fraser, J.D. Full functional activity of SSL7 requires binding of both complement C5 and IgA. *Immunol. Cell Biol.* **2013**, *91*, 469–476. [CrossRef] [PubMed]

17. Wines, B.D.; Ramsland, P.A.; Trist, H.M.; Gardam, S.; Brink, R.; Fraser, J.D.; Hogarth, P.M. Interaction of human, rat, and mouse immunoglobin A (IgA) with staphylococcal superantigen-like 7 (SSL7) decoy protein and leukocyte IgA receptor. *J. Biol. Chem.* **2011**, *286*, 33118–33124. [CrossRef] [PubMed]

18. Hermans, S.J.; Baker, H.M.; Sequeira, R.P.; Langley, R.J.; Baker, E.N.; Fraser, J.D. Structural and functional properties of staphylococcal superantigen-like protein 4. *Infect. Immun.* **2012**, *80*, 4004–4013. [CrossRef] [PubMed]

19. Itoh, S.; Yamaoka, N.; Kamoshida, G.; Takii, T.; Tsuji, T.; Hayashi, H.; Onozaki, K. Staphylococcal superantigen-like protein 8 (SSL8) binds to tenascin C and inhibits tenascin C-fibronectin interaction and cell motility of keratinocytes. *J. Biol. Chem.* **2013**, *288*, 21569–21580. [CrossRef]

20. Koymans, K.J.; Feitsma, L.J.; Brondijk, T.H.; Aerts, P.C.; Lukkien, E.; Lossl, P.; van Kessel, K.P.; de Haas, C.J.; van Strijp, J.A.; Huizinga, E.G. Structural basis for inhibition of TLR2 by staphylococcal superantigen-like protein 3 (SSL3). *Proc. Natl. Acad. Sci. USA* **2015**, *112*, 11018–11023. [CrossRef]

21. Yokoyama, R.; Itoh, S.; Kamoshida, G.; Takii, T.; Fujii, S.; Tsuji, T.; Onozaki, K. Staphylococcal superantigen-like protein 3 binds to the Toll-like receptor 2 extracellular domain and inhibits cytokine production induced by *Staphylococcus aureus*, cell wall component, or lipopeptides in murine macrophages. *Infect. Immun.* **2012**, *80*, 2816–2825. [CrossRef] [PubMed]

22. Kuroda, M.; Ohta, T.; Uchiyama, I.; Baba, T.; Yuzawa, H.; Kobayashi, I.; Cui, L.; Oguchi, A.; Aoki, K.; Nagai, Y.; et al. Whole genome sequencing of methicillin-resistant *Staphylococcus aureus*. *Lancet* **2001**, *357*, 1225–1240. [CrossRef]

23. Al-Shangiti, A.M.; Nair, S.P.; Chain, B.M. The interaction between staphylococcal superantigen-like proteins and dendritic cells. *Clin. Exp. Immun.* **2005**, *140*, 461–469. [CrossRef] [PubMed]

24. Tenover, F.C.; McDougal, L.K.; Goering, R.V.; Killgore, G.; Projan, S.J.; Patel, J.B.; Dunman, P.M. Characterization of a strain of community-associated methicillin-resistant *Staphylococcus aureus* widely disseminated in the Untied States. *J. Clin. Microbiol.* **2006**, *44*, 108–118. [CrossRef] [PubMed]

25. Shukla, S.K.; Karow, M.E.; Brady, J.M.; Stemper, M.E.; Kislow, J.; Moore, N.; Wroblewski, K.; Chyou, P.H.; Warshauer, D.M.; Reed, K.D.; et al. Virulence genes and genotypic associations in nasal carriage, community-associated methicillin-susceptible and methicillin-resistant USA400 *Staphylococcus aureus* isolates. *J. Clin. Microbiol.* **2010**, *48*, 3582–3592. [CrossRef]

26. Francis, K.P.; Yu, J.; Bellinger-Kawahara, C.; Joh, D.; Hawkinson, M.J.; Xiao, G.; Purchio, T.F.; Caparon, M.G.; Lipitsch, M.; Contag, P.R. Visualizing pneumococcal infections in the lungs of live mice using bioluminescent *Streptococcus pneumoniae* transformed with a novel gram-positive lux transposon. *Infect. Immun.* **2001**, *69*, 3350–3358. [CrossRef] [PubMed]

27. Novick, R.R. The *Staphylococcus* as a molecular genetic system. In *Molecular Biology of the Staphylococci*; Novick, R.P., Ed.; VCH Publishers: New York, NY, USA, 1990; pp. 1–40.

28. Recsei, P.; Kreiswirth, B.; O'Reilly, M.; Schlievert, P.; Gruss, A.; Novick, R.P. Regulation of exoprotein gene expression in *Staphylococcus aureus* by *agr*. *Mol. Gen. Genet.* **1986**, *202*, 58–62. [CrossRef]

29. Yarwood, J.M.; Schlievert, P.M. Quorum sensing in *Staphylococcus* infections. *J. Clin. Investig.* **2003**, *112*, 1620–1625. [CrossRef]

30. Nygaard, T.K.; Pallister, K.B.; Ruzevich, P.; Griffith, S.; Vunong, C.; Voyich, J.M. SaeR binds a consensus sequence within virulence promoters to advance USA300 pathogenesis. *J. Infect. Dis.* **2010**, *201*, 241–254. [CrossRef]

31. Benson, M.A.; Lilo, S.; Wassaerman, G.A.; Thoendel, M.; Smith, A.; Horswill, A.R.; Fraser, J.; Novick, R.P.; Shopsin, B.; Torres, V.J. *Staphylococcus aureus* regulates the expression and production of the staphylococcal superantigen-like secreted proteins in a Rot-dependent manner. *Mol. Microbiol.* **2011**, *81*, 659–675. [CrossRef]

32. Sun, F.; Li, C.; Jeong, D.; Sohn, C.; He, C.; Bae, T. In the *Staphylococcus aureus* two-component sae, the response regulator SaeR binds to a direct repeat sequence and DNA binding requires phosphorylation by the sensor kinase SaeS. *J. Bacteriol.* **2010**, *192*, 2111–2127. [CrossRef] [PubMed]

33. Pantrangi, M.; Singh, V.K.; Wolz, C.; Shukla, S.K. Staphylococcal superantigen-like genes, *ssl5* and *ssl8*, are positively regulated by Sae and negatively by Agr in the Newman strain. *FEMS Microbiol. Lett.* **2010**, *308*, 175–184. [CrossRef] [PubMed]

34. Dutta, D.; Dutta, A.; Bhattacharajee, A.; Basak, A.; Das, A.K. Cloning, expression, crystallization and preliminary X-ray diffraction studies of staphylococcal superantigen-like protein 1 (SSL1). *Acta Crystallogr. F Struct. Biol. Commun.* **2014**, *70*, 600–603. [CrossRef] [PubMed]

35. Koymans, K.J.; Bisschop, A.; Vughs, M.M.; van Kessel, K.P.; de Haas, C.J.; van Strijp, J.A. Staphylococcal superantigen-like protein 1 and 5 (SSL1 & SSL5) limit neutrophil chemotaxis and migration through MMP-inhibition. *Int. J. Mol. Sci.* **2016**, *17*, E1072. [PubMed]

36. Tang, A.; Caballero, A.R.; Bierdeman, M.A.; Marquart, M.E.; Foster, T.J.; Monk, I.R.; O'Callaghan, R.J. *Staphylococcus aureus* superantigen-like protein SSL1: A toxic protease. *Pathogens* **2019**, *8*, 2. [CrossRef] [PubMed]

37. Dunman, P.M.; Murphy, E.; Haney, S.; Palacios, D.; Tucker-Kellogg, G.; Wu, S.; Brown, E.L.; Zagursky, R.J.; Shlaes, D.; Projan, S.J. Transcription profiling-based identification of *Staphylococcus aureus* genes regulated by the *agr* and/or *sarA* loci. *J. Bacteriol.* **2001**, *183*, 7341–7353. [CrossRef]

38. Foster, T.J.; O'Reilly, M.; Phonimdaeng, P.; Cooney, J.; Patel, A.H.; Bramley, A.J. Genetic studies of virulence factors of *Staphylococcus aureus*. Properties of coagulase and gamma-toxin, alpha-toxin, beta-toxin and protein A in the pathogenesis of *S. aureus* infections. In *Molecular Biology of the Staphylococci*; Novick, R.P., Ed.; VCH Publishers: New York, NY, USA, 1990; pp. 403–420.

39. Laughton, J.M.; Devillard, E.; Heinrichs, D.E.; Reid, G.; McCormick, J.K. Inhibition of expression of a staphylococcal superantigen-like protein by a soluble factor from *Lactobacillus reuteri*. *Microbiology* **2006**, *152*, 1155–1167. [CrossRef]

40. Tremaine, M.T.; Brockman, D.K.; Betley, M.J. Staphylococcal enterotoxin A gene (*sea*) expression is not affected by the accessory gene regulator (*agr*). *Infect. Immun.* **1993**, *61*, 356–359.

41. Zhang, S.; Iandolo, J.J.; Stewart, G.C. The enterotoxin D plasmid of *Staphylococcus aureus* encodes a second enterotoxin determinant (*sej*). *FEMS Microbiol. Lett.* **1998**, *168*, 227–233. [CrossRef]

42. Mekalanos, J.J. Environmental signals controlling expression of virulence determinants in bacteria. *J. Bacteriol.* **1992**, *174*, 1–7. [CrossRef]

43. McNamara, P.J.; Milligan-Monroe, K.C.; Khalili, S.; Proctor, R.A. Identification, cloning, and initial characterization of rot, a locus encoding a regulator of virulence factor expression in *Staphylococcus aureus*. *J. Bacteriol.* **2000**, *182*, 3197–3203. [CrossRef] [PubMed]

44. Giraudo, A.T.; Calzolari, A.; Cataldi, A.A.; Bogni, C.; Nagel, R. The *sae* locus of *Staphylococcus aureus* encodes a two-component regulatory system. *FEMS Microbiol. Lett.* **1999**, *177*, 15–22. [CrossRef] [PubMed]

45. Montgomery, C.P.; Boyle-Vavra, S.; Daum, R.S. Importance of the global regulators Agr and SaeRS in the pathogenesis of CA-MRSA USA300 infection. *PLoS ONE* **2010**, *5*, e15177. [CrossRef] [PubMed]

46. Voyich, J.M.; Vuong, C.; DeWald, M.; Nygaard, T.K.; Kocianova, S.; Griffith, S.; Jones, J.; Iverson, C.; Sturdevant, D.E.; Braughton, K.R.; et al. The SaeR/S gene regulatory system is essential for innate immune evasion by *Staphylococcus aureus*. *J. Infect. Dis.* **2009**, *199*, 1698–1706. [CrossRef] [PubMed]

47. Benson, M.A.; Lilo, S.; Nygaard, T.; Voyich, J.M.; Torres, V.J. Rot and SaeRS cooperate to activate expression of the staphylococcal superantigen-like exoproteins. *J. Bacteriol.* **2012**, *194*, 4355–4365. [CrossRef] [PubMed]

48. Bestebroer, J.; Poppelier, M.J.J.G.; Ulfman, L.H.; Lenting, P.J.; Denis, C.V.; van Kessel, K.P.M.; van Strijp, J.A.G.; de Haas, C.J.C. Staphylococcal superantigen-like 5 binds PSGL-1 and inhibits P-selectin-mediated neutrophil rolling. *Blood* **2007**, *109*, 2936–2943.

49. Chung, M.C.; Wines, B.D.; Baker, H.; Langley, R.J.; Baker, E.N.; Fraser, J.D. The crystal structure of staphylococcal superantigen-like toxin 11 (SSL11) in complex with sialyl Lewis X reveals the mechanism for cell binding and immune inhibition. *Mol. Microbiol.* **2007**, *66*, 1342–1355. [CrossRef]

50. Swierstra, J.; Debets, S.; de Vogel, C.; Lemmens-den Toom, N.; Verkaik, N.; Ramdani-Bouguessa, N.; Jonkman, M.F.; van Dijl, J.M.; Fahal, A.; van Belkum, A.; et al. IgG4 subclass-specific responses to *Staphylococcus aureus* antigens shed new light on host-pathogen interaction. *Infect. Immun.* **2015**, *83*, 492–501. [CrossRef]

51. Fey, P.D.; Endres, J.L.; Yajjala, V.K.; Widhelm, T.J.; Boissy, R.J.; Bose, J.L.; Bayles, K.W. A genetic resource for rapid and comprehensive phenotype screening of nonessential *Staphylococcus aureus* genes. *mBio* **2013**, *4*, e00537012. [CrossRef]

52. Rudin, L.; Sjostrom, J.E.; Lindberg, M.; Philipson, L. Factors affecting competence for transformation in *Staphylococcus aureus*. *J. Bacteriol.* **1974**, *118*, 155–164.

53. Schwan, W.R.; Wetzel, K.J.; Gomez, T.S.; Stiles, M.A.; Beitlich, B.D.; Grunwald, S. Low-proline environments impair growth, proline transport and in vivo survival of *Staphylococcus aureus* strain-specific *putP* mutants. *Microbiology* **2004**, *150*, 1055–1061. [CrossRef] [PubMed]

54. Novick, R.P.; Ross, H.F.; Projan, S.J.; Kornblum, J.; Kreiswirth, B.; Mohgazeh, S. Synthesis of staphylococcal virulence factors is controlled by a regulatory RNA molecule. *EMBO J.* **1993**, *12*, 3967–3975. [CrossRef] [PubMed]

55. Duthie, E.S.; Lorenz, L.L. Staphylococcal coagulase; mode of action and antigenicity. *J. Gen. Microbiol.* **1952**, *6*, 95–107. [CrossRef] [PubMed]

56. Iandolo, J.; Kraemer, G.R. High frequency transformation of *Staphylococcus aureus* by electroporation. *Curr. Microbiol.* **1990**, *21*, 373–376.

57. Kloos, W.E.; Pattee, P.A. Transduction analysis of the histidine region in *Staphylococcus aureus*. *J. Gen. Microbiol.* **1965**, *39*, 195–207. [CrossRef] [PubMed]

58. Schwan, W.R.; Polanowski, R.; Dunman, P.M.; Medina-Bielski, S.; Lane, M.; Rott, M.; Lipker, L.; Wescott, A.; Monte, A.; Cook, J.M.; et al. Identification of *Staphylococcus aureus* cellular pathways affected by the stilbenoid lead drug SK-03-92 using a microarray. *Antibiotics* **2017**, *6*, 17. [CrossRef] [PubMed]

59. Bustin, S.S.; Benes, V.; Garson, J.A.; Hellemans, J.; Huggett, J.; Kubista, M.; Mueller, R.; Nolan, T.; Pfaffl, M.W.; Shipley, G.L.; et al. The MIQE guidelines: Minimum information for publication of quantitative real-time PCR experiments. *Clin. Chem.* **2009**, *55*, 611–622. [CrossRef] [PubMed]

60. Livak, K.J.; Schmittgen, T.D. Analysis of relative gene expression data using real-time quantitative PCR and the $2^{-\Delta\Delta CT}$ method. *Methods* **2001**, *25*, 402–408. [CrossRef]

61. Beonton, B.M.; Zhang, J.P.; Pope, C.; Christian, T.; Lee, L.; Winterberg, K.M.; Schmid, M.B.; Buysse, J.M. Large-scale identification of genes required for full virulence of *Staphylococcus aureus*. *J. Bacteriol.* **2004**, *186*, 8478–8489. [CrossRef]

Article

Novel Regulation of Alpha-Toxin and the Phenol-Soluble Modulins by Peptidyl-Prolyl *cis/trans* Isomerase Enzymes in *Staphylococcus aureus*

Rebecca A. Keogh [1], Rachel L. Zapf [1], Emily Trzeciak [1], Gillian G. Null [1], Richard E. Wiemels [1] and Ronan K. Carroll [1,2,*]

[1] Department of Biological Sciences, Ohio University, Athens, OH 45701, USA; rk145815@ohio.edu (R.A.K.); rz537816@ohio.edu (R.L.Z.); et884814@ohio.edu (E.T.); gn454115@ohio.edu (G.G.N.); wiemels@ohio.edu (R.E.W.)

[2] The Infectious and Tropical Disease Institute, Ohio University, Athens, OH 45701, USA

* Correspondence: carrolr3@ohio.edu; Tel.: +1-740-593-2201

Received: 30 April 2019; Accepted: 12 June 2019; Published: 16 June 2019

Abstract: Peptidyl-prolyl *cis/trans* isomerases (PPIases) are enzymes that catalyze the *cis*-to-*trans* isomerization around proline bonds, allowing proteins to fold into their correct confirmation. Previously, we identified two PPIase enzymes in *Staphylococcus aureus* (PpiB and PrsA) that are involved in the regulation of virulence determinants and have shown that PpiB contributes to *S. aureus* virulence in a murine abscess model of infection. Here, we further examine the role of these PPIases in *S. aureus* virulence and, in particular, their regulation of hemolytic toxins. Using murine abscess and systemic models of infection, we show that a *ppiB* mutant in a USA300 background is attenuated for virulence but that a *prsA* mutant is not. Deletion of the *ppiB* gene leads to decreased bacterial survival in macrophages and nasal epithelial cells, while there is no significant difference when *prsA* is deleted. Analysis of culture supernatants reveals that a *ppiB* mutant strain has reduced levels of the phenol-soluble modulins and that both *ppiB* and *prsA* mutants have reduced alpha-toxin activity. Finally, we perform immunoprecipitation to identify cellular targets of PpiB and PrsA. Results suggest a novel role for PpiB in *S. aureus* protein secretion. Collectively, our results demonstrate that PpiB and PrsA influence *S. aureus* toxins via distinct mechanisms, and that PpiB but not PrsA contributes to disease.

Keywords: PPIase; *S. aureus*; toxins; PpiB; PrsA; alpha-toxin; PSMs

Key Contribution: This work demonstrates novel regulation of *S. aureus* toxins by two PPIase enzymes with distinct mechanisms of regulation.

1. Introduction

Staphylococcus aureus is a Gram-positive bacterium that resides in the anterior nares of approximately one-third of the population. Diseases caused by *S. aureus* range in severity from minor skin and soft tissue infections, to life-threatening infections such as endocarditis, necrotizing fasciitis, and sepsis [1]. This incredible diversity in diseases is largely due to the multitude of virulence factors that *S. aureus* produces, such as exoenzymes that assist in the degradation of host molecules, adhesins that aid in attachment to surfaces, and toxins that lyse host cells [2–5]. Two of the best characterized toxins for their role in infection are α-toxin (encoded by the *hla* gene) and the phenol-soluble modulins (PSMs) [6,7].

Hla is a receptor-mediated pore-forming toxin, which binds the sheddase ADAM10 on the surface of host cells and disrupts their cellular membranes. Hla has been implicated as the primary toxin responsible for the lysis of rabbit erythrocytes, which have high amounts of ADAM10 coating their surface [6,8,9]. Interestingly, human red blood cells have very little ADAM10 on their surface and consequently, it takes high levels of accumulated Hla to lyse human erythrocytes. Lysis of human erythrocytes is more efficiently accomplished by the alpha phenol-soluble modulins (αPSMs) [10]. *S. aureus* encodes four αPSMs, each approximately 20–25 amino acids in size, on a single polycistronic transcript, called the αPSM transcript. A fifth αPSM (Hld or the delta-toxin) is encoded within the regulatory RNA molecule RNAIII. Studies on the regulation of both Hla and the αPSMs have been largely centered around the *agr* system. AgrA has been shown to bind directly to the promoter region of the αPSM transcript where it activates transcription, while *hla* translation is regulated by the *agr* effector RNAIII [11]. Interestingly, it has also been demonstrated that the αPSMs can regulate Hla production in murine skin and lung models of infection [12], although the exact mechanism is unclear. Investigating the regulation of these toxins will improve our understanding of their relative contribution during human infection and may help in the development of "anti-virulence" approaches to combat infections caused by *S. aureus*.

Peptidyl-prolyl *cis/trans* isomerases (PPIases) are a family of enzymes that catalyze the *cis*-to-*trans* isomerization of proline peptide bonds. Proline is a unique amino acid in that it can exist in both the *cis* and *trans* isomerization state in vivo. Correct protein folding is often not possible when a proline peptide bond is in the incorrect configuration and therefore the isomerization rate of proline peptide bonds can be the rate-limiting step in protein folding [13]. PPIase enzymes accelerate this isomerization and therefore assist in the regulation of proteins via a post-translational mechanism. Numerous studies have identified bacterial PPIases that contribute to virulence [14–18]. In addition to acting as foldases proteins, PPIases often have moonlighting roles and functions not limited to prolyl isomerization in the cell [19–21]. *S. aureus* encodes three PPIase enzymes—trigger factor (Tig), PrsA, and PpiB. Recent work in our lab (and others) has demonstrated that both PrsA and PpiB influence the activity of secreted virulence factors [21–23].

Initial work in our lab showed that a Δ*prsA* mutant has reduced phospholipase C (PI-PLC) activity and decreased protease activity, and that a Δ*ppiB* mutant has reduced hemolysis and nuclease activity [24]. A follow-up study demonstrated that PpiB contributes to virulence in a murine abscess model of infection, independently of its PPIase activity [21]. Recently, work by Lin et al. on the methicillin-sensitive *S. aureus* strain HG001, demonstrated that a Δ*prsA* mutant had better survival than wild-type (WT) in a murine systemic model of infection [22]. They also concluded that there was altered abundance of 67 exoproteins and 163 cell wall-associated proteins in a Δ*prsA* mutant, including the virulence factors surface protein A (SpA), immunodominant staphylococcal antigen B (IsaB) and the αPSMs.

Due to the contribution of these two PPIases to the regulation of multiple virulence determinants, we hypothesized that PrsA and PpiB would contribute to virulence in a murine systemic model of infection using a community acquired methicillin-resistant *S. aureus* (CA-MRSA) USA300 strain. In this study, we demonstrate that a USA300 Δ*ppiB* mutant is attenuated in both an abscess and systemic model of infection but there is no significant attenuation with a Δ*prsA* mutant. The same virulence trend was observed during intracellular survival assays using macrophage and human nasal-epithelial cells. To understand the molecular mechanism underlying the attenuation of virulence in the Δ*ppiB* mutant, we examined toxin production by measuring the hemolytic activity of culture supernatants against human and rabbit erythrocytes. We identify a significant reduction in the activity and secretion of the αPSMs in a Δ*ppiB* mutant as well as a reduction in the activity of Hla in both the Δ*prsA* and Δ*ppiB* mutants. Immunoprecipitation analysis of PpiB and PrsA suggests a role for PpiB in the Sec secretion pathway and that PrsA is involved in cell-wall processing. Together, these data suggest a role for two PPIase proteins in the regulation of *S. aureus* hemolytic toxins via two distinct mechanisms.

2. Results

2.1. PpiB Is Required for Virulence in a Murine Abscess and Systemic Model of Infection

Previously, we demonstrated that a Δ*ppiB* mutant was attenuated for virulence in an abscess model of infection using the CA-MRSA strain USA300 TCH1516 [21]. Recently, Lin et al. demonstrated increased survival of a Δ*prsA* mutant in a sepsis model using the methicillin-sensitive *S. aureus* strain HG001 [22]. To better understand the role of each PPIase protein, we performed both localized (abscess) and systemic (sepsis) murine infection models using the USA300 strain TCH1516 and isogenic PPIase deletion mutants. USA300 strains are the most frequent cause of community-acquired MRSA infection in North America and are also a frequent cause of hospital-associated MRSA (HA-MRSA).

The murine abscess model of infection was performed as described previously [25]. Briefly, 6-week-old BALB/c mice were injected in the lower right flank with 10^6 CFU of either, W.T.; a Δ*ppiB* mutant or a Δ*prsA* mutant strain. Following a 7-day infection, animals were sacrificed, abscesses were excised and number of bacteria present was determined. The Δ*ppiB* mutant displayed a reduced bacterial load in a skin abscess, confirming our previous finding. However, a Δ*prsA* mutant showed no apparent alteration in recovered bacteria (Figure 1A).

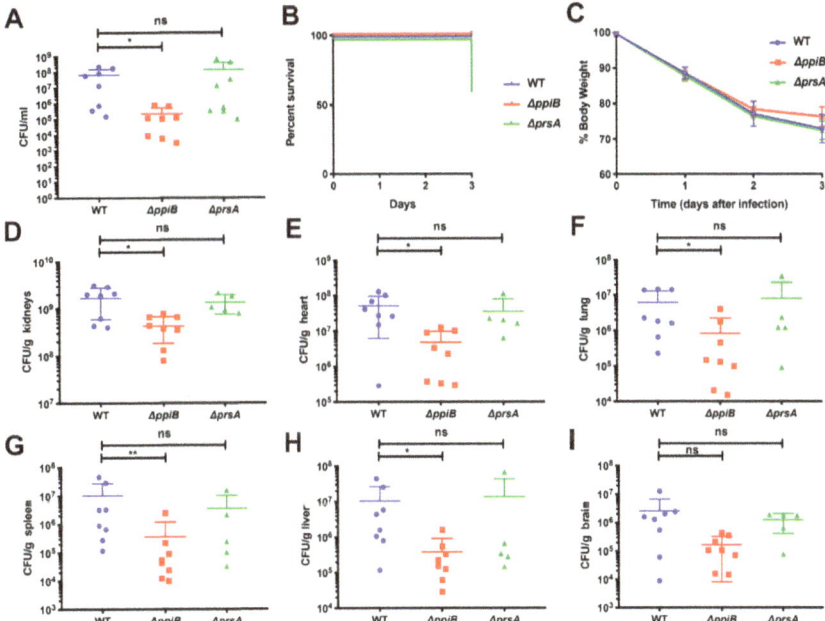

Figure 1. PpiB is required for virulence in a murine abscess (**A**) and systemic (**B–I**) model of infection. (**A**) Female 6-week-old BALB/c mice were injected subcutaneously with WT *S. aureus*, a Δ*ppiB* mutant or a Δ*prsA* mutant strain. The infection was allowed to proceed for 7 days. Mice were then sacrificed before abscesses were excised, homogenized, and diluted and plated to enumerate bacteria present in the abscesses. (**B–H**) Female 6-week-old BALB/c mice were injected retro orbitally with WT *S. aureus*, a Δ*ppiB* mutant or a Δ*prsA* mutant strain. The infection was allowed to proceed for 3 days. Mice were then sacrificed before organs were excised, weighed, diluted and plated to enumerate bacteria present in the organs. (**B**) Kaplan–Meyer curve for mouse survival during 3-day systemic infection. Three mice infected with the Δ*prsA* mutant strain died. (**C**) Percent initial body weight during 3-day systemic infection. The reduction in body weight in Δ*ppiB*-infected mice was less than that observed in WT or Δ*prsA*-infected mice, although the difference was not statistically significant. (**D–I**) CFU/g recovered from kidneys, heart, lung, spleen, liver, and brains of infected mice. In mice infected with the Δ*ppiB* mutant strain, a significant reduction in bacterial numbers was detected in the kidneys, heart, and lungs of infected mice. Experiments were performed with an *n* = 8 for each strain. Error bars represent standard deviation. Significance was determined by Student's t test for the abscess model of infection (**A**). * *p* < 0.05; and by a Mann–Whitney *U*-test for systemic organ data, * *p* < 0.05.

We next performed a murine systemic model of infection as described by Alonzo et al. [26]. Six-week-old BALB/c mice were injected with 10^7 CFU of each strain via the retro-orbital venous plexus and monitored for 3 days. Three of the eight $\Delta prsA$-infected mice died over the course of infection (Figure 1B) and the surviving mice were monitored for change in weight (Figure 1C). Following 3 days of infection, mice were sacrificed, and the brain, lungs, heart, liver, spleen and kidneys were excised for processing. Each organ was homogenized, diluted, and plated to quantify the number of bacteria per organ. For $\Delta ppiB$-infected mice, a significant reduction in bacterial burden was identified in the kidneys, heart, spleen, liver, and lungs (Figure 1D–I). A non-significant reduction in bacterial burden was also observed in the brain of the $\Delta ppiB$-infected mice ($p = 0.065$), and overall a modest non-significant reduction in weight loss was also observed in the $\Delta ppiB$-infected mice (Figure 1I,C). No difference in bacterial burden was observed in any of the organs from mice infected with the $\Delta prsA$ mutant (Figure 1D–I). This is in contrast to the findings of Lin et al. which showed increased murine survival when infected with a $\Delta prsA$ mutant [22].

2.2. PpiB is Required for Survival inside Macrophages and Human Nasal-Epithelial Cells

Although long considered an extracellular pathogen, abundant evidence now exists to demonstrate that *S. aureus* has the ability to reside inside of both professional phagocytic cells and non-professional phagocytic cells during infection. To further characterize the contribution of PrsA and PpiB to virulence, we examined the ability of isogenic mutants to survive in the intracellular environment. These experiments were performed in professional phagocytic cells Tohoku Hospital Pediatrics (THP1) human monocytes/macrophage), and non-professional phagocytic cells Roswell Park Memorial Institute (RPMI) 2650, a human nasal epithelial cell line). Cells were infected at a multiplicity of infection (MOI) of 10 with either, W.T.; the $\Delta ppiB$ or the $\Delta prsA$ mutant strains. The three strains were used to infect THP-1 macrophages and RPMI 2650 cells and the number of intracellular bacteria determined at 2 h and 48 h. The $\Delta ppiB$ mutant displayed a small but significant decrease ($p = 0.0365$) in recovery compared to WT in macrophages at 2 h (Figure 2B). The $\Delta ppiB$ mutant also had a significant decrease ($p = 0.0018$) in recovery compared to WT in RPMI 2650 cells at 48 h (Figure 2C). These results are consistent with a $\Delta ppiB$ mutant being attenuated for virulence in murine models of infection. The $\Delta prsA$ mutant showed no significant differences in recovery at any of the selected timepoints in both cell types (Figure 2A–D).

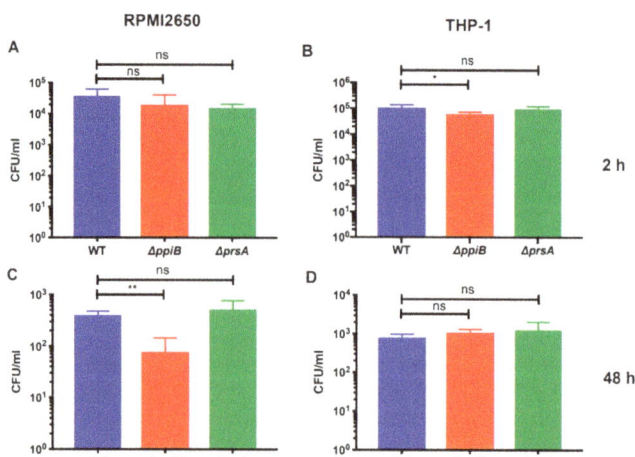

Figure 2. PpiB is required for survival in RPMI2650 nasal epithelial (**A,C**) and THP-1 macrophage cells (**B,D**). Gentamycin protection assays were performed at a multiplicity of infection (MOI) of 10 with WT, the $\Delta ppiB$ or the $\Delta prsA$ mutant strains. After 2 and 48 h of infection, bacterial cells were diluted and plated to enumerate surviving bacteria. Significance was determined by Student's *t* test (A). * $p < 0.05$; ** $p < 0.01$.

2.3. Exoproteome Analysis Reveals Greater Alterations in Secreted Protein Abundance in a ΔppiB Mutant Than a ΔprsA Mutant

To investigate further why attenuation of virulence was observed in the ΔppiB mutant but not the ΔprsA mutant, we analyzed the secreted proteome of the, W.T.; ΔppiB and ΔprsA mutant strains to identify proteins with altered abundance that may contribute to infection. Secreted proteins were detected via label-free mass spectrometry and data was analyzed by the Scaffold program and compared based on average normalized abundance. Proteins with a minimum of a 2-fold change in abundance when compared to WT and a *p*-value < 0.05 (based on Student's *t*-test) were considered for analysis. Based on these parameters, 86 proteins displayed altered abundance in culture supernatants of the ΔppiB mutant (Table 1, Table S1). Sixteen were found in greater abundance in the ΔppiB mutant, while 60 were less abundant compared to WT. In contrast, only 16 proteins were identified as having significantly altered abundance in the secreted fraction of a ΔprsA mutant compared to WT. Of these, 6 were found at a higher level and 10 were less abundant in culture supernatants of the ΔprsA mutant (Table 2, Table S2). Of note is that PpiB and PrsA peptides were detected in small amounts at 0.01 and 0.03, respectively, in their isogenic mutant strains when compared to WT. We attribute this to artifact based on the technique detecting small peptides and not full-length proteins.

Table 1. Proteins with altered abundance in ΔppiB culture supernatants.

Gene Designation	Protein Name	Fold Change [1]	Functional Grouping
SAUSA300_1533	YdfA	25.51	Conserved hypothetical protein
SAUSA300_0062	ArcB	23.74	Amino acid biosynthesis
SAUSA300_0962	QoxB	18.74	Energy metabolism
SAUSA300_0226	FadB	13.36	Fatty acid and phospholipid metabolism
SAUSA300_2581	SasA	7.36	Cell envelope
SAUSA300_1305	OdhB	4.93	Energy metabolism
SAUSA300_0912	FabI	4.78	Fatty acid and phospholipid metabolism
SAUSA300_1594	YajC	4.71	Protein fate
SAUSA300_2061	AtpH	3.70	Energy metabolism
SAUSA300_2060	AtpA	3.44	Energy metabolism
SAUSA300_0547	SdrD	3.37	Cell envelope
SAUSA300_2143		3.32	Conserved hypothetical protein
SAUSA300_2170	RpoA	3.23	Transcription
SAUSA300_2058	AtpD	3.13	Energy metabolism
SAUSA300_0527	RpoB	2.87	Transcription
SAUSA300_2059	AtpG	2.80	Energy metabolism
SAUSA300_0528	RpoC	2.63	Transcription
SAUSA300_0194	MurP	2.49	Cellular processes (includes toxins and virulence factors)
SAUSA300_1565		2.40	Central intermediary metabolism
SAUSA300_1685		2.28	Conserved hypothetical protein
SAUSA300_2453	NcaC	2.23	Transport and binding proteins
SAUSA300_2440	FnbB	2.17	Cell envelope
SAUSA300_0724		2.10	Cell envelope
SAUSA300_0514	CysE	2.08	Amino acid biosynthesis
SAUSA300_0963	QoxA	2.05	Energy metabolism
SAUSA300_2573	IsaB	2.00	Unknown function
SAUSA300_1881	GatA	0.50	Protein synthesis
SAUSA300_0691	SaeR	0.49	Regulatory functions
SAUSA300_2469	SdaAA	0.49	Energy metabolism
SAUSA300_0386	Xpt	0.48	Purines, pyrimidines, nucleosides, and nucleotides
SAUSA300_1258		0.48	Energy metabolism
SAUSA300_1293	LysA	0.48	Amino acid biosynthesis
SAUSA300_0325	GcvH	0.48	Energy metabolism
SAUSA300_1360	UbiE	0.48	Protein synthesis
SAUSA300_0480	Pth	0.48	Protein synthesis
SAUSA300_0716		0.48	Purines, pyrimidines, nucleosides, and nucleotides
SAUSA300_2066	Upp	0.47	Purines, pyrimidines, nucleosides, and nucleotides
SAUSA300_0492	FolP	0.47	Biosynthesis of cofactors, prosthetic groups, and carriers

Table 1. *Cont.*

Gene Designation	Protein Name	Fold Change [1]	Functional Grouping
SAUSA300_2076		0.46	Central intermediary metabolism
SAUSA300_1640	Icd	0.46	Energy metabolism
SAUSA300_0841		0.46	Conserved hypothetical protein
SAUSA300_2197	RplP	0.45	Protein synthesis
SAUSA300_1443	RluB	0.45	Protein synthesis
SAUSA300_1159	NusA	0.45	Transcription
SAUSA300_1530	YbeY	0.44	Conserved hypothetical protein
SAUSA300_0820	SufS	0.44	Biosynthesis of cofactors, prosthetic groups, and carriers
SAUSA300_1937		0.44	Mobile and extrachromosomal element functions
SAUSA300_1049	MurI	0.43	Cell envelope
SAUSA300_1679	AcsA	0.43	Central intermediary metabolism
SAUSA300_1178	RecA	0.42	DNA metabolism
SAUSA300_1269	FemA	0.42	Cellular processes (includes toxins and virulence factors)
SAUSA300_1882	GatC	0.41	Signal transduction
SAUSA300_1614	HemL1	0.41	Biosynthesis of cofactors, prosthetic groups, and carriers
SAUSA300_0067		0.41	Unknown function
SAUSA300_1634	CoaE	0.40	Biosynthesis of cofactors, prosthetic groups, and carriers
SAUSA300_1288	DapA	0.40	Amino acid biosynthesis
SAUSA300_1478		0.37	Cell envelope
SAUSA300_1285		0.35	Transport and binding proteins
SAUSA300_0971	PurL	0.35	Purines, pyrimidines, nucleosides, and nucleotides
SAUSA300_0368	RpsR	0.33	Protein synthesis
SAUSA300_1357	AroC	0.33	Amino acid biosynthesis
SAUSA300_0919	MurE	0.32	Cell envelope
SAUSA300_1156	ProS	0.32	Protein synthesis
SAUSA300_0753		0.30	Conserved hypothetical protein
SAUSA300_0741	UvrB	0.29	DNA metabolism
SAUSA300_0692	SaeQ	0.27	Conserved hypothetical protein
SAUSA300_1523		0.27	Conserved hypothetical protein
SAUSA300_2526	PyrD	0.26	Purines, pyrimidines, nucleosides, and nucleotides
SAUSA300_0364	YchF	0.26	Unknown function
SAUSA300_1144	TrmFO	0.24	Unknown function
SAUSA300_1861		0.24	Conserved hypothetical protein
SAUSA300_1007		0.24	Unknown function
SAUSA300_0329		0.24	Unknown function
SAUSA300_0732		0.23	Conserved hypothetical protein
SAUSA300_0538		0.23	Energy metabolism
SAUSA300_2251		0.22	Energy metabolism
SAUSA300_2025	RsbU	0.22	Cellular processes (includes toxins and virulence factors)
SAUSA300_2525		0.21	Conserved hypothetical protein
SAUSA300_2510		0.20	Conserved hypothetical protein
SAUSA300_2312	Mqo	0.18	Energy metabolism
SAUSA300_2296		0.17	Unknown function
SAUSA300_0737	SecA1	0.10	Protein fate
SAUSA300_2477	CidC	0.09	Energy metabolism
SAUSA300_1711	PutA	0.05	Energy metabolism
SAUSA300_2125		0.05	Transport and binding proteins
SAUSA300_0857	PpiB	0.01	Conserved hypothetical protein

[1] Fold change is based on comparing abundance of proteins in Δ*ppiB*/WT. Fold change >1 is indicative of an increase in abundance in Δ*ppiB* culture supernatants. Fold change <1 is indicative of a decrease in abundance in Δ*ppiB* culture supernatants.

Table 2. Proteins with altered abundance in Δ*prsA* culture supernatants.

Gene Designation	Protein Name	Fold Change [1]	Functional Grouping
SAUSA300_1018		11.13	Conserved hypothetical protein
SAUSA300_0062	ArcB	7.92	Amino acid biosynthesis
SAUSA300_2052		2.96	DNA metabolism
SAUSA300_1606		2.63	Conserved hypothetical protein
SAUSA300_0963	QoxA	2.23	Energy metabolism
SAUSA300_1341	Pbp2	2.06	Cell envelope
SAUSA300_0318	NanE	0.50	Central intermediary metabolism
SAUSA300_1763	EpiP	0.44	Protein fate
SAUSA300_1937		0.44	Mobile and extrachromosomal element functions
SAUSA300_2082	RpoE	0.42	Transcription
SAUSA300_0923	HtrA2	0.38	Protein fate
SAUSA300_0279	EsaA	0.37	Cell envelope
SAUSA300_2032	KdpC	0.32	Transport and binding proteins
SAUSA300_0226	FadB	0.31	Fatty acid and phospholipid metabolism
SAUSA300_1934		0.30	Mobile and extrachromosomal element functions
SAUSA300_1790	PrsA	0.03	Protein fate

[1] Fold change is based on comparing abundance of proteins in Δ*prsA*/WT. Fold change >1 is indicative of an increase in abundance in Δ*prsA* culture supernatants. Fold change <1 is indicative of a decrease in abundance in Δ*prsA* culture supernatants.

Proteins that made our cutoffs were grouped into categories based on known roles in the cell to identify common pathways or regulons with altered abundance (Figure 3). Notably, 2 of the 16 proteins with altered abundance in a Δ*prsA* mutant were cell envelope proteins. One of these, penicillin binding protein 2 (Pbp2a), was found in greater abundance in culture supernatants of the Δ*prsA* mutant. Work by Jousselin et al. demonstrated that *S. aureus* PrsA was involved in oxacillin resistance via regulation of *pbp2a*, and that a Δ*prsA* mutant had a reduced amount of Pbp2a in the cell wall [23]. We hypothesize that in the Δ*prsA* mutant, Pbp2a may have a defect in cell wall anchoring, which could account for the greater levels in the supernatant.

A large number of cytoplasmic proteins had decreased abundance in the exoproteome of a Δ*ppiB* mutant. Interestingly, we did not see the same trend in a Δ*prsA* mutant. Work by Wang et al. has recently demonstrated that *S. aureus* secretes numerous cytoplasmic proteins in extracellular vesicles to allow intracellular communication [27]. Notably, they show that this secretion is mediated by the αPSMs, which promote the release of extracellular vesicles from the bacterial cell. This finding, coupled with our previous work that shows a defect in human erythrocyte lysis in a Δ*ppiB* mutant [21,24], led us to hypothesize that PpiB might be regulating secretion of the αPSMs. A reduction in the amount of αPSMs could explain why cytoplasmic proteins were found in less abundance in a Δ*ppiB* mutant.

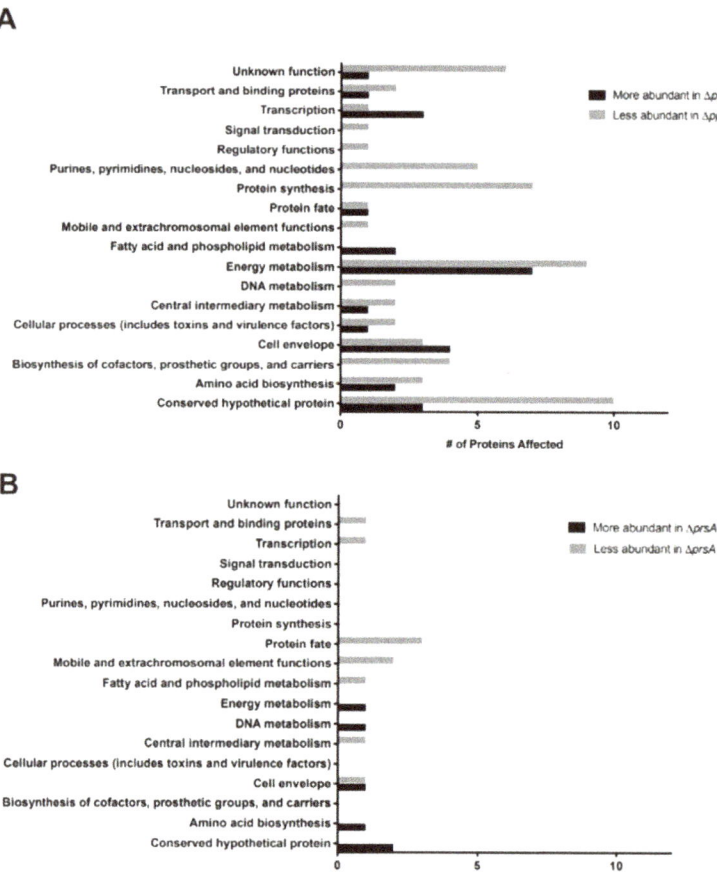

Figure 3. Exoproteome analysis reveals greater alterations in secreted protein abundance in a Δ*ppiB* mutant than a Δ*prsA* mutant. (**A**) Culture supernatants from a Δ*ppiB* mutant reveal that there are 86 proteins with altered abundance when *ppiB* is deleted. (**B**) Culture supernatants from a Δ*prsA* mutant reveal that there are 16 proteins with altered abundance when *prsA* is deleted. Proteins were grouped according to general function and the number of proteins falling into each functional category were plotted.

2.4. A ΔppiB Mutant Has Reduced PSM Production

Our previous publications demonstrated that culture supernatants from a Δ*ppiB* mutant were less hemolytic for human blood than supernatants from the WT [21,24]. Although the reduction in hemolytic activity was clear, the specific toxin responsible for this phenotype was not identified. Based on the results of the exoproteome analysis above, and the facts that the primary toxins responsible for human erythrocyte lysis are the αPSMs and that Hla activity against human erythrocytes has been shown to be negligible (Figure S1, [28]), we hypothesized that there is decreased abundance (or activity) of αPSMs in a Δ*ppiB* mutant. To test this hypothesis, we performed a butanol extraction to isolate the αPSM peptides from *S. aureus* culture supernatants [29]. This experiment allows visualization of lysis mediated exclusively via the PSMs, as they are the only toxin that will be extracted from culture supernatants, and thus, no other hemolysins will be present. The resulting extracted peptides were used in a hemolysis assay and analyzed by SDS-PAGE to visualize peptide abundance (Figure 4). Extracts from a Δ*ppiB* mutant displayed a modest (1.5-fold), yet significant reduction in hemolysis when compared to WT (Figure 4A). No reduction in hemolytic activity was observed in peptide extracts taken from the Δ*prsA* mutant or a *hla* mutant (negative control). SDS-PAGE analysis of extracts revealed a band corresponding to the size to the PSM peptides in each of the samples that was reduced in a Δ*ppiB* mutant (Figure 4B). Densitometry analysis was then performed on the SDS-PAGE gel with duplicate samples. Densitometry values were determined using Image J and normalized to WT (Figure 4C). Results show reduced PSM abundance in the Δ*ppiB* mutant (68.5% compared to the WT), mirroring the decrease in PSM activity observed in Figure 4A. Together, these results demonstrate a reduction in PSM abundance when *ppiB* is absent from the cell, confirming that PpiB influences PSM production.

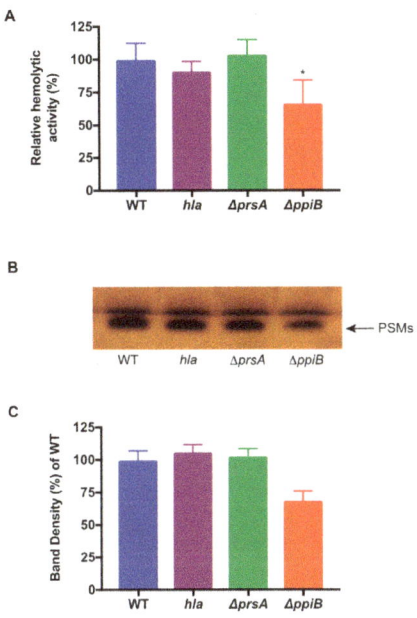

Figure 4. Decreased αPSM production in a Δ*ppiB* mutant. (**A**) Hemolytic activity of butanol-extracted peptides. Butanol-extracted samples were resuspended in water and incubated with whole human blood. The deletion of *ppiB* results in a significant decrease in hemolytic activity in comparison to WT. (**B**) SDS-PAGE analysis of butanol-extracted samples from panel A. PSM levels are reduced in a Δ*ppiB* mutant. (**C**) Densitometry analysis of PSMs on SDS-PAGE gel. Densitometry analysis was performed on duplicate samples and normalized to WT. The PSM band showed decreased abundance in the *ppiB* mutant strain (68.5% of WT). No differences in PSM abundance were observed in the *hla* or *prsA* mutant strains. Significance was determined by Student's *t* test. * $p < 0.05$.

2.5. The Δ*ppiB* and Δ*prsA* Mutants Both Display Reduced Hemolysis of Rabbit Erythrocytes

The αPSMs have been shown to regulate Hla by delaying transcription and subsequently, Hla secretion [12]. Consequently, we hypothesized that alpha-toxin activity would also be altered in a Δ*ppiB* mutant. Previously, we demonstrated that the Δ*ppiB* mutant displayed a slight reduction in hemolytic activity against rabbit erythrocytes, however, results were not deemed significant [21]. To better test the activity of Hla in a Δ*ppiB* mutant, we repeated the hemolysis assay over a time course of 20 min, with measurements taken every 5 min to determine the degree of lysis. Interestingly, culture supernatants from both a Δ*ppiB* and a Δ*prsA* mutant displayed reduced hemolysis at earlier time points (Figure 5A) when compared to WT. A representative time point at 10 min displayed a significant reduction in rabbit erythrocyte lysis in both the Δ*ppiB* and Δ*prsA* mutants, although not to the levels of a *hla* mutant strain (Figure 5B). At later timepoints (>20 min), the degree of hemolytic activity in the Δ*ppiB* and Δ*prsA* mutant samples increased to a level similar to that of the WT. This result explains our previous observation of a modest, non-significant reduction in hemolytic activity in the Δ*ppiB* mutant [21].

Taken collectively, these data suggest that both PpiB and PrsA are involved in the regulation of Hla. To investigate if this regulation occurred at the level of transcription or post-trancriptionally, RT-qPCR was performed to examine *hla* and αPSM transcript levels in the, W.T.; Δ*ppiB* and Δ*prsA* strains. No significant differences were detected in either transcript, indicating that the regulation is likely post transcriptional (Figure S2). We postulate that the reduction in Hla in a Δ*ppiB* mutant could be due to the reduction of αPSMs in culture supernatants. This supports the findings of Berube et al. in which they demonstrate that the αPSMs regulate Hla production during infection [12]. Since no significant reduction in PSM activity was observed in a Δ*prsA* mutant, we hypothesize that PrsA is regulating Hla independently of the PSMs. One explanation for why a Δ*prsA* mutant may have reduced cytolytic activity against rabbit erythrocytes is if the PPIase activity of PrsA is required to help assist in the folding of Hla. If this is the case, deleting *prsA* could lead to less active Hla in culture supernatants.

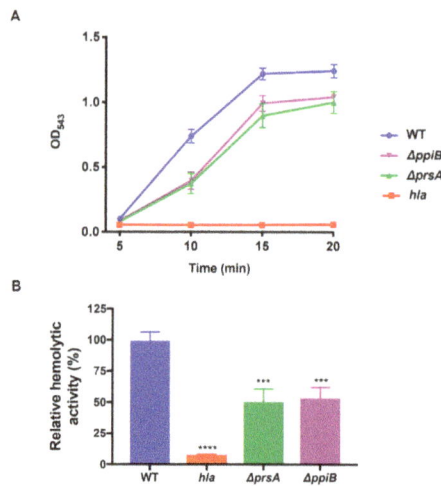

Figure 5. A *ppiB* and *prsA* mutant display reduced hemolysis in rabbit blood. Rabbit erythrocyte lysis assays were performed using *S. aureus* culture supernatants and whole rabbit blood. (**A**) A decrease in hemolytic activity against rabbit erythrocytes over time was observed using culture supernatants from *ppiB*, *prsA* and *hla* mutant strains. (**B**) A representative time point at 10 min reveals a significant reduction in hemolysis in the *ppiB* and *prsA* mutant strains. The data presented in A and B are the averages of four replicates. Significance was determined by Student's t test. ****, $p < 0.001$; ***, $p < 0.005$.

2.6. In Vivo Immunoprecipitation Identifies Greater Abundance of PpiB Target Proteins Than PrsA

Previous work in our lab has demonstrated that PpiB contributes to *S. aureus* virulence independently of its PPIase activity, suggesting an alternative function/activity for this protein [21]. To further investigate this alternative function, and to elucidate how PpiB and PrsA independently regulate *S. aureus* toxins, we performed an immunoprecipitation experiment to identify proteins that interact with PpiB and PrsA. The *ppiB* gene was amplified, HA-tagged, and cloned into the multicopy plasmid pMK4 under the control of its native promoter (plasmid pRKC0131). To identify proteins interacting with PpiB, pRKC0131 (*ppiB*-HA) was transduced into the Δ*ppiB* mutant strain, and immunoprecipitation performed using anti-HA magnetic beads. Immunoprecipitation experiments were performed by formaldehyde crosslinking proteins in cultures that had been grown for 3 h to mid-exponential phase. A negative control strain containing the empty pMK4 vector in the Δ*ppiB* mutant was used for comparison. Magnetic beads containing PpiB-HA (and associated proteins) were collected and analyzed via mass-spectrometry to identify all peptides associated with PpiB. A similar analysis was performed for PrsA using a His-tagged *prsA* construct in the same pMK4 plasmid (plasmid pRKC0126). Anti-His magnetic beads were analyzed and compared to an empty vector control in the Δ*prsA* mutant strain.

Data were analyzed using the Scaffold program and compared via the exclusive spectrum count of each protein identified in the samples. Proteins with a minimum of 3-fold increase in abundance (in the tagged strain compared to the empty vector control) were selected for further analysis. We also excluded any data where the average spectrum count was >10 for both the overexpresser and empty vector strains. Surprisingly, neither the αPSMs nor Hla were identified in the immunoprecipitation assay, suggesting that the role of PpiB and PrsA in regulating these toxins is indirect (see below).

As predicted (based on its cellular location), PrsA was found to interact with a large number of cell-wall anchored proteins (Table 3, Table S3). This included the signal peptidase I (SpsB) which assists in the anchoring of cell-wall proteins by cleaving the N-terminal signal peptide in the general secretory pathway. Work by Schallenberger et al. demonstrated that inhibition of SpsB with arylomycin led to increased abundance of the proteins PrsA, HtrA and SAOUHSC_01761 [30]. They suggested the reason for increased secretion of these proteins could have been as a response to the inhibition of SpsB (28). Notably, both *htrA* and SAUSA300_1606 (SAOUHSC_01761 homologue in TCH1516) both have altered abundance in the secreted fraction of a Δ*prsA* mutant (Table. 3, Table S3). HtrA is a highly conserved protein in Gram-positives where it functions as a chaperone and a protease [31,32]. SAUSA300_1606 is an uncharacterized protein and relatively little is known about it other than its induction during vancomycin treatment [33]. How these proteins might be interacting in the cell remains unknown, but our findings support a connection between the anchoring of PrsA to the cell wall and its connection to antibiotic resistance.

Table 3. Proteins interacting with PrsA.

Gene Designation	Protein Name	Fold Change [1]	Protein Description
SAUSA300_1790	PrsA	15.61748634	Foldase protein PrsA
SAUSA300_1512	Pbp3	9.75	Penicillin-binding protein 3
SAUSA300_0838	DltD	7.25	D-alanine-activating enzyme/D-alanine-D-alanyl, dltD protein
SAUSA300_0958	LcpB	6	Transcriptional regulator
SAUSA300_1588	LytH	6	Probable cell wall amidase LytH
SAUSA300_0963	QoxA	5.636363636	Probable quinol oxidase subunit 2
SAUSA300 1974	LukB	4.833333333	Uncharacterized leukocidin-like protein 1
SAUSA300_0032	MecA	4.5	Penicillin-binding protein 2
SAUSA300_2259	LcpC	4.454545455	Putative transcriptional regulator
SAUSA300_0274		4.285714286	Uncharacterized protein
SAUSA300_0419		3.8125	Uncharacterized lipoprotein
SAUSA300_2136	HtsA	3.705882353	Iron compound ABC transporter, iron compound-binding protein
SAUSA300_1982	GroL	3.702702703	60 kDa chaperonin
SAUSA300_0962	QoxB	3.608695652	Probable quinol oxidase subunit 1
SAUSA300_2578		3.6	Putative phage infection protein
SAUSA300_2213		3.5	AcrB/AcrD/AcrF family protein
SAUSA300_2328		3.333333333	Uncharacterized protein
SAUSA300_2092	Dps	3.142857143	General stress protein 20U
SAUSA300_2100		3.133333333	Lytic regulatory protein
SAUSA300_0992		3.090909091	Putative lipoprotein
SAUSA300_2144		3.083333333	Uncharacterized protein
SAUSA300_0868	SpsB	3	Signal peptidase I
SAUSA300_0279	EsaA	∞	Putative membrane protein
SAUSA300_2579	LytZ	∞	N-acetylmuramoyl-L-alanine amidase domain protein

[1] Fold change is based on proteins found in association with PrsA in comparison to an empty-vector control. Fold chance >1 indicates greater abundance in PrsA immunoprecipitation samples. Fold change = ∞ indicates protein not detected in negative control.

The largest fold enrichment for a protein identified in the PpiB immunoprecipitation analysis (after PpiB itself) was for the Sec translocase subunit SecA1 (8.5-fold enrichment, Table 4, Table S4). SecA is the ATPase that facilitates protein translocation through the SecYEG translocon, in the general secretory pathway [34]. In *Escherichia coli*, the chaperone protein SecB transports newly synthesized proteins from the ribosome to SecA for secretion [35]. Interestingly, *S. aureus* does not encode a SecB homologoue, and to date it is unknown if any alternative chaperone substitutes for SecB in the secretion process [34].

As previously stated, many PPIases also function as chaperone proteins [36–38]. This, in addition to our recent finding that PpiB possesses a PPIase-independent activity [21], and the data shown here of an association between PpiB and SecA, suggests that PpiB may function as a chaperone in the general secretion system. If so, this could account for the reduced activity of secreted virulence factors, such as Hla and nuclease (Nuc), in a Δ*ppiB* mutant [24].

Table 4. Proteins interacting with PpiB.

SAUSA300 Gene Number	Protein Name	Fold Change [1]	Protein Description
SAUSA300_0857	PpiB	31.5	Putative peptidyl-prolyl cis-trans isomerase
SAUSA300_0737	SecA1	8.5	Protein translocase subunit SecA 1
SAUSA300_1027	RpmF	7.8	50S ribosomal protein L32
SAUSA300_2364	Sbi	7.5	Immunoglobulin-binding protein sbi
SAUSA300_1535	RpsU	6.25	30S ribosomal protein S21
SAUSA300_1178	RecA	5.75	Protein RecA
SAUSA300_0220	PflB	5.5	Formate acetyltransferase
SAUSA300_1193	GlpD	4.88	Aerobic glycerol-3-phosphate dehydrogenase
SAUSA300_1645	PfkA	4.83	ATP-dependent 6-phosphofructokinase
SAUSA300_0757	Pgk	4.83	Phosphoglycerate kinase
SAUSA300_0798		4.63	Lipoprotein
SAUSA300_0388	GuaB	4.5	Inosine-5′-monophosphate dehydrogenase
SAUSA300_1525	GlyQS	4.18	Glycine–tRNA ligase
SAUSA300_0489	FtsH	3.86	ATP-dependent zinc metalloprotease FtsH
SAUSA300_2067	GlyA	3.75	Serine hydroxymethyltransferase
SAUSA300_1880	GatB	3.63	Aspartyl/glutamyl-tRNA (Asn/Gln) amidotransferase subunit B
SAUSA300_0491	CysK	3.63	Cysteine synthase
SAUSA300_1150	Tsf	3.62	Elongation factor Ts
SAUSA300_0389	GuaA	3.56	GMP synthase [glutamine-hydrolyzing]
SAUSA300_1684		3.25	Uncharacterized protein
SAUSA300_0533	Tuf	3.09	Elongation factor Tu
SAUSA300_0693	SaeP	3.07	Putative lipoprotein
SAUSA300_1459	Gnd	3.06	6-phosphogluconate dehydrogenase, decarboxylating
SAUSA300_0496	LysS	3	Lysine–tRNA ligase
SAUSA300_0716	NrdE	∞	Ribonucleoside-diphosphate reductase
SAUSA300_2251		∞	Dehydrogenase family protein
SAUSA300_1881	GatA	∞	Glutamyl-tRNA (Gln) amidotransferase subunit A
SAUSA300_1586	AspS	∞	Aspartate–tRNA ligase
SAUSA300_1167	Pnp	∞	Polyribonucleotide nucleotidyltransferase
SAUSA300_0009	SerS	∞	Serine–tRNA ligase
SAUSA300_1629	ThrS	∞	Threonine–tRNA ligase
SAUSA300_2214	FemX	∞	Lipid II:glycine glycyltransferase

[1] Fold change is based on proteins found in association with PpiB in comparison to an empty-vector control. Fold chance >1 indicates greater abundance in PpiB immunoprecipitation samples. Fold change = ∞ indicates protein not detected in negative control.

3. Discussion

In bacteria, PPIase enzymes are traditionally studied for their roles in protein (re)folding. In Gram-positive bacteria, parvulin-type PPIases (including PrsA) are often anchored on the external leaflet of the cell membrane, and are thought to fold secreted proteins after they have been translocated out of the cell. *S. aureus* PrsA appears to function in a similar manner. A *prsA* mutant strain has decreased proteolytic and phospholipase activity [24], increased sensitivity to β-lactam antibiotics [23] and decreased hemolytic activity (this study). We hypothesize that all of these defects arise as a result of defective protein folding in the absence of PrsA. This idea is supported by a number of observations in our proteomic and immunoprecipitation data analysis. First, the number of secreted proteins displaying altered levels in *prsA* mutant culture supernatants was relatively low (16 proteins, Table 2, Table S2). This suggests that PrsA targets are present in these supernatants, however, they may exhibit reduced activity due to incorrect folding (Figure 6A). Second, the majority of proteins found to interact with PrsA by immunoprecipitation are localized to the cell envelope and play important roles in cell wall synthesis and stability. Of particular interest was the identification of Pbp2a in the immunoprecipitation assay. This result supports the findings of Jousselin et al. and validates the approach used in this study to identify proteins that interact with PrsA [23]. Interestingly, EsaA, a component of the recently identified *S. aureus* type 7 secretion system, was identified in the immunoprecipitation assay as interacting with

PrsA and was also found at lower levels in the culture supernatant. This may indicate a potential role for PrsA in type 7 secretion.

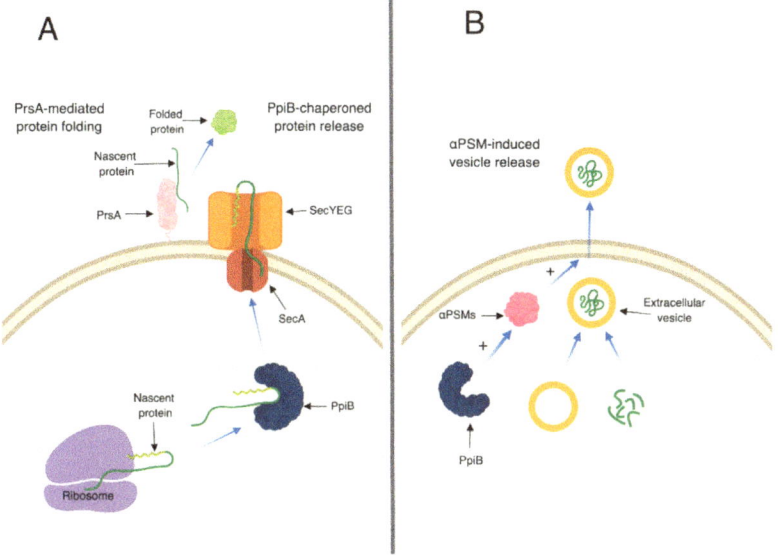

Figure 6. Schematic diagram of proposed mechanisms of PpiB and PrsA. (**A**) PpiB functioning as a chaperone to deliver nascent proteins to the Sec secretion machinery. Once secreted, proteins begin to fold and the membrane-anchored PrsA, which acts as a PPIase, assists this process. (**B**) PpiB positively regulates the αPSMs. When active αPSMs are within the cell, they promote the release of EVs from the membrane. Image created with BioRender.

An unexpected finding in this study was that PrsA did not contribute to virulence in either a murine abscess or sepsis model of infection. This result conflicts with the results obtained by Lin et al., who demonstrate attenuation of virulence in a HG001 *prsA* mutant [22]. One possible explanation for the discrepancy in results observed is the strain background used in each study. HG001, the strain used by Lin et al., is a methicillin-sensitive strain derived from NCTC8325 (with the *rsbU* gene repaired) [22]. The strain used in this study, TCH1516, is a methicillin-resistant USA300 strain. USA300 strains typically produce high levels of toxins and previously it was shown that a USA300 strain is more hemolytic than HG001 [39]. We speculate that, in USA300 strains, the high level of toxin production and overlapping activities of *S. aureus* toxins may negate the requirement for PrsA activity when it comes to causing disease in a mouse model of infection. In strains with relatively lower levels of toxin production (such as HG001), the loss of PrsA activity may have a more dramatic effect and result in attenuation. To test this hypothesis, we are currently broadening our investigations and exploring the role of PrsA in a variety of *S. aureus* strains.

While the role of PrsA in *S. aureus* appears to be similar to that of homologues in other Gram-positive bacteria (such as *Listeria monocytogenes*), the mechanism of action of PpiB remains elusive. PpiB clearly plays an important role in infection, as a *ppiB* mutant is attenuated in abscess and sepsis models of infection, and displays reduced intracellular survival. The activity and abundance of several virulence factors (including nuclease, alpha toxin, and the PSMs) is reduced in *ppiB* culture supernatants, which explains the attenuation of virulence, however the mechanism through which PpiB functions is not clear. We previously demonstrated that PpiB (i) is found in the cytoplasm, (ii) possesses PPIase activity, and (iii) its PPIase activity was dispensable during infection [21,24]. Taken together, this implies that PpiB functions inside the cell, in a PPIase-independent manner, to influence virulence factor production. Our first indication of a potential biological role for PpiB comes from the secreted proteomic and immunoprecipitation data presented herein. SecA, a component of the general secretion pathway in *S. aureus*, was found to interact with PpiB and was also found at significantly reduced levels in *ppiB* mutant culture supernatants. In *E. coli*, during general secretion, a chaperone protein called SecB binds to newly synthesized proteins and delivers them to SecA for translocation. No SecB homologue has been identified in *S. aureus*, therefore, based on our data, we hypothesize that one potential role for PpiB could be to functionally compensate for SecB and chaperone proteins prior to secretion (Figure 6A). Thus, the loss of PpiB could result in general secretion defects and/or the secretion of misfolded proteins. This might explain why so many proteins had altered abundance in the culture supernatants of *ppiB* mutant strains compared to *prsA* mutants (86 proteins altered in *ppiB* mutant, 16 in *prsA* mutant). Another role of chaperone proteins is to prevent the aggregation of misfolded proteins under stress [40]. If PpiB is functioning as a chaperone, it could additionally function to protect the αPSM peptides within the *S. aureus* cell. If this is the case, it could explain why there are less active αPSMs in culture supernatants of a Δ*ppiB* mutant.

S. aureus, like many other cells, has the ability to produce and secrete extracellular vesicles (EVs) [27]. Recently, the formation of EVs in *S. aureus* has been shown to be enhanced by the αPSMs [27]. EVs are loaded with proteins, including many proteins typically thought of as being cytoplasmic. The presence of EVs in culture supernatants may explain why cytoplasmic proteins are commonly identified during proteomic studies of secreted fractions, which we also observed in our proteomic data set. Interestingly, of the 86 proteins that were altered in abundance in *ppiB* mutant supernatants, a large proportion of them are considered cytoplasmic. These proteins were present in WT supernatants, but decreased in *ppiB* mutant supernatants. It is possible that PpiB plays a role in EV biogenesis and that in the mutant there is reduced EV production, which accounts for the variation in cytoplasmic proteins. Since a *ppiB* mutant produces less αPSMs which are necessary for EV release, it is also possible that the same number of EVs are being generated in a Δ*ppiB* mutant but they are unable to be released from the cell due to the decrease in αPSMs (Figure 6B). We are currently investigating EV production in a *ppiB* mutant.

In conclusion, while the role of PpiB in pathogenesis is clear, its molecular mechanism of action remains uncertain. The results generated in this study further our understanding of *S. aureus* virulence factors regulated by PpiB and suggest that PpiB may be functioning in a secretion-related manner. Further studies are being conducted to investigate this novel protein to better understand how it regulates virulence factor production and virulence.

4. Materials and Methods

4.1. Strains and Strain Construction

All bacterial strains and plasmids are listed in Table 5 and oligonucleotides in Table 6. The Δ*ppiB* and Δ*prsA* mutant strains were constructed via allelic exchange as previously described [21,24,41]. For overexpresser strains used in the immunoprecipitation, plasmids pRKC0131, pRKC0126 and pMK4 (empty vector) were transduced into RKC0323 (Δ*ppiB*) and RKC0085 (Δ*prsA*) respectively. A Δ*apsm* strain was constructed by allelic exchange using plasmid pJB38 [42]. DNA sequences flanking the αPSM transcript were amplified using primer pairs 490/644 and 645/493 and a third sequence containing the erythromycin cassette from the *bursa aurealis* transposon was amplified using primers 638 and 639 [42]. These three fragments were cloned to generate plasmid pRKC0674. This plasmid was recombined onto the *S. aureus* chromosome and excised to make the deletion strain according to previously published protocol [41].

Table 5. Strains and plasmids used in this study.

Name	Characteristics	Source
Strains		
S. aureus		
RN4220	Restriction-deficient transformation recipient	[43]
TCH1516	Community-associated USA300 MRSA isolate	[44]
RKC0323	TCH1516 Δ*ppiB*	[21]
RKC0085	TCH1516 Δ*prsA*	[24]
RKC0183	TCH1516 *hla*::Bursa, *hla* mutant	(21)
RKC0374	TCH1516 Δ*ppiB* pMK4_*ppiB*-HA	This work
RKC0536	TCH1516 Δ*ppiB* pMK4	This work
RKC0283	TCH1516 Δ*prsA* pMK4_*prsA*-his	(24)
RKC0129	TCH1516 Δ*prsA* pMK4	(24)
JE2	USA300 LAC isolate cured of plasmids LAC-p01 and LAC-p03	[45]
NE1354	USA300 JE2 *hla*::Bursa, *hla* NTML transposon mutant	[45]
AH1263	USA300 Lac isolate cured of plasmid Lac-p03	[46]
RKC0521	AH1263 *hla*::Bursa, *hla* mutant	[47]
RKC0753	AH1263 Δ*αPSMs*	This work
Plasmids		
pMK4	Gram-positive shuttle vector (Cm^r)	[48]
pRKC0131	pMK4_*ppiB*-HA (vector overexpressing *ppiB* with an HA tag)	[24]
pRKC0126	pMK4_prsA-His (vector overexpressing prsA with a poly-histidine tag)	[24]
pRKC0674	pJB38 containing DNA flanking αPSM transcript with ery cassette	This work

Table 6. Oligonucleotides used in this study.

Name	Sequence
#0273	GGTGCTGGGCAAATACAAGT (*gyrB*)
#0274	TCCCACACTAAATGGTGCAA (*gyrB*)
#0263	TGCAAATGTTTCGATTGGTC (*hla*)
#0264	CCCCAATTTTGATTCACCATA (*hla*)
#0271	ACAGGAGGACAAAACGATGG (*psmα*)
#0272	CCCTATTGGTATAGTGGCCTGA (*psmα*)
#0490	CAAGACGTCCGTCGGTCTACCTTTCCATGC
#0493	GGGGTACCACGTGGCACTTTCCAAAAAC
#0638	CCGGAATTCGCTCCTTGGAAGCTGTCAGT
#0639	AAAACTGCAGGAAGCAAACTTAAGAGTGTGTTGA
#0644	CCGGAATTCGATGTGAGGTGAGTCTTGTTAGTTTG
#0645	AAAACTGCAGAGATTACCTCCTTTGCTTATGAGT

4.2. Bacterial Growth Conditions

S. aureus cultures were grown in tryptic soy broth (TSB) shaking at 37 °C. Where appropriate, the antibiotic chloramphenicol was used at the concentration of 5 μg mL^{-1}. For analysis of culture supernatants, *S. aureus* cultures were synchronized as follows. Replicate overnight cultures were grown in 5 mL of TSB for 15 h. The next day, cultures were diluted 1:100 in 10 mL of pre-warmed TSB and grown for 3 h to mid-exponential phase. The 3-h cultures were then diluted into 25 mL of TSB in 250 flasks and normalized to an optical density at 600 nm (OD$_{600}$) of 0.05. Resulting flasks were grown overnight for 15 h.

4.3. Murine Abscess Model of Infection

A subcutaneous abscess infection was performed as described by us previously [21]. Cultures were grown for 2.5 h in TSB to an OD$_{600}$ of 0.75. Resulting bacterial cells were pelleted via centrifugation and resuspended in sterile phosphate-buffered saline (PBS) to prepare an inocula of 10^6 CFU/50 μL. Each inoculum was then confirmed via serial diluting and plating. Six-week-old female BALB/c mice were anesthetized by isoflourane inhalation before being shaved at the right flank and treated with

Nair to remove fur. Mice were then injected with 50 μL of their respective strains and the infection was allowed to proceed for 7 days. Following 7 days of infection, mice were euthanized with CO_2 and abscesses were excised and homogenized. Homogenates were serial diluted and plated onto TSB to count recovered CFU/ abscess.

4.4. Murine Systemic Model of Infection

A systemic model of dissemination was chosen to mimic septic infection and cultures were prepared as previously described by Spaan et al. [49]. Cultures were grown for approximately 2.5 h in TSB to an OD_{600} of 0.75. Resulting bacterial cells were pelleted via centrifugation and resuspended in sterile phosphate-buffered saline (PBS) to prepare an inocula of 10^7 CFU/100 μL. Six-week-old BALB/c mice were injected retro-orbitally with 100 μL of bacterial cultures and the infection was allowed to proceed for 3 days. Following 3 days of infection, mice were euthanized with CO_2 and the brain, lungs, heart, liver, kidneys and spleen were harvested. Each organ was weighed and homogenized before serial dilution and plating was conducted to quantify recovered bacteria/ organ.

4.5. Macrophage Infection and Cell Differentiation

THP-1 infection assays were performed as outlined in Carroll et al. [50]. Macrophages were seeded at a density of 2×10^5 macrophages/well in a volume of 500 μL of RPMI 1640 with 10% FBS and 1% penicillin/streptomycin. A multiplicity of infection (MOI) of 10 was used to infect each well (2×10^6 bacteria/well). Strains were prepared by taking 250 μL of overnight culture grown at 37 °C and inoculating it into a 250 mL flask containing 25 mL TSB. The bacteria were then grown in a shaking 37 °C incubator until their optical density (OD_{600}) reached 0.4–0.6. The volume of bacteria needed to perform the infection was then transferred into a 1.5 mL centrifuge tube and pelleted for 20 min at 3000 rpm. The supernatant was removed and discarded. The pellet was then washed with 500 μL of phosphate buffered saline (PBS) and pelleted for 20 min at 3000 rpm. The pellet was resuspended in 20% human serum and 80% PBS. This solution was then incubated in a 37 °C water bath for 30 min to opsonize the bacteria. Opsonized bacteria were resuspended into the correct volume of 1640 RPMI medium with 10% FBS.

The macrophages were washed twice with 37 °C prewarmed PBS. Each well received 500 μL of the opsonized bacteria resuspended in 1640 RPMI with 10% FBS. Plates were centrifuged at 1000 rpm for 10 min at room temperature and placed into a 37 °C incubator supplemented with 5% CO_2 for 60 min. After 60 min, the medium was aspirated, and the wells were washed once with prewarmed PBS as described above. To each well, 500 μL of RPMI 1640 with 10% FBS and 30 μg/mL of gentamicin was added. Cells were incubated in a 37 °C incubator supplemented with 5% CO_2 for 60 min. The end of this 60 min incubation marked hour 0 of intracellular. After this incubation, the medium was aspirated and replaced with 500 μL of RPMI 1640 with 10% FBS and 5 μg/mL of gentamicin for the rest of the infection.

Intracellular bacteria were harvested from the macrophages at 2 h and 48 h, representing initial and late stages of infection. Medium was aspirated, and the cells were washed twice with prewarmed PBS as described above. A 500 μL aliquot of 0.5% Triton X-100 in PBS was added to each well to lyse the macrophages and release the intracellular bacteria. The lysates for each time point were serially diluted in PBS in sterile 96-well plates. A volume of 100 μL of each of the diluted lysate was plated on TSA plates. Plates were incubated in a 37 °C incubator and colonies were counted to determine the number of recovered intracellular bacteria and calculated to determine CFU/mL.

4.6. Nasal Epithelial Cell Infection

Strains were prepared and washed in the same way as described above except the bacteria were not opsonized. After washing the bacterial pellet with PBS, the bacteria were resuspended into the correct volume of EMEM with 10% FBS. RPMI 2650 cells were seeded at the same density and volume as described above except using EMEM with 10% FBS and 1% penicillin/streptomycin. Cells were washed twice with prewarmed PBS and infected with bacteria at an MOI of 10 (2 x 10^6 bacteria/well).

4.7. Exoproteome Analysis

Bacterial strains were grown in triplicate in 5 mL overnight cultures shaking at 37 °C. The next day, cultures were diluted 1:100 in 10 mL of TSB and incubated for 3 h shaking at 37 °C. After 3 h, the OD_{600} of each culture was measured and each sample was normalized to an OD_{600} of 0.05 in 100 mL of TSB. After 15 h of growth in the flask, samples were split into 2 × 50 mL tubes and centrifuged at 3000 rpm for 15 min. Cell-free supernatants were harvested and filter sterilized through a 0.45 μm filter disk to ensure all bacterial cells were removed from the sample. An amount of 10 mL of TCA was added to the culture supernatants and incubated overnight at 4 °C. The following day, samples were centrifuged at 11,000 RPM for 10 min and the supernatant was removed. Resulting pellets containing concentrated proteins were washed with ice-cold acetone and sent to the University of Nebraska Lincoln Proteomics Facility for mass spectrometry analysis.

TCA precipitated samples were dissolved in a solution of 7M urea, 2M thiourea, 5 mM DTT, 0.1 M Tris, pH 8 and 1× PhosStop by gentle shaking for 1.5 h at 24 °C. After reduction, the samples were alkylated for 30 min using a 3-fold molar excess of iodoacetamide. The protein concentration was determined using the CB-X protein assay kit. An amount of 100 μg of protein in 10 μL of the urea solution was diluted and digested with 2 μg trypsin (1:50 enzyme:substrate ratio) for 16 h overnight, then an additional 1 μg of trypsin was added and digestion continued for a further 4 h. An amount of 200 ng of each of the 9 samples was run by nanoLC-MS/MS using a 2 h gradient on a 0.075 mm × 250 mm C18 Waters CSH column feeding into a Q-Exactive HF mass spectrometer.

Raw datafiles were loaded directly in to Progenesis QI for proteomics version 2.0 and alignment of the chromatograms showed a ≥95% match. Features (peaks) were extracted across all runs and areas under the peaks calculated. Runs were normalized using the "normalize to all proteins" setting. Data were exported from Progenesis and analyzed using the search engine Mascot (Matrix Science, London, UK; version 2.6.1, 2016. Mascot was set up to search 2 databases: the common contaminants database cRAP_20150130 database (117 entries) and a combined database containing 2 uniprot reference proteomes for *S. aureus* NCTC8325 & C0673_20170607 (5705 entries) assuming the digestion enzyme trypsin. Mascot was searched with a fragment ion mass tolerance of 0.060 Da and a parent ion tolerance of 10.0 PPM. Oxidation of methionine; deamidated of asparagine and glutamine; and carbamidomethyl of cysteine were specified in Mascot as variable modifications.

4.8. Butanol Extraction of PSMs and Densitometry Analysis

PSM peptides were isolated as previously described [29]. An amount of 5 mL of synchronized, cell-free culture supernatants was incubated shaking at 37 °C with 3 mL of *n*-butanol for 2 h. Cultures were then centrifuged and 1 mL of the organic layer was removed for vacuum centrifugation for 12 h at 5000 rpm. Samples containing the PSMs were then resuspended in water and utilized for human erythrocyte hemolysis assays. The same set of samples was run on SDS-PAGE gels in duplicate and silver stained. Densitometry analysis of the duplicate extract samples was performed using Image J software. Samples were averaged and normalized to the percent density of WT.

4.9. Human-Erythrocyte Hemolysis Assay

Synchronized bacterial cultures were grown in quadruplicate as described in bacterial growth conditions for 15 h. Samples were centrifuged and cell-free culture supernatants were harvested and diluted 1:2 in reaction buffer containing 40 mM $CaCl_2$, and 1.7% NaCl. An amount of 25 μL of whole-human blood was added to samples and they were incubated at 37 °C while rotating. After 10 min, samples were centrifuged at 5500× *g* and the resulting supernatant was transferred to a 96-well plate. The degree of erythrocyte lysis was determined by reading sample absorbance at an OD_{543}.

4.10. Rabbit Erythrocyte Hemolysis Assay

Bacterial strains were grown in quadruplicate in a 25 mL flask of TSB as described above. Cultures were then centrifuged to obtain the cell-free supernatant. Supernatants were diluted 1:10 before being further diluted 1:2 in reaction buffer containing 40 mM $CaCl_2$, and 1.7% NaCl and incubated in a 37 °C water bath with 125 μL of rabbit blood. An amount of 200 μL of samples was removed every 5 min after incubation and centrifuged at 5500× *g*. An amount of 100 μl of supernatant was transferred to a 96-well plate and erythrocyte lysis was determined by reading the absorbance of the samples at OD_{543}.

4.11. Protein Immunoprecipitation Assay

Duplicate bacterial cultures were inoculated into 100 mL of TSB with appropriate antibiotics in 250 mL flasks for 3 h to mid-exponential phase. Samples were centrifuged at 3000 rpm for 20 min and pellets were resuspended in sterile PBS. In total, 37% formaldehyde was added to cultures to a final concentration of 1% and a 20 min shaking incubation was performed at room temperature. The reaction was quenched with the addition of glycine to a final concentration of 200 mM. Cells were centrifuged and the remaining pellets were washed with sterile PBS and then re-suspended in sterile water. Cells were then lysed with lysostaphin at a concentration of 10 mg/mL and incubated at 37 °C for 30 min. Following this, cells were treated with DNase I and incubated at 37 °C for 30 min. Resulting lysates were then sonicated at 20% amplitude, 2 × 15 s. Samples were then centrifuged at 13,000 rpm for 1 min and supernatants were used for immunoprecipitation.

For immunoprecipitation, anti-HA or anti-His magnetic beads were added to respective samples and incubated at 4 °C for 1 h. Beads were collected with a magnetic rack and washed 10 times with 1× PBS. Final washed beads were resuspended in 20 μL of water and sent to the University of Nebraska Lincoln Proteomics Facility for analysis.

The 2 magnetic bead samples were resuspended in ammonium bicarbonate containing 5 mM DTT and reduced at 37 °C for 1 h. Samples were then alkylated (10 mM IAM for 20 min at 22 °C in the dark). The IAM was quenched with a molar equivalent of DTT. Trypsin was added and digestion carried out overnight at 37 °C. The digest was dried down and dissolved in 2.5% acetonitrile, 0.1% formic acid. An amount of 5 uL of one digest was run by nanoLC-MS/MS using a 2 h gradient on a 0.075 mm × 250 mm C18 Waters CSH column feeding into a Q-Exactive HF mass spectrometer.

All MS/MS samples were analyzed using Mascot (Matrix Science, London, UK; version 2.6.1, 2016). Mascot was set up to search the cRAP_20150130.fasta (117 entries); uniprot_S_aureus_USA300_20170823 database (2607 entries) assuming the digestion enzyme trypsin. Mascot was searched with a fragment ion mass tolerance of 0.060 Da and a parent ion tolerance of 10.0 PPM. Deamidated of asparagine and glutamine, oxidation of methionine and carbamidomethyl of cysteine were specified in Mascot as variable modifications.

Scaffold (Proteome Software Inc., Portland, OR, USA, version Scaffold_4.8.4, 2017) was used to validate MS/MS-based peptide and protein identifications. Peptide identifications were accepted if they could be established at greater than 80.0% probability by the Peptide Prophet algorithm [51] with Scaffold delta-mass correction. Protein identifications were accepted if they could be established at greater than 99.0% probability and contained at least 2 identified peptides. Protein probabilities were assigned by the Protein Prophet algorithm [52]. Proteins that contained similar peptides and could not be differentiated based on MS/MS analysis alone were grouped to satisfy the principles of parsimony. Proteins sharing significant peptide evidence were grouped into clusters.

4.12. Reverse Transcriptase-Quantitative PCR (RT-qPCR)

RT-qPCR was performed as described previously [47]. Briefly, bacterial pellets were collected 6 h after subculture and total RNA was isolated. Complimentary DNA (cDNA) was synthesized from 1 μg of total RNA using iScript reverse transcriptase (Bio-Rad) according to the manufacturer's directions. The cDNA was diluted 10 times and used in SYBR Green reactions in technical duplicates to analyze the expression of *αPSM*, and *hla*. Transcription of the housekeeping gene *gyrB* was used as the endogenous control in each strain. Relative expression was determined by first comparing the amount of each individual gene transcript to *gyrB* within the same strain, followed by expression of these values in comparison to each respective gene in WT strain TCH1516.

4.13. Ethics Statement

Whole human-blood was isolated from donors in agreement with the Ohio University Institutional Review Board (Identification code; 17-X-79 date of approval: 22 February 2019). Rabbit blood was purchased from Hemostat-laboratories. Six-week-old female BALB/c mice were purchased from Envigo and held at the Ohio University Office of Laboratory Animal Resources. Animal work was performed under approval of the Institutional Animal Care and Use Committee (Identification code; 18-H-029 date of approval: 04 September 2018) at Ohio University and performed by trained lab personnel.

Supplementary Materials: The following are available online at http://www.mdpi.com/2072-6651/11/6/343/s1, Figure S1: The αPSMs are the primary toxins responsible for human erythrocyte lysis while Hla is the primary toxin active against rabbit erythrocytes. Figure S2: RT-qPCR analysis of *hla* and *α*PSM transcript levels in, W.T.; *ΔppiB*, and *ΔprsA* strains. Table S1: PpiB secreted proteome analysis. Table S2: PrsA secreted proteome analysis. Table S3: PrsA immunoprecipitation. Table S4: PpiB immunoprecipitation.

Author Contributions: Conceptualization, R.A.K.; Data curation, R.A.K.; Funding acquisition, R.K.C.; Investigation, R.A.K., R.L.Z., E.T., G.G.N. and R.E.W.; Supervision, R.K.C.; Writing–original draft, R.A.K.; Writing-review and editing, R.K.C.

Funding: This study was supported by a grant from the National Institute of Allergy and Infectious Diseases (grant AI128376).

Acknowledgments: Special thank you to Michael Naldrett and Sophie Alvarez at the University of Nebraska Lincoln proteomics and metabolomics facility for their assistance with the secreted proteomics and immunoprecipitation.

Conflicts of Interest: The authors declare no conflict of interest.

References

1. Lu, T.; DeLeo, F.R. Pathogenesis of staphylococcus aureus in Humans. In *Human Emerging and Re emerging Infections*; John Wiley & Sons, Inc.: Hoboken, NJ, USA, 2015; pp. 711–748.
2. Otto, M. Staphylococcus aureus toxins. *Curr. Opin. Microbiol.* **2014**. [CrossRef] [PubMed]
3. Jenkins, A.; Diep, B.A.; Mai, T.T.; Vo, N.H.; Warrener, P.; Suzich, J.; Stover, C.K.; Sellman, B.R. Differential expression and roles of staphylococcus aureus virulence determinants during colonization and disease. *MBio* **2015**, *6*, e02272-14. [CrossRef] [PubMed]
4. Mann, E.E.; Rice, K.C.; Boles, B.R.; Endres, J.L.; Ranjit, D.; Chandramohan, L.; Tsang, L.H.; Smeltzer, M.S.; Horswill, A.R.; Bayles, K.W. Modulation of eDNA release and degradation affects staphylococcus aureus biofilm maturation. *PLoS ONE* **2009**, *4*, e5822. [CrossRef] [PubMed]
5. Williams, R.J.; Henderson, B.; Nair, S.P. Staphylococcus aureus fibronectin binding proteins A and B possess a second fibronectin binding region that may have biological relevance to bone tissues. *Calcif. Tissue Int.* **2002**, *70*, 416–421. [CrossRef] [PubMed]
6. Bhakdil, S. Alpha-Toxin of Staphylococcus aureus. *Microbiol. Mol. Biol. Rev.* **1991**, *55*, 733–751.
7. Joo, H.S.; Chatterjee, S.S.; Villaruz, A.E.; Dickey, S.W.; Tan, V.Y.; Chen, Y.; Sturdevant, D.E.; Ricklefs, S.M.; Otto, M. Mechanism of gene regulation by a staphylococcus aureus toxin. *MBio* **2016**, *7*, e01579-16. [CrossRef] [PubMed]
8. Kobayashi, S.D.; Malachowa, N.; Whitney, A.R.; Braughton, K.R.; Gardner, D.J.; Long, D.; Wardenburg, J.B.; Schneewind, O.; Otto, M.; DeLeo, F.R. Comparative analysis of USA300 virulence determinants in a rabbit model of skin and soft tissue infection. *J. Infect. Dis.* **2011**, *204*, 937–941. [CrossRef]
9. Wilke, G.A.; Wardenburg, J.B. Role of a disintegrin and metalloprotease 10 in Staphylococcus aureus—Hemolysin-mediated cellular injury. *Proc. Natl. Acad. Sci. USA* **2010**, *107*, 13473–13478. [CrossRef]
10. Cheung, G.Y.C.; Joo, H.S.; Chatterjee, S.S.; Otto, M. Phenol-soluble modulins—Critical determinants of staphylococcal virulence. *FEMS Microbiol. Rev.* **2014**, *38*, 698–719. [CrossRef]

11. Queck, S.Y.; Jameson-Lee, M.; Villaruz, A.E.; Bach, T.-H.L.; Khan, B.A.; Sturdevant, D.E.; Ricklefs, S.M.; Li, M.; Otto, M. RNAIII-independent target gene control by the agr quorum-sensing system: Insight into the evolution of virulence regulation in staphylococcus aureus. *Mol. Cell* **2008**, *32*, 150–158. [CrossRef]

12. Berube, B.J.; Sampedro, G.R.; Otto, M.; Wardenburg, J.B. The psmα locus regulates production of staphylococcus aureus alpha-toxin during infection. *Infect. Immun.* **2014**, *82*, 3350–3358. [CrossRef] [PubMed]

13. Kofron, J.L.; Kuzmic, P.; Kishore, V.; Colon-Bonilla, E.; Rich, D.H. Determination of kinetic constants for peptidyl prolyl cis-trans isomerases by an improved spectrophotometric assay. *Biochemistry* **1991**, *30*, 6127–6134. [CrossRef] [PubMed]

14. Unal, C.M.; Steinert, M. Microbial peptidyl-prolyl cis/trans isomerases (PPIases): Virulence factors and potential alternative drug targets. *Microbiol. Mol. Biol. Rev.* **2014**, *78*, 544–571. [CrossRef] [PubMed]

15. Alonzo, F.; Freitag, N.E. Listeria monocytogenes PrsA2 is required for virulence factor secretion and bacterial viability within the host cell cytosol. *Infect. Immun.* **2010**, *78*, 4944–4957. [CrossRef] [PubMed]

16. Alonzo, F.; Port, G.C.; Cao, M.; Freitag, N.E. The posttranslocation chaperone PrsA2 contributes to multiple facets of Listeria monocytogenes pathogenesis. *Infect. Immun.* **2009**, *77*, 2612–2623. [CrossRef]

17. Reffuveille, F.; Connil, N.; Sanguinetti, M.; Posteraro, B.; Chevalier, S.; Auffray, Y.; Rince, A. Involvement of peptidylprolyl cis/trans isomerases in enterococcus faecalis virulence. *Infect. Immun.* **2012**, *80*, 1728–1735. [CrossRef] [PubMed]

18. Roset, M.S.; Fernández, L.G.; Delvecchio, V.G.; Briones, G. Intracellularly induced cyclophilins play an important role in stress adaptation and virulence of brucella abortus. *Infect. Immun.* **2013**, *81*, 521–530. [CrossRef]

19. Pandey, S.; Sharma, A.; Tripathi, D.; Kumar, A.; Khubaib, M.; Bhuwan, M.; Chaudhuri, T.K.; Hasnain, S.E.; Ehtesham, N.Z. Mycobacterium tuberculosis Peptidyl-Prolyl Isomerases Also Exhibit Chaperone like Activity In-Vitro and In-Vivo. *PLoS ONE* **2016**, *11*, e0150288. [CrossRef]

20. Skagia, A.; Zografou, C.; Vezyri, E.; Venieraki, A.; Katinakis, P.; Dimou, M. Cyclophilin PpiB is involved in motility and biofilm formation via its functional association with certain proteins. *Genes Cells* **2016**, *21*, 833–851. [CrossRef]

21. Keogh, R.A.; Zapf, R.L.; Wiemels, R.E.; Wittekind, M.A.; Carroll, R.K. The intracellular cyclophilin PpiB contributes to the virulence of staphylococcus aureus independently of Its peptidyl-prolyl cis/trans isomerase activity. *Infect. Immun.* **2018**, *86*, 1–13. [CrossRef]

22. Lin, M.-H.; Li, C.-C.; Shu, J.-C.; Chu, H.-W.; Liu, C.-C.; Wu, C.-C. Exoproteome profiling reveals the involvement of the foldase PrsA in the cell surface properties and pathogenesis of staphylococcus aureus. *Proteomics* **2018**, *18*, 1700195. [CrossRef] [PubMed]

23. Jousselin, A.; Manzano, C.; Biette, A.; Reed, P.; Pinho, M.G.; Rosato, A.E.; Kelley, W.L.; Renzoni, A. The Staphylococcus aureus Chaperone PrsA Is a New Auxiliary Factor of Oxacillin Resistance Affecting Penicillin-Binding Protein 2A. *Antimicrob. Agents Chemother.* **2016**, *60*, 1656–1666. [CrossRef] [PubMed]

24. Wiemels, R.E.; Cech, S.M.; Meyer, N.M.; Burke, C.A.; Weiss, A.; Parks, A.R.; Shaw, L.N.; Carroll, R.K. An intracellular peptidyl-prolyl cis/trans isomerase is required for folding and activity of the Staphylococcus aureus secreted virulence factor nuclease. *J. Bacteriol.* **2017**, *199*, e00453-16. [CrossRef] [PubMed]

25. Malachowa, N.; Kobayashi, S.D.; Braughton, K.R.; DeLeo, F.R. *Mouse Model of Staphylococcus Aureus Skin Infection*; Humana Press: Totowa, NJ, USA, 2013; pp. 109–116.

26. Alonzo, F., III; Benson, M.A.; Chen, J.; Novick, R.P.; Shopsin, B.; Torres, V.J. Staphylococcus aureus leucocidin ED contributes to systemic infection by targeting neutrophils and promoting bacterial growth in vivo. *Mol. Microbiol.* **2012**, *83*, 423–435. [CrossRef] [PubMed]

27. Wang, X.; Thompson, C.D.; Weidenmaier, C.; Lee, J.C. Release of staphylococcus aureus extracellular vesicles and their application as a vaccine platform. *Nat. Commun.* **2018**, *9*, 1379. [CrossRef] [PubMed]

28. Berube, B.J.; Wardenburg, J.B. Staphylococcus aureus α-toxin: Nearly a century of intrigue. *Toxins* **2013**, *5*, 1140–1166. [CrossRef] [PubMed]

29. Otto, H.-S.J. *The isolation and Analysis of Phenol-Soluble Modulins of Staphylococcus Epidermidis*; Humana Press: Totowa, NJ, USA, 2014; pp. 93–100.

30. Schallenberger, M.A.; Niessen, S.; Shao, C.; Fowler, B.J.; Romesberg, F.E. Type I signal peptidase and protein secretion in staphylococcus aureus. *J. Bacteriol.* **2012**, *194*, 2677–2686. [CrossRef] [PubMed]

31. Antelmann, H.; Darmon, E.; Noone, D.; Veening, J.W.; Westers, H.; Bron, S.; Kuipers, O.P.; Devine, K.M.; Hecker, M.; Van Dijl, J.M. The extracellular proteome of Bacillus subtilis under secretion stress conditions. *Mol. Microbiol.* **2003**, *49*, 143–156. [CrossRef] [PubMed]

32. Rigoulay, C.; Entenza, J.M.; Halpern, D.; Widmer, E.; Moreillon, P.; Poquet, I.; Gruss, A. Comparative analysis of the roles of HtrA-like surface proteases in two virulent staphylococcus aureus strains. *Infect. Immun.* **2005**, *73*, 563–572. [CrossRef]

33. McCallum, N.; Spehar, G.; Bischoff, M.; Berger-Bächi, B. Strain dependence of the cell wall-damage induced stimulon in staphylococcus aureus. *Biochim. Biophys. Acta (BBA)-Gen. Subj.* **2006**, *1760*, 1475–1481. [CrossRef]

34. Schneewind, O.; Missiakas, D. Sec-secretion and sortase-mediated anchoring of proteins in Gram-positive bacteria. *Biochim. Biophys. Acta (BBA)-Gen. Subj.* **2014**, *1843*, 1687–1697. [CrossRef] [PubMed]

35. Pugsley, A.P. The complete general secretory pathway in gram-negative bacteria. *Microbiol. Rev.* **1993**, *57*, 50–108. [PubMed]

36. Jousselin, A.; Renzoni, A.; Andrey, D.O.; Monod, A.; Lew, D.P.; Kelley, W.L. The posttranslocational chaperone lipoprotein PrsA Is Involved in both glycopeptide and oxacillin resistance in staphylococcus aureus. *Antimicrob. Agents Chemother.* **2012**, *56*, 3629–3640. [CrossRef] [PubMed]

37. Skagia, A.; Vezyri, E.; Sigala, M.; Kokkinou, A.; Karpusas, M.; Venieraki, A.; Katinakis, P.; Dimou, M. Structural and functional analysis of cyclophilin PpiB mutants supports an in vivo function not limited to prolyl isomerization activity. *Genes Cells* **2017**, *22*, 32–44. [CrossRef] [PubMed]

38. Göthel, S.F.; Marahiel, M.A. Peptidyl-prolyl cis-trans isomerases, a superfamily of ubiquitous folding catalysts. *Cells Mol. Life Sci.* **1999**, *55*, 423–436. [CrossRef] [PubMed]

39. Herbert, S.; Ziebandt, A.K.; Ohlsen, K.; Schäfer, T.; Hecker, M.; Albrecht, D.; Novick, R.; Götz, F. Repair of global regulators in staphylococcus aureus 8325 and comparative analysis with other clinical isolates. *Infect. Immun.* **2010**, *78*, 2877–2889. [CrossRef] [PubMed]

40. Buchner, J. Supervising the fold: Functional principles of molecular chaperones. *FASEB J.* **1996**, *10*, 10–19. [CrossRef] [PubMed]

41. Bose, J.L. Genetic manipulation of staphylococci. In *Methods in Molecular Biology*; Humana Press: Clifton, NJ, USA, 2014; pp. 101–111.

42. Bose, J.L.; Fey, P.D.; Bayles, K.W. Genetic tools to enhance the study of gene function and regulation in staphylococcus aureus. *Appl. Environ. Microbiol.* **2013**, *79*, 2218–2224. [CrossRef]

43. Kreiswirth, B.N.; Löfdahl, S.; Betley, M.J.; O'Reilly, M.; Schlievert, P.M.; Bergdoll, M.S.; Novick, R.P. The toxic shock syndrome exotoxin structural gene is not detectably transmitted by a prophage. *Nature* **1983**, *305*, 709. [CrossRef]

44. Highlander, S.K.; Hultén, K.G.; Qin, X.; Jiang, H.; Yerrapragada, S.; Mason, E.O.; Shang, Y.; Williams, T.M.; Fortunov, R.M.; Liu, Y.; et al. Subtle genetic changes enhance virulence of methicillin resistant and sensitive Staphylococcus aureus. *BMC Microbiol.* **2007**, *7*, 1–14. [CrossRef]

45. Fey, P.D.; Endres, J.L.; Yajjala, V.K.; Widhelm, T.J.; Boissy, R.J.; Bose, J.L.; Bayles, K.W. A genetic resource for rapid and comprehensive phenotype screening of nonessential staphylococcus aureus genes. *MBio* **2013**, *4*, 1–8. [CrossRef] [PubMed]

46. Boles, B.R.; Thoende, M.; Roth, A.J.; Horswill, A.R. Identification of genes involved in polysaccharide-independent staphylococcus aureus biofilm formation. *PLoS ONE* **2010**, *5*, e10146. [CrossRef] [PubMed]

47. Zapf, R.L.; Wiemels, R.E.; Keogh, R.A.; Holzschu, D.L.; Howell, K.M.; Trzeciak, E.; Caillet, A.R.; King, K.A.; Selhorst, S.A.; Naldrett, M.J.; et al. The small RNA Teg41 regulates expression of the alpha phenol-soluble modulins and Is required for Virulence in staphylococcus aureus. *MBio* **2019**, *10*, e02484-18. [CrossRef] [PubMed]

48. Sullivan, M.A.; Yasbin, R.E.; Young, F.E. New shuttle vectors for bacillus subtilis and escherichia coli which allow rapid detection of inserted fragments. *Gene* **1984**, *29*, 21–26. [CrossRef]

49. Spaan, A.N.; Reyes-Robles, T.; Badiou, C.; Cochet, S.; Boguslawski, K.M.; Yoong, P.; Day, C.J.; De Haas, C.J.C.; Van Kessel, K.P.M.; Vandenesch, F.; et al. Staphylococcus aureus targets the duffy antigen receptor for chemokines (DARC) to lyse erythrocytes. *Cell Host Microbe* **2015**, *18*, 363–370. [CrossRef] [PubMed]

50. Carroll, R.K.; Rivera, F.E.; Cavaco, C.K.; Johnson, G.M.; Martin, D.; Shaw, L.N. The lone S41 family C-terminal processing protease in Staphylococcus aureus is localized to the cell wall and contributes to virulence. *Microbiology* **2014**, *160*, 1737–1748. [CrossRef] [PubMed]

51. Pandey, A.; Mann, M.; Aebersold, R.; Goodlett, D.R.; Link, A.J.; Eng, J.; Schieltz, D.M.; Carmack, E. Empirical statistical model to estimate the accuracy of peptide identifications made by MS/MS and database search. *Anal. Chem.* **2002**, *74*, 5383–5392.
52. Nesvizhskii, A.I.; Keller, A.; Kolker, E.; Aebersold, R. A statistical model for identifying proteins by tandem mass spectrometry. *Anal. Chem.* **2003**, *75*, 4646–4658. [CrossRef]

 *toxins*

Article

Rational Design of Toxoid Vaccine Candidates for *Staphylococcus aureus* Leukocidin AB (LukAB)

Shweta Kailasan [†], Thomas Kort [†], Ipsita Mukherjee, Grant C. Liao, Tulasikumari Kanipakala, Nils Williston, Nader Ganjbaksh, Arundhathi Venkatasubramaniam, Frederick W. Holtsberg, Hatice Karauzum, Rajan P. Adhikari * and M. Javad Aman *

Integrated Biotherapeutics Inc., Rockville, MD 20850, USA; skailasan@IntegratedBiotherapeutics.com (S.K.);
Tom@IntegratedBiotherapeutics.com (T.K.); imukher@integratedbio.onmicrosoft.com (I.M.);
gliao@IntegratedBiotherapeutics.com (G.C.L.); Tula@IntegratedBiotherapeutics.com (T.K.);
nils.williston@gmail.com (N.W.); nader@IntegratedBiotherapeutics.com (N.G.);
arundathi@IntegratedBiotherapeutics.com (A.V.); rick@IntegratedBiotherapeutics.com (F.W.H.);
hkarauzum@IntegratedBiotherapeutics.com (H.K.)
* Correspondence: rajan@integratedbiotherapeutics.com (R.P.A.); javad@integratedbiotherapeutics.com (M.J.A.)
† Equally contributed.

Received: 1 May 2019; Accepted: 12 June 2019; Published: 14 June 2019

Abstract: *Staphylococcus aureus* (SA) infections cause high mortality and morbidity in humans. Being central to its pathogenesis, *S. aureus* thwarts the host defense by secreting a myriad of virulence factors, including bicomponent, pore-forming leukotoxins. While all vaccine development efforts that aimed at achieving opsonophagocytic killing have failed, targeting virulence by toxoid vaccines represents a novel approach to preventing mortality and morbidity that are caused by SA. The recently discovered leukotoxin LukAB kills human phagocytes and monocytes and it is present in all known *S. aureus* clinical isolates. While using a structure-guided approach, we generated a library of mutations that targeted functional domains within the LukAB heterodimer to identify attenuated toxoids as potential vaccine candidates. The mutants were evaluated based on expression, solubility, yield, biophysical properties, cytotoxicity, and immunogenicity, and several fully attenuated LukAB toxoids that were capable of eliciting high neutralizing antibody titers were identified. Rabbit polyclonal antibodies against the lead toxoid candidate provided potent neutralization of LukAB. While the neutralization of LukAB alone was not sufficient to fully suppress leukotoxicity in supernatants of *S. aureus* USA300 isolates, a combination of antibodies against LukAB, α-toxin, and Panton-Valentine leukocidin completely neutralized the cytotoxicity of these strains. These data strongly support the inclusion of LukAB toxoids in a multivalent toxoid vaccine for the prevention of *S. aureus* disease.

Keywords: Leukocidin; *Staphylococcus aureus*; LukAB; LukGH; toxin neutralization; polyclonal antibody; toxoid vaccine

Key Contribution: The present study described the first fully characterized vaccine candidate for LukAB and provides evidence in support of its inclusion in a multivalent toxoid vaccine for *S. aureus* disease.

---

1. Introduction

Staphylococcus aureus, which is a pervasive human pathogen, is a leading cause of life-threatening community and hospital-acquired infections world-wide. This Gram-positive bacterium is often associated with a range of diseases, from mild skin and soft tissue infections to invasive bacteremia, septic arthritis, endocarditis, and osteomyelitis [1]. Moreover, recent widespread emergence of multi-drug resistant strains, specifically methicillin and vancomycin-resistance, have not only complicated the use of available treatment options, but have also considerably raised the economic burden that is associated with staphylococcal infections [2–4]. Methicillin-resistant *S. aureus* (MRSA) causes ~80,000 invasive infections and 11,000 deaths per year in the United States alone [3].

S. aureus expresses a myriad of virulence factors, including cell surface attachment factors, capsular polysaccharides, enzymes, immune modulatory molecules, pore-forming toxins, and superantigens that aimed at establishing the infection or colonization as well as immune evasion [5]. Among the pore-forming toxins, *S. aureus* produces single component alpha hemolysin (Hla or α-toxin) and bicomponent pore-forming toxins (BCPFTs), Panton-Valentine leukocidin (PVL; composed of LukS-PV and LukF-PV), Leukocidin AB (LukAB), Leukocidin ED (LukED), and γ-hemolysins (HlgAB and HlgCB) [6]. While Hla is secreted as a monomer and oligomerizes on the plasma membrane of target cells upon interaction with its specific cellular receptor ADAM10 [7], the BCPFTs are produced from two distinct polypeptides, S (~32.4 kDa) and F (~34.6 kDa), which have β-barrel structures and hetero-oligomerize in a stepwise fashion with alternating S (LukS-PV, LukE, HlgA, HlgC, and LukA) and F (LukF-PV, LukD, HlgB, and LukB) subunits on the cell surface [8]. Following oligomerization, structural rearrangements within the C-terminal stem domain promote membrane insertion, resulting in ion efflux, disruption of the host cell lipid bilayer, and ultimately cell death.

Among the BCPFTs, LukAB is the most recently identified and it is one of the most potent members of the leukocidin family that kills human neutrophils, macrophages, monocytes, and dendritic cells [9,10]. While the majority of the different BCPFTs exhibit a high sequence identity of 70–80% among the S and F components, LukA and LukB share a low sequence identity of 30–40% with the other leukotoxins [6]. LukAB is also unique, in that it is secreted as a dimer in solution and it requires both S and F components for cell surface engagement in contrast to other BCPFTs [11]. The crystal structure of LukAB from USA300, which is a predominant methicillin-resistant *S. aureus* (MRSA) strain circulating in the United States (US), revealed an octameric arrangement of four LukAB dimers with two unique interfaces 1 (intra-protomeric) and 2 (inter-protomeric), which primarily govern the cap and rim domains [11]. Highly conserved residues among known LukAB variants that form salt bridges, a characteristic that is not found in other BCPFTs, hold these interfaces together. Of the two interfaces, residues within interface 1 were reported to be important for dimer formation, whilst interface 2 is important for octamerization. A surface-exposed residue E323 within the rim domain of LukA is critical to LukAB-mediated cytotoxicity by directly interacting with the integrin αM/β2 receptor (CD11b/CD18) on the surface of human neutrophils and monocytes (THP-1) [11–14]. In addition to the receptor binding properties, the rim domain harbors a high-affinity antibody epitope, which is also conserved among different LukAB sequence variants, and is suggestive of its role as an antigenic determinant as well as a site important for neutralization [15]. LukAB production not only allows the bacteria to escape from phagocytic killing by human neutrophils, but also lyse monocytes in a CD11b-targeted manner [9,12,13,16]. Furthermore, LukAB, along with α-toxin, were identified as key players in host macrophage dysfunction enabling USA 300 biofilm formation, highlighting yet another role of LukAB in circumventing the immune-mediated clearance in the host [17].

Highly virulent MRSA strains, like USA300, contribute to the clinical severity of SA infections by secreting a higher load of toxins to evade the innate immune system. Amongst other leukocidins, high titers to LukAB have been reported in acute and convalescent patients when compared to healthy controls [18]. Additionally, a study reported high levels of functional antibodies against LukAB in patients with invasive *S. aureus* infections as compared to healthy individuals or commercially available intravenous immunoglobulin (IVIG) [19]. Altogether, these studies suggest LukAB as a key protagonist in SA-related disease pathogenesis, bacterial survival, and persistence, and therefore a potential vaccine target.

Here, iterative rounds of targeted single-amino acid and combination mutations within different functional domains of LukAB were designed in a structure-guided manner to identify the attenuated forms of LukAB. To this end, we first developed a novel production strategy that was based on co-expression from a single vector and a multistep purification process to generate tag-free, soluble dimers, and then functionally characterized them based on toxicity, thermostability, immunogenicity, and the ability to compete with cytotoxicity of wild type (WT) LukAB to identify several attenuated toxoids. Furthermore, we generated a polyclonal antibody against a selected attenuated LukAB mutant that demonstrated neutralizing activity towards supernatants from several laboratory and clinical strains, including USA300, which is responsible for the current CA-MRSA outbreak in the United States. Our findings further indicate that a multivalent toxoid vaccine targeting all the pore-forming toxins is needed to fully neutralize the cytotoxicity of different clinical strains of *S. aureus* toward neutrophils and monocytes.

2. Results

2.1. Expression of LukAB using pET Duet versus pET24a(+) Dual Plasmid System

Unlike other staphylococcal leukotoxins, the LukAB proteins, when expressed as separate subunits, do not fold properly, and expression is directed to inclusion bodies [12]. In the past, we also noted that, when these proteins are refolded together from solubilized inclusion bodies, the resulting yield and toxicity of the product is lower than that of the co-expressed [12] and co-purified LukAB proteins [20]. Therefore, we compared purified LukAB proteins from two expression vectors, pET-Duet expressing the subunit from the same plasmid, and pET24a (+) dual plasmid expression systems and observed reasonably good and equivalent yields of LukAB$_{WT}$ proteins (Table S1). Both of the systems have their merits. With pET-Duet, as both subunits are expressed from a single plasmid while using the same antibiotic marker, and therefore easier to scale up from a manufacturing prospective. On the other hand, with pET24a (+) dual system, as LukA and LukB are co-expressed within the same cell by two different plasmids, it is a better system for introduction of mutations while screening a mutation library. Additionally, we found that the toxicity and physicochemical properties of LukAB$_{WT}$ from the two systems are comparable (Table S1). Hence, we used a pET24a (+) dual system during mutant screening and pET-Duet for production.

2.2. Rational Design of Attenuated LukAB Toxoids

The objective of this study was to generate attenuated forms of LukAB as the vaccine candidates. To this end, we used the available crystal structure of LukAB heterodimer (PDB: 5K59) and octamer (PDB: 4TW1, Figure 1) as the structural framework to design mutations that could potentially reduce the cytotoxicity, while maintaining structural integrity and other indispensable functional characteristics, such as dimer formation, receptor binding, and immunogenicity. The mutations were predominantly made within four domains of the LukAB dimer: interface 1 (protomer-protomer interface), interface 2 (dimer-dimer interface), membrane binding cleft that interacts with the lipid bilayer, and the stem domain, which forms a β-barrel pore inserted into the lipid bilayer (Figure 1). All of the 39 mutants designed, including combinations, target residues that are highly conserved among more than 40 LukAB$_{WT}$ sequences from different *S. aureus* strains for which annotated genomic data are available (Figure S1). Additionally, the majority of these mutants were expressed as soluble proteins and they could be purified at reasonable yield, unless otherwise indicated (Table 1). All of the mutants were analyzed for (1) purity measured by SDS-PAGE (data not shown) and yield (Table 1), (2) toxicity levels tested using a cytotoxicity assay in induced HL-60 cells differentiated into polymononuclear neutrophils (PMNs), as described in the Methods (Figure 2A–C and Table 1), (3) thermostability measured by differential scanning fluorimetry (DSF) (Figure 2D), and (4) immunogenicity levels in mice (Figure 3 and Table 1). The non-toxic LukAB vaccine candidates formulated in Alhydrogel®(Al(OH)$_3$)) as an adjuvant were examined for the induction of neutralizing antibodies. Groups of five female ICR (CD-1) mice were intramuscularly immunized (IM) with 10μg of LukAB mutants or LukAB$_{WT}$ in 50μg of Al(OH)$_3$ three times at two-week intervals and sera was collected 10 days after final immunization. Antibody-binding and -neutralizing titers were measured either for individual mice or from pooled sera for each mutant group by ELISA and toxin neutralization assays (TNA), respectively, as described in the Methods.

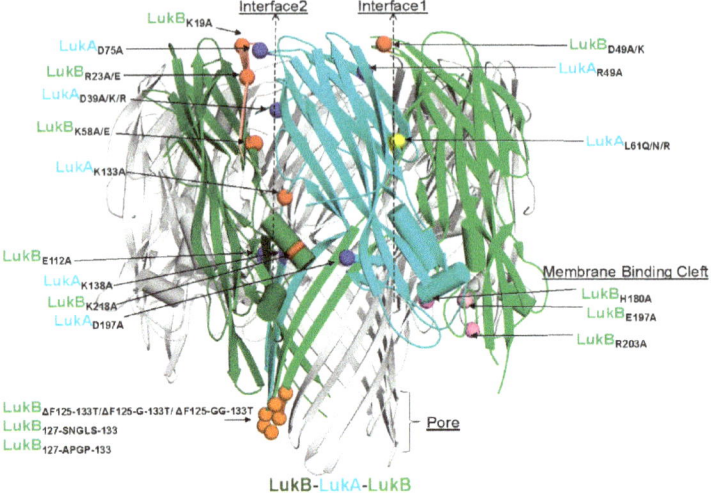

Figure 1. The repertoire of selected Leukocidin AB (LukAB) residues for mutagenesis. Cartoon representation of the crystal structure of LukAB (PDB: 4TWI) with polypeptides LukA and LukB shown in blue and green, respectively, along with the neighboring LukB polypeptide within the octameric ring shown in dark green. The α-carbon atoms of residues selected for mutation are labeled and depicted as spheres. Of the selected mutations, basic residues are shown in red, acidic in blue, hydrophobic in yellow, surface-loop residues near the pore in orange and those part of the membrane binding cleft in pink. As the complete LukB N terminal sequence was not entirely resolved in this crystal structure, mutated residue LukB$_{K12}$ is not shown in this depiction.

Table 1. Characteristics of the LukAB mutants, including cytotoxicity, production yield, and immunogenicity.

LukA	LukB	Location	Tox. EC$_{50}$ (nM)	Relative Attenuation (Mutant/WT)	Expression (mg/L)	ELISA EC$_{50}$	TNA NT$_{50}$	EC$_{50}$ Mut/WT	NT$_{50}$ Mut/WT
WT	WT	WT	0.03	1.00	20.574	17800	2260	1	1
L61R	WT	Interface1	>1000	>36,752	4.6	9000	1290	0.51	0.57
L61N	WT	Interface1	0.72	26.36	8.82	ND	ND		
L61Q	WT	Interface1	0.11	4.06	8.98	ND	ND		
WT	D49A	Interface1	0.11	4.04	3.765	ND	ND		
WT	D49K	Interface1	0.01	0.35	9.6	ND	ND		
WT	R23A	Interface2	>1000	>36,752	8.616	1320	597.5	0.07	0.26
WT	R23E	Interface2	>1000	>36,752	3.384	16,500	1540	0.93	0.68
WT	K58E	Interface2	>1000	>36,752	7.056	ND	ND		
WT	K58A	Interface2	>1000	>36,752	4.384	14,700	5900	0.83	2.61
WT	E112A	Interface2	0.12	4.35	40.72	5070	3010	0.28	1.33
D39A	WT	Interface2	0.04	1.36	9.648	16,500	3170	0.93	1.40
D39A	R23E	Interface2	>1000	>36,752	23.24	14,000	2640	0.79	1.17
D39R	R23E	Interface2	>1000	>36,752	21.64	682	334	0.04	0.15
D39R	K218A	Interface2	0.95	35.01	20.84	ND	ND		
D39A	E112A	Interface2	0.06	2.04	29.6	4770	4060	0.27	1.80
K133A	K218A	Interface2	0.05	1.89	42.56	2320	1570	0.13	0.69
D39R	E112A	Interface2	0.00	0.13	55.2	1800	765	0.10	0.34
L61R	R23E	Interface1&2	>1000	>36,752	31.6	1560	219	0.09	0.10
L61R	R23A	Interface1&2	>1000	>36,752	30.2	863	164.5	0.05	0.07
L61R	K218A	Interface1&2	>1000	>36,752	21.28	2640	162	0.15	0.07
D197K	R23A	Interface1&2	>1000	>36,752	34.04	1610	764	0.09	0.34
D39A	K218A	Interface1&2	0.00	0.02	50.4	2740	748	0.15	0.33
L61R	E112A	Interface1&2	99.00	3638	39.8	12,700	2100	0.71	0.93

Table 1. *Cont.*

LukA	LukB	Location	Tox. EC$_{50}$ (nM)	Relative Attenuation (Mutant/WT)	Expression (mg/L)	ELISA EC$_{50}$	TNA NT$_{50}$	EC$_{50}$ Mut/WT	NT$_{50}$ Mut/WT
WT	HlgB	Interface 2 (N Term)	>1000	>36,752	28	15,600	2110	0.88	0.93
WT	K12/K19/R23A	Interface 2 (N Term)	15.61	573.7	17.72	3120	1050	0.18	0.46
L61R	K12/K19/R23A	Interface1 & 2(N Term)	>1000	>36,752	35.8	2310	613	0.13	0.27
L61R	HlgB	Interface1 & 2(N Term)	>1000	>36,752	14.33	22,900	611	1.29	0.27
D39A	K12/K19/R23A	Interface2+(N Term)	>1000	>36,752	23.52	965	1030	0.05	0.46
D39R	K12/K19/R23A	Interface2+(N Term)	>1000	>36,752	26.8	942	770	0.05	0.34
D197A	WT	MBC	0.16	5.94	2.96	ND	ND		
WT	E197A	MBC	0.12	4.51	ND	ND	ND		
WT	H180A	MBC	0.00	0.11	4.18	ND	ND		
WT	R203A	MBC	0.00	0.10	27.16	4840	1100	0.27	0.49
D197K	WT	MBC	0.00	0.06	14.74	ND	ND		
L61R	R203A	Interface1& MBC	>1000	>36,752	27.4	ND	ND		
D39R	R23E/R203A	Interface1, 2 & MBC	>1000	>36,752	15.88	ND	ND		
WT	125-133_1G	Pore	33.16	1218.7	did not elute	47,100	2570	2.65	1.14
WT	127-133_APGP	Pore	0.26	9.38	19.4	ND	ND		
WT	127-33_SNGLS	Pore	0.00	0.01	20.52	5000	3380	0.28	1.50

* MBC: Membrane Binding Cleft; ND: Not determined.

2.2.1. Mutations within Interface 1 of LukAB

We targeted L61 on Interface 1, a key residue that aligns with H35 of alpha hemolysin (α-toxin; Hla) and T28 of Panton-Valentine leukocidin (PVL) S subunit (LukS-PV) [21]. N terminal residues H35 and H48 of Hla are imperative to protomer-protomer interactions and their substitution results in the loss of hemolytic activity due to the formation of incomplete, larger heptameric rings, as observed by electron microscopy [22]. The substitution of T28 of LukS-PV with a bulky residue, such as phenylalanine, is also known to affect toxicity due to steric or electrostatic repulsions that might interfere with oligomerization [21,23]. We designed three substitutions—L61N, L61Q, and L61R—to incrementally disrupt the hydrophobic pocket found within LukAB interface 1 (Figure 1). The second mutagenesis site targeted within interface 1 was the salt bridge comprised of LukA$_{R49}$ and LukB$_{D49}$ (Figure 1). We disrupted the electrostatic interactions by alanine and lysine substitutions at LukB$_{D49}$. Of these mutants, LukA$_{L61Q/N}$LukB$_{wt}$, and LukA$_{wt}$LukB$_{D49A/K}$, exhibited similar or enhanced cytotoxicity, respectively, in PMNs with EC$_{50}$s in the range of 0.01–0.7 nM as compared to WT (EC$_{50}$ = 0.027 nM) (Table 1, Figure 2A,B). In contrast, the introduction of a bulky, positively charged residue, like arginine (LukA$_{L61R}$LukB$_{wt}$), reduced the toxicity by >30,000-fold as compared to WT (Table 1, Figure 2A). Additionally, thermostability measurements by DSF indicated that the mutation of L61 of LukA increases protein stability, because mutants LukA$_{L61R/N/Q}$LukB$_{wt}$ recorded melting temperatures (Tm) that were in the range of 51.3–52.2 °C, representing a positive thermal shift of 7.3–8 °C as compared to LukAB$_{WT}$ (Tm = 44 °C) (Figure 2D). However, when tested for immunogenicity in mice, the highly attenuated mutant LukA$_{L61R}$LukB$_{wt}$ resulted in ~50% reduction of total IgG and TNA titers as compared to LukAB$_{WT}$ (Table 1). Therefore, this mutant was not further considered as a vaccine candidate.

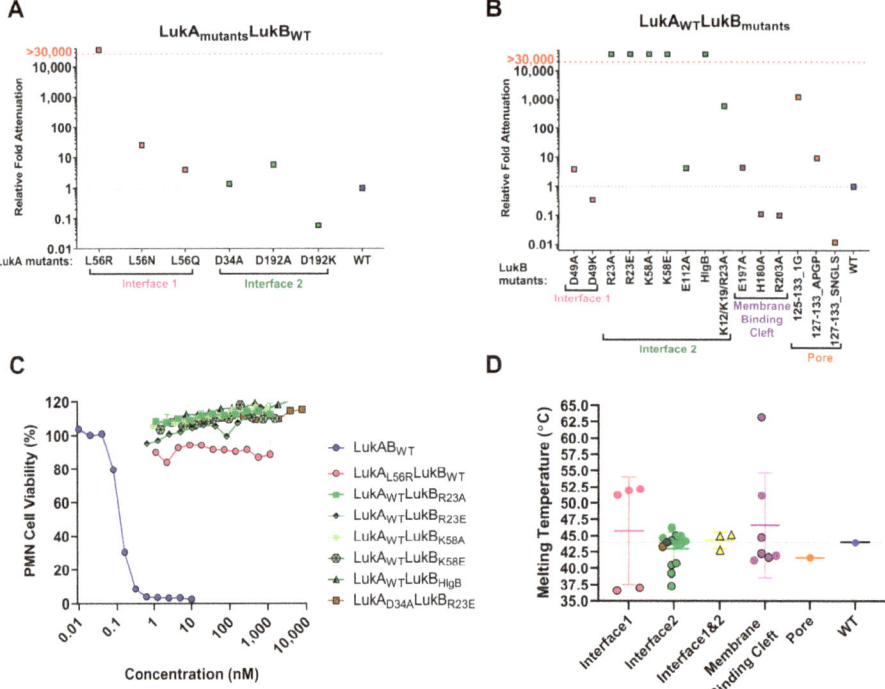

Figure 2. Characterization of LukAB mutants. Polymononuclear neutrophils (PMNs) were treated with increasing concentrations of LukAB$_{WT}$ or mutants within the concentration range of 1000 to 0.001 nM to calculate toxicity EC$_{50}$s. Results represent the fold attenuation (mutant EC$_{50}$/WT EC$_{50}$) and shown in (**A**) for mutants of LukA combined with wild type (WT) LukB and in (**B**) for mutants of LukB combined with WT LukA mean. Calculated EC50 values are listed in Table 1 and were measured in 3 independent experiments. (**C**) Toxicity profiles of LukAB$_{WT}$ or the most highly attenuated mutant proteins at the indicated concentration range were tested in PMNs as % cell viability. (**D**) Thermostability measurements using differential scanning fluorimetry (DSF) to calculate melting temperatures (T$_m$s) are plotted for the different, individual LukAB mutants within the different domains. Average values and spread of recorded Tms indicated for each domain. For Interface 1, outlined circles indicate non-LukA$_{L61}$ mutants. For interface 2, mutants LukA$_{D39A}$LukB$_{R23E}$ and LukA$_{wt}$LukB$_{HlgB}$ have been shown as outlined circles in brown and dark green, respectively.

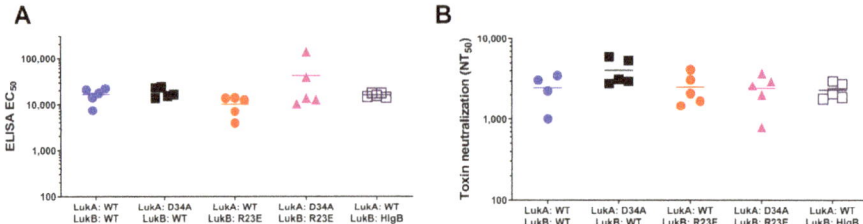

Figure 3. Immunogenicity titers for LukAB mutants. (**A**) Enzyme-Linked Immunosorbent Assay (ELISA) and (**B**) toxin neutralization assays (TNA) titers calculated for select mutants with titers for each individual mouse per group and average titers represented by the appropriate symbol and horizontal line, respectively. The Y axis is shown in log scale.

2.2.2. Pore and Membrane Binding Domain LukAB Mutants

It has been reported that the membrane binding cleft in the rim domain of both Hla and LukF-PV binds to polar head groups on the lipid bilayer [23]. Hence, we targeted three LukB residues within the membrane binding cleft, namely, H180A, E197A, and R203A, with the anticipation of reducing toxicity as membrane lipid binding is a preliminary step in pore formation (Figure 1). When combined with $LukA_{WT}$, the $LukB_{E197A}$ mutation modestly attenuated toxicity (~4.5-fold), while the other two mutations failed to attenuate cytolytic activity in PMNs with EC_{50}s in the range of $LukAB_{wt}$ (Figure 2B; Table 1). The mutations were also designed within a surface loop on LukB between residues 125-133 (FSINRGGLT) in the β-barrel pore in order to obturate the cytoplasmic edge of the pore (Figure 1, Figure S1). To this end, we made deletion mutants $LukB_{Δ125-133}$, $LukB_{Δ125-G-133}$, and $LukB_{Δ125-GG-133}$, supplemented with or without additional glycine(s) to introduce flexibility between residues D124 and G134 and to maintain structure stability. Of these mutants, only $LukA_{wt}LukB_{Δ125-G-133}$ could be expressed, although poorly, and demonstrated moderate 1218-fold attenuation with TNA titers similar to wild type (Figure 2B, Table 1). This mutant was discarded due to expression and purification difficulties. We also substituted residues 125–133 with a short Type-1 β-turn sequence (APGP) as $LukA_{wt}LukB_{127-APGP-133}$ or with the analogous loop sequence of HlgB (SNGLS) as $LukA_{wt}LukB_{127-SNGLS-133}$. While $LukA_{wt}LukB_{127-APGP-133}$ diminished the cytotoxicity by ~10-fold, the EC_{50} for $LukA_{wt}LukB_{127-SNGLS-133}$ toxicity was enhanced by ~100 fold (Figure 2B). Overall, the mutagenesis of the targeted residues within the membrane binding domain and pore region failed to provide viable vaccine candidates for either a lack of proper attenuation or poor yield.

2.2.3. Mutations within Interface 2 of LukAB

We made several mutations on LukB to target the LukAB dimer-dimer interface and to potentially disrupt its ability to oligomerize on the cell surface (Figure 1). First, we targeted three salt bridges that were observed in the crystal structure on interface 2, namely, $LukA_{D75}$-$LukB_{R23}$ and $LukA_{D39}$-$LukB_{K58}$, which lie within the cap domain and $LukA_{K133}$-$LukB_{E112}$ that lies closer to the rim domain (Figure 1). Mutations that were made to disrupt the latter salt bridge, $LukA_{wt}LukB_{E112A}$, failed to significantly attenuate toxicity (Table 1, Figure 2B), and we therefore focused our efforts on those on the apical side (cap). The disruption of electrostatic interactions at position R23 in the form of mutant $LukA_{wt}LukB_{R23A}$ or the reversal of positive charge with mutant $LukA_{wt}B_{R23E}$ resulted in an EC_{50} greater than 1000 nM representing >30,000-fold attenuation of cytotoxicity in PMNs (Table 1; Figure 2B). Visual inspection of the crystal structure (PDB 5K59) highlighted two additional positively charged residues in the vicinity of $LukB_{R23}$, K12, and K19, which may also contribute to a positively charged surface juxtaposing $LukA_{D75}$. K12 and K19 line the inner face of the β-barrel core. Therefore, we generated a triple mutation at the N terminus of LukB, $LukA_{wt}LukB_{K12A/K19A/R23A}$ to disrupt this basic amino acid patch. However, it resulted in an EC_{50} of 15.6 nM with a moderate attenuation of 573-fold as compared to $LukAB_{wt}$ (Table 1, Figure 2B). Altogether, these data indicated that $LukB_{R23}$ at the N terminus plays the most critical role in maintaining the ionic interactions at the cap domain of LukB and it is central to its cytotoxicity, but a more disruptive mutation, like R23E, is required for optimal attenuation (Table 1). However, a larger area of disrupted positive charges does not augment the level of cytotoxic attenuation. Moreover, relative antibody and neutralizing titers for $LukA_{wt}LukB_{K12A/K19A/R23A}$ and $LukA_{wt}LukB_{R23A}$ were considerably lower than $LukA_{wt}LukB_{R23E}$ and $LukAB_{wt}$ (Table 1), and therefore these mutants were no longer pursued as potential vaccine candidates. The average immunogenicity titers for $LukB_{R23E}$-containing mutants were comparable to WT (Figure 3).

Similarly, we generated $LukA_{wt}LukB_{K58A}$ for salt bridge $LukA_{D39}$-$LukB_{K58}$, which exhibited a >30,000-fold reduction of toxicity, suggesting a critical functional role for K58 (Figure 2B; Table 1). Similar results were obtained with glutamic acid substitution at K58, indicating the importance of a positive charge at this position (Figure 2B; Table 1). Protein stability and IgG titers were found to be analogous to WT, while neutralizing titers for $LukA_{wt}LukB_{K58A}$ were ~2.61-fold better than $LukAB_{wt}$ (Table 1). However, the overall protein yields that were obtained for $LukA_{wt}LukB_{K58A/E}$

were consistently five-times lower than LukAB$_{wt}$ and therefore not pursued (Table 1). Surprisingly, despite the indispensable role of K58, the reciprocal mutant LukA$_{D39A}$LukB$_{wt}$ showed toxicity levels that were comparable to WT with an EC$_{50}$ of 0.037 nM (Table 1). The IgG titers that were elicited by LukA$_{D39A}$LukB$_{wt}$ were as high as LukAB$_{WT}$, but the neutralizing titers were higher at 1.4-fold better as compared to LukAB$_{wt}$ (Figure 3, Table 1). By combining LukA$_{D39A}$ with attenuating LukB$_{R23E}$ mutation within the same interface, we observed an enhancement of the overall neutralizing titers of LukA$_{wt}$LukB$_{R23E}$ by a factor of 2, while retaining >30,000-fold attenuation in toxicity (Table 1, Figure 2C). Altogether, our observations suggest that disruption of the positively charged residues on LukB are more effective for designing LukAB toxoids than the corresponding residues on LukA.

When considering our observations, where the LukB N terminal residues, K12, K19, and R23 reduced toxicity, we wanted to examine whether the LukB N terminus is directly involved in mediating cytotoxicity. Both available crystals structures of LukAB are missing the first 10–15 residues of the LukB N terminus, which suggests that this region may be highly unstructured or disordered in the crystal form and it may be stabilized upon binding or oligomerization. Moreover, the superposition of the crystal structures of LukB (4TWI) and HlgB suggested high structural similarity, except for the loops at the rim domain. Taking these observations together, we further investigated the role of the first 30 N terminal residues of LukB by either deleting them (LukA$_{wt}$LukB$_{\Delta30AA}$) or replacing with the analogous sequence of HlgB (LukA$_{wt}$LukB$_{HlgB}$) (Figure S1). HlgB has ~70% sequence divergence from LukAB in this region and it does not contain the analogous bulky, positively-charged residues at the same location as LukB. While we were unable to purify a meaningful amount of LukA$_{wt}$LukB$_{\Delta30aa}$ due to the poor expression of LukB$_{\Delta30aa}$ (data not shown), mutant LukA$_{wt}$LukB$_{HlgB}$ was well expressed and it exhibited a cytotoxicity EC$_{50}$ of >1000 nM, representing >30,000-fold attenuation (Figure 2B, Table 1). The protein stability (Figure 2D) and relative immunogenicity levels were also comparable to LukAB$_{WT}$ (Figure 3, Table 1). We also tested whether the N terminal HlgB sequence mediates any HlgB specific neutralizing responses; however, no such response was observed, which suggested a lack of neutralizing epitopes within this region of HlgB.

2.2.4. Fully Attenuated Interface 1&2 Combination Mutants

We tested the cytotoxicity of several attenuated mutations on HL-60-derived PMNs at the highest concentration possible and detected no residual toxicity for mutants LukA$_{wt}$LukB$_{R23A/E}$, LukA$_{wt}$LukB$_{K58/E}$, LukA$_{D39A}$LukB$_{R23E}$, and LukA$_{wt}$LukB$_{HlgB}$. Mutant LukA$_{L61R}$LukB$_{wt}$ retained 6–4% toxicity at very high concentrations (Figure 2C). Altogether, based on the data from expression, toxicity, stability, and immunogenicity experiments, we concluded that mutants LukA$_{wt}$LukB$_{R23E}$, LukA$_{D39A}$LukB$_{R23E}$, and LukA$_{wt}$LukB$_{HlgB}$ show the most significant attenuation in PMN cytotoxicity, and LukA$_{D39A}$LukB$_{R23E}$ and LukA$_{wt}$LukB$_{HlgB}$ also retain the ability to produce high-titer, neutralizing antibodies in ICR mice. Therefore, we focused our efforts to further characterize these two mutants.

2.3. Biophysical Characterization of Selected LukAB Vaccine Candidates

Next, we wanted to confirm whether the selected mutations within the attenuated LukAB candidates affect protein dimerization in solution. This is important, as the elicitation of effective neutralizing antibodies is dependent on maintaining the heterodimer structure [15]. Initial efforts to detect the LukAB dimer by size exclusion chromatography were unsuccessful, as LukAB could not be resolved in SEC-HPLC while using the hydrophilic polymer-coated AdvanceBio SEC column (PL1180-5301) or silica-based Agilent ProSEC300 (PL1147-6501). Additionally, no peaks were detected on the A$_{280}$ chromatograph when increasing the salt conditions (up to 500mM NaCl) and/or varying pH 6–7.0 were implemented. Therefore, we sought to fractionate the proteins by reverse-phase (RP) HPLC. We first fractionated LukAB$_{WT}$, LukA$_{D39A}$LukB$_{R23E}$, and LukA$_{wt}$LukB$_{HlgB}$ while using acetonitrile that resolved into two prominent peaks (Figure 4A). We analyzed the two RP-HPLC fractions by Western blot while using two isolated rabbit polyclonal antibodies—one that cross-reacts to LukA and LukB (referred to as αLukAB) and another LukB-specific (referred to as αLukB-specific) antibody generated, as described in the Methods. As seen in Figure 4B, the results suggested that the first and second peak consist of LukA and LukB, respectively, and the mutants are comparable to WT. Once we confirmed that our expressed form of LukAB is composed of both components, we wanted to verify its oligomerization level in solution. Wild-type and mutant LukAB proteins were incubated with glutaraldehyde, as described in the Methods section to cross-link the LukA and LukB components. LukS-PV and LukF-PV were used as the controls, as they are known to exist as monomers in solution and only dimerize in the presence of the host cellular receptor. As shown in Figure 4C,D, the examination of the glutaraldehyde cross-linked products on SDS-PAGE and Western blot indicated that LukAB$_{WT}$ exists as a dimer (~77 kDa). In contrast, LukS-PV (MW = 70.6 KDa) and LukF-PV (MW = 73.9 KDa) remained largely monomeric in the presence of glutaraldehyde. Mutants LukA$_{D39A}$LukB$_{R23E}$ and LukA$_{WT}$LukB$_{HlgB}$ are similar to LukAB$_{WT}$ as dimers, indicating that they retain the heterodimeric structure that is important for eliciting neutralizing antibodies. Additionally, we also ran the poorly expressed LukA$_{WT}$LukB$_{\Delta30aa}$ mutant, in which the LukA component fails to associate with LukB$_{\Delta30AA}$, suggesting a previously unknown role for the N terminus of LukB in protomer-protomer interaction (Figure 4C,D).

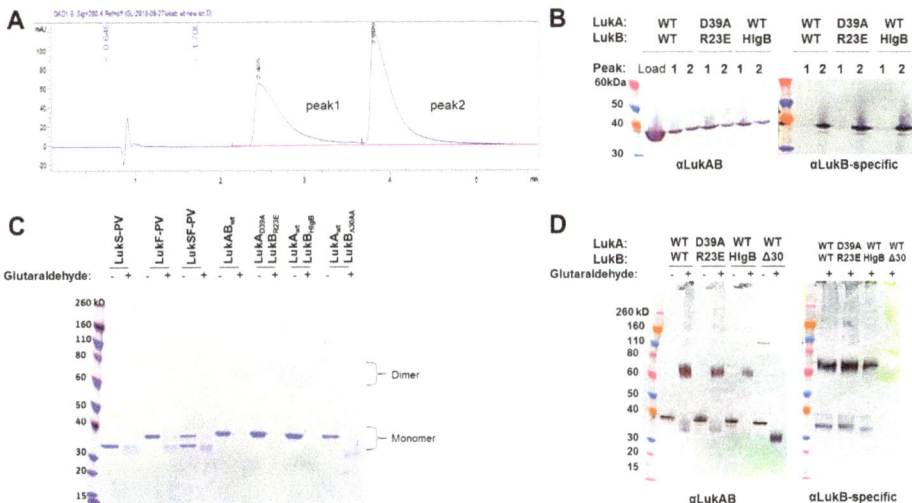

Figure 4. Biophysical characterization of selected LukAB mutants. (**A**) Example of the reverse-phase HPLC chromatograph of LukAB$_{WT}$ highlighting the two fractionated peaks, 1 and 2. (**B**) Immunoblots treated with either 1 µg/mL of αLukAB or 0.1 µg/mL of αLukB-specific pAB highlighting peak 1 and 2 fractionated samples of LukAB$_{WT}$, LukA$_{D39A}$LukB$_{R23E}$, and LukA$_{wt}$LuK$_{HlgB}$. Molecular weights are also indicated. Individual proteins as indicated were incubated with or without 0.25% glutaraldehyde and analyzed by (**C**) SDS-PAGE without boiling and (**D**) Western blot using αLukAB or 0.1 µg/mL of αLukB-specific pAB.

2.4. Functional Characterization of Selected LukAB Vaccine Candidates

The ability of mutants to bind to the cellular receptor CD11b would indicate an intact receptor binding site, which could be important for the induction of neutralizing antibodies. Therefore, we sought to determine whether the selected mutants compete with LukAB$_{WT}$ for receptor binding. Direct competition assay between the mutant and WT, followed by the detection of bound protein by flow cytometry is not possible because of the rapid cytotoxic effect of LukAB. We reasoned that competition between WT and mutant for receptor binding should result in the reversal of LukAB toxicity when the toxicity is measured in the presence of excess amount of mutant. To this end, various concentrations (0.4 pM-13 nM) of the mutants LukA$_{D39A}$LukB$_{R23E}$, LukA$_{wt}$LukB$_{HlgB}$, LukA$_{wt}$LukB$_{R23E}$, LukA$_{wt}$LukB$_{125-133\_1G}$, LukA$_{D39R}$LukB$_{R23E}$, and LukA$_{wt}$LukB$_{K12/19/R23A}$, which are fully attenuated, or

LukA$_{D39A}$LukB$_{wt}$, which retains toxicity, were incubated with PMNs along with a fixed concentration of 0.42 nM LukAB$_{WT}$. Relative toxicity of LukAB in the presence of various mutants was then plotted against the molar ratio of mutant to wild type (Figure 5A). Consistent with our hypothesis, the mutants suppressed LukAB toxicity in a dose dependent manner. Of these mutants, LukA$_{D39A}$LukB$_{R23E}$ and LukA$_{wt}$LukB$_{HlgB}$, followed by LukA$_{wt}$LukB$_{R23E}$, exhibited the most dramatic effect, with 50% reversal of LukAB$_{WT}$ toxicity at a mutant/WT ratio of 0.17, 0.18, and 0.64, respectively. Even the full inhibition of LukAB toxicity was observed with the first two mutants at molar ratios below 1. These data suggested that, at least for these three mutants, the reversal of toxicity cannot be entirely explained by the competition for receptor binding. A potential explanation for these data is that these mutants may be able to form mixed, defective oligomeric structures on the cell surface with wild type LukAB dimers. Mutants LukA$_{D39R}$LukB$_{R23E}$ and LukA$_{wt}$LukB$_{K12/19/R23A}$ were less efficient in competing with WT with 50% reversal only being observed at the highest mutant/WT molar ratio of >30 (high mutant excess). On the other hand, attenuated LukA$_{WT}$LukB$_{125-133\_1G}$, having a deletion in the loop that is involved in pore formation, failed to reverse the lytic effect of LukAB, even at high concentrations, which suggested that this mutation can neither compete with receptor binding nor form mixed defective oligomers. As expected, LukA$_{D39A}$LukB$_{wt}$, a mutant that retains substantial toxicity, had no reverting effect on LukAB activity.

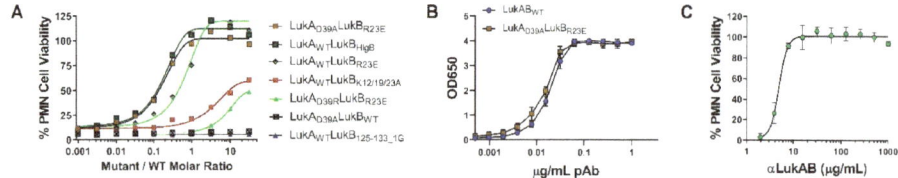

Figure 5. Functional characterization of selected LukAB mutants in PMNs. (**A**) LukAB$_{WT}$ and mutant competition binding to PMNs. (**B**) ELISA and (**C**) TNA results testing the quality of binding and neutralization of αLukAB polyclonal antibodies to LukAB$_{WT}$ and LukA$_{D39A}$LukB$_{R23E}$. All the error bars represent measurements made in duplicates.

Out of the two best attenuated candidates, LukA$_{D39A}$LukB$_{R23E}$ and LukA$_{wt}$LukB$_{HlgB}$, we tested the ability of LukA$_{D39A}$LukB$_{R23E}$ for its ability to generate neutralizing antibodies in rabbits. The total IgG was purified from the sera of rabbits immunized with the LukAB mutant showing high and equal binding to both wild type and mutant LukA$_{D39A}$LukB$_{R23E}$ (Figure 5B), in addition to being able to effectively neutralize LukAB$_{WT}$ at an NT$_{50}$ of 4.9 µg/mL (Figure 5C).

Published work using SA Newman strain or individual isogenic mutants that were deficient of LukAB, α-toxin, HlgABC, or LukED demonstrated that SA directly kills human derived monocytes by promoting necrotic cell death in primary CD14$^+$ human monocytes in a LukAB-CD11b-dependent manner [14]. Therefore, we wanted test whether our best mutants are also attenuated in the THP-1 cells. To test this, while using flow cytometry, we first confirmed that CD11b expression increased from ~13% to 85% upon the differentiation of THP-1 cells using PMA, which results in the macrophage-like phenotype, mimicking primary human macrophages (Figure S2). Consistently, the measured CD14$^+$ levels were also high for these cells. Cytotoxicity assay with purified LukAB and mutants, LukA$_{D39A}$LukB$_{R23E}$ and LukA$_{wt}$LukB$_{HlgB}$, showed an EC$_{50}$ of 0.04 nM for LukAB and the complete loss of toxicity for LukA$_{D39A}$LukB$_{R23E}$ at concentrations as high as 160 nM (Figure 6A). LukA$_{WT}$LukB$_{HlgB}$ exhibited ~33% residual toxicity at the highest concentration tested. Similar to HL-60 cells, polyclonal antibodies that were generated against LukA$_{D39A}$LukB$_{R23E}$ were able to neutralize LukAB$_{WT}$-mediated toxicity in differentiated THP-1 cells (Figure 6B). Together, these data show that LukA$_{D39A}$LukB$_{R23E}$, as a vaccine candidate, can generate potent, neutralizing antibodies that are capable of reversing LukAB$_{WT}$-mediated toxicity and preventing monocyte lysis.

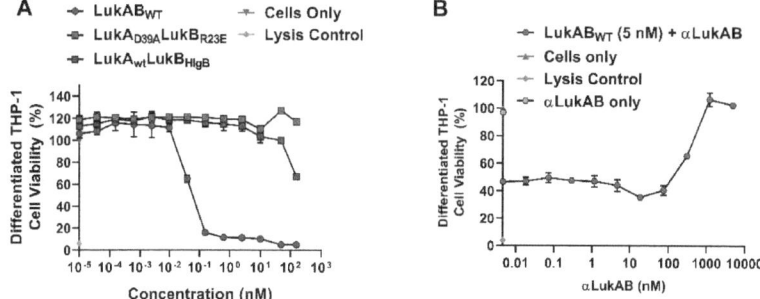

Figure 6. LukAB-mediated cytotoxicity and TNA in THP-1. (**A**) Differentiated THP-1 cells intoxicated with purified LukAB$_{WT}$, LukA$_{D39A}$LukB$_{R23E}$, or LukALukB$_{HlgB}$ proteins in a dose-dependent manner to measure cell viability and cytotoxicity. (**B**) Cells intoxicated with a constant concentration of LukAB$_{WT}$ (5 nM) in the presence of titrated αLukAB polyclonal antibody starting at a concentration of 5000 nM and diluted 2-fold. For all graphs, percent cell viability is plotted as a function of log concentration shown in nM and error bars represent measurements made in duplicates. Readout using CTG luminescence. Lysis control shown contains cells incubated with 0.2% TritonX-100.

2.5. Attenuated LukAB Candidate as a Component of a Multivalent Toxoid Vaccine

Multiple pore-forming toxins, including Hla, PVL, LukED, HlgAB, HlgCB, and LukAB, mediate the cytotoxic activity of SA. While LukAB and PVL are most lytic to PMNs, this activity is shared by several other leukotoxins [6]. We measured the ability of antibodies against LukAB and other leukotoxins to neutralize leukotoxicity in HL-60 and THP-1 cells induced by bacterial culture supernatants from different standard clinical SA strains, as listed in Table S2, to establish the potential contribution of a LukAB toxoid to a multivalent toxoid vaccine. Bacterial supernatants were produced by growing the cells to stationary phase in BHI or TSB media, sterile-filtered, and used in PMN cytotoxicity assay in the presence of rabbit polyclonal antibodies raised against LukA$_{D39A}$LukB$_{R23E}$ (αLukAB), two PVL subunit toxoids LukS$_{mut9}$ (αLukS) and LukF$_{mut1}$ (αLukF), which we previously reported [21], or the toxoid Hla$_{I135LI148L}$ (αHla). Besides three isolates of USA300, we also used a LukAB isogenic knock out of the USA300 strain SF8300 (SF8300ΔlukAB). As shown in Figure 7A, supernatants from these three isolates, as well as SF8300ΔlukAB, were highly toxic toward PMNs. Either a combination of antibodies against PVL and α-toxin or LukAB alone had modest neutralizing activity towards supernatants of the three wild-type strains, while a combination of the antibodies against all of these toxins was able to fully neutralize the cytotoxicity (Figure 7A). In contrast, when the lukAB gene was deleted, the cytotoxicity was entirely neutralized by antibodies to PVL and Hla. These results were also confirmed using TSB media for bacterial growth (Figure 7B). These data indicated that efficient neutralization of leukotoxicity of USA300 requires a broadly neutralizing antibody response, including LukAB. In contrast to USA300, LukAB did not appear to significantly contribute to leukotoxicity mediated by USA100, ST80, USA1000, and COL strains (Figure 7A). Total cytotoxicity induced by USA200, MNHoCH, and MRSA252 was too low to determine the role of LukAB.

We next tested whether αLukAB also protects differentiated THP-1 cells against cytotoxic activity in the bacterial culture supernatants. For this, we examined the cell viability in presence of supernatants from SF8300 and SF8300ΔlukAB grown in TSB or BHI media to gauge the level of LukAB-mediated toxicity in differentiated THP-1 cells. The cytotoxicity of WT strain was 2.6-fold or 1.8-fold fold higher than SF8300ΔlukAB when grown in BHI or TSB, respectively (Figure 7C), which suggested that LukAB, despite its potent cytotoxicity towards THP-1 cells (Figure 6A), only partially contributes to monocyte killing by SF8300 culture supernatant and other pore-forming toxins may also contribute to overall cytotoxicity. To this end, we evaluated the cytotoxicity of other pore-forming toxins Hla, PVL, HlgAB, HlgCB, and LukED. As shown in Figure 7D, of these toxins, HlgCB and PVL, followed by Hla, displayed the strongest cytotoxicity toward differentiated THP-1 cells, while HlgAB and LukED showed poor activity. However, all of these toxins were far less potent than LukAB (EC$_{50}$ 0.04 nM, Figure 6A). We next evaluated the ability of antibodies against PVL, Hla, and LukAB to neutralize the toxicity of SF8300 supernatant toward THP-1 cells. At a 1:10 dilution of the supernatant that caused >90% cytotoxicity in THP-1 cells, only a combination of all these antibodies was able to fully neutralize the toxic activity of the supernatant (Figure 7E). These data, again, point to the importance of neutralizing multiple pore-forming toxins for the complete protection of monocytic cells.

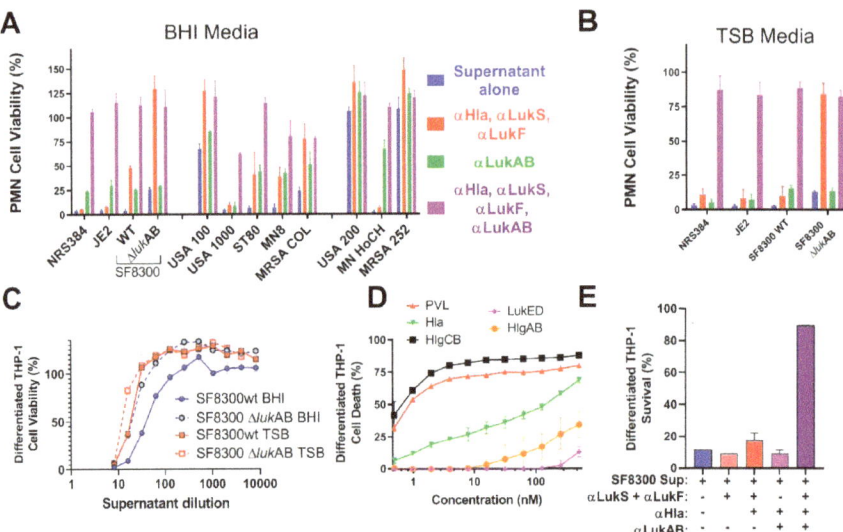

Figure 7. Toxicity and TNA of bacterial culture supernatants. Bacterial culture supernatants grown in (**A**) BHI or (**B**) TSB media incubated with indicated polyclonal antibodies to measure relative cytotoxicity and neutralization post a 3 h incubation in PMN cells. Error bars indicate measurements made in triplicates. (**C**) Toxicity profiles of SF8300 wild-type and LukAB-deficient isogenic mutant grown in BHI or TSB media on differentiated THP-1 cells measured 4 h post incubation. Error bars represent measurements made in triplicates. (**D**) Measurement of toxicity of purified toxoids such as α-toxin and BCPFTs on differentiated THP-1 cells. CTG readout was carried out 1 h post incubation with toxoids and error bars indicate measurements made in duplicates. (**E**) TNA results of supernatants from SF8300WT or SF8300ΔlukAB strains grown in TSB and BHI media and incubated with αLukAB, αLukS, αLukF, or αHla, or a combination as labeled for 4 h to measure extent of neutralization indicated by cell viability measurements. Error bars represent measurements made in duplicates.

3. Discussion

The evolution and emergence of MRSA strains has become a major challenge to global health. Among the diverse virulence factors that were secreted by SA to subvert the encountering host defense system, α-toxin/Hla and the bicomponent leukotoxins, such as LukAB, HlgAB, HlgCB, LukED, and PVL, are the most potent, because they specifically target and kill innate immune cells and disrupt biological barriers by lysing epithelial and endothelial cells [24], as well as keratinocytes [25]. Prior efforts toward vaccine development for *S. aureus* have myopically focused on promoting the opsonophagocytic uptake of the bacteria and the subsequent killing by phagocytic cells, an approach that has been successful for several pathogens, such as *S. pneumoniae*, *N. meningitidis*, and *H. influenzae* B. However, these efforts have failed to deliver an effective vaccine for *S. aureus* and at least one of the experimental vaccines, Merck V710, led to increased mortality in vaccinated individuals who developed SA infection [26], which suggested possible immunopathology. Several epidemiological studies indicate that the SA toxins are important vaccine targets [27–29]. We have previously developed vaccine candidates for PVL subunits that elicit cross-reactivity to HlgAB, HlgCB, and LukED [21]. LukAB plays a significant role in mediating SA virulence and its potency is most comparable to PVL [8]. Unlike PVL, which is carried by phages and is only found in 5–15% clinical isolates, LukAB is chromosomally encoded and found in the majority of the isolates [6]. However, LukAB is phylogenetically distant to these bicomponent toxins and anti-PVL antibodies are unable to neutralize this toxin. Here, we took a systematic and rational approach to design vaccine candidates for LukAB that are attenuated, stable, and highly immunogenic, and we identified several candidates for inclusion in a multivalent toxoid vaccine.

All of the bicomponent leukotoxins, except for LukAB, are secreted as S and F subunit monomers. Cell receptor binding is initiated by S subunit followed by the binding of F and oligomerization, which leads to pore formation [6]. In contrast, LukAB is secreted as a stable dimer before engaging with the CD11b receptor molecules on the target cells following, which it hetero-oligomerizes into octomeric pore [16]. In this report, we developed a process to express LukAB from a single vector (pETDuet) to purify the protein at high yield without the use of an affinity tag. Additionally, we were able to confirm that the produced LukAB contains both S and F components and are present as dimers in solution by using reverse-phase HPLC and glutaraldehyde cross-linking.

Badarau et al. solved the crystal structure of LukAB heterodimer in complex with a potent neutralizing antibody (ASN102) [15], as well as the octameric LukAB [11]. These studies identified 56 hydrogen bonds and four electrostatic interactions that hold the heterodimer together, as well as 34 hydrogen bonds and three electrostatic interactions that govern the formation of LukAB octamers. The authors also demonstrated that maintaining a stable dimer is critical for binding to neutralizing antibodies [15]. Badarau et al. also generated several attenuated mutants, including double mutations in LukB consisting of R23A/E and K218A or LukA D75A and D197A in interface 2 (referred to as interface 1 in [11]), which appear to interfere with octamerization. We undertook a broad screening strategy that is based on these important findings, and on the basis of LukAB sequence from USA300, to identify potential vaccine candidates that maintain LukAB structural integrity and immunogenicity, but lack toxicity. By making single and combination mutations, we were able to highlight residues in different, functional domains of the LukAB dimer that are imperative to mediating LukAB cytotoxicity. Additionally, targeting different domains on LukAB also allowed for biochemical, biophysical, and functional characterization, and delineated additional "hotspots" within LukAB dimer that act as key determinants of solubility, cytotoxicity, stability, and immunogenicity.

Of the electrostatic charges mediating protomer-protomer interactions (mediating heterodimer formation) and dimer-dimer interactions (mediating octamer formation), referred to here as interface 1 and 2, respectively, our data indicates that the residues within the salt bridges on interface 2 along with the bulky positively charged arginine and lysine residues at LukB N terminus are the key to mediating LukAB cytotoxicity. The electrostatic interactions that are part of the salt bridges within Interface 2, particularly participating LukB residues, which form a basic patch near the apical side of the cap domain, are crucial to cytotoxicity, as seen in this study. These residues have been previously shown to be important for LukAB octamerization and are known to be fully conserved between the LukAB sequence variants [11]. In contrast, the salt bridge residues mutated within Interface 1 did not contribute to cytotoxicity as much. In our observation, substituting charged residues (K12/K19 and R23) or swapping of the first 30 residues of LukB N terminus with those of HlgB exhibited the highest impact on cytotoxic function. A previous report by DuMont et al. showed that the deletion of the first 33 residues of LukA does not affect copurification with LukB or heterodimer formation, but increases pore formation, thereby moderately increasing cytotoxicity [12]. Our data indicates that, unlike LukA, the N terminus of LukB is indispensable for both heterodimer formation and cytotoxicity. While the N-terminal 30 residues of HlgB transplanted to LukB allow for heterodimer formation, it does not restore cytotoxicity, which is likely because of the loss of R23 that is critical for the salt bridge that is involved in dimer-dimer interaction. In contrast, the complete removal of those 30 residues resulted in the failure to express LukB. These observations are reminiscent of the functional properties of the amino latch that are found at the N terminus of α-toxin, where the residues control hemolytic activity and remain central to protomer-protomer interactions [30]. Interestingly, the N terminal residues of LukB are unique to LukAB and they are not conserved among other S and F components or α-toxin alluding to a possibility that LukB N terminus may be functionally unique with putative roles in oligomerization and toxicity, requiring further investigation.

Mutations that were made to disrupt the buried hydrophobic pocket found within Interface 1, particularly L61, was found to be crucial to LukAB thermostability and attenuating toxicity; however, drastically hampering immunogenicity. As this residue is not surface exposed, the pocket near L61 is not an antigenic target, but it lowers immunogenicity, which is likely due to altered oligomeric status. The structures of LukAB and α-toxin show that L61 is nestled between β-strands that are part of the β-sandwich within the cap domain and the introduction of bulky mutations perturbs the hydrophobicity within this pocket affecting crucial protomer-protomer interactions that are sacred to stability, thereby reducing its ability to generate neutralizing antibodies.

In our efforts, we have been successful in identifying a strong vaccine candidate in LukA$_{D39A}$LukB$_{R23E}$ that satisfied biochemical and biophysical characterization, in addition to showing complete attenuation in PMNs. It is interesting that the second-best candidate, LukA$_{wt}$LukB$_{HlgB}$ mutant, which exhibited the complete attenuation in PMN lytic activity, retained some toxicity, albeit at high concentrations in the THP-1 cells, suggesting that the substituted HlgB residues may be responsible for the relapsed cytotoxic effects. Interestingly, both of these mutants were able to compete with WT LukAB toxicity at a low molar ratio of below 1, which indicated that the competition cannot be entirely due to receptor binding by the mutant. The reversal of LukAB toxicity by these mutants may relate to the formation of defective mixed oligomeric structures, an observation that warrants further investigation. Moreover, with this study, we can show that the identified vaccine candidate generates antibodies that can reverse the toxicity and show protection in PMNs, as well THP-1 monocytes.

Previous studies showed the important role of LukAB in cell specific lysis of monocytes [14] and dendritic cells (DC) [31], as well as its role in macrophage dysfunction [17]. However, these activities are also shared by some of the other pore-forming toxins. Most of the strains that were tested in our study showed clear synergism between polyclonal antibodies against LukAB and other pore-forming toxins. Cocktail polyclonal: αHla, LukS-PV, and LukF-PV (generated against three toxoid proteins) alone were unable to neutralize the toxicity of culture supernatants from most of the virulent strains in PMN lysis study. However, when this cocktail polyclonal was mixed with anti-LukAB polyclonal antibodies, 100% neutralization of culture supernatants was achieved, indicating the importance of neutralizing all the pore-forming toxins.

In summary, in this study, we have developed at least two vaccine candidates for LukAB, which is an important virulence factor of *S. aureus*. Our findings indicate that a multivalent approach targeting the related leukotoxins PVL, LukED, HlgAB, HlgCB, and the divergent LukAB, as well as the single component Hla is critical for protection against cytolytic activity of the most prevalent *S. aureus* strains. These data strongly support the development of a multivalent toxoid vaccine for *S. aureus*, which covers all major pore-forming toxins.

4. Materials and Methods

4.1. Generation of LukAB Wild-Type (WT) and Mutants in pET Duet and pET24a (+)

General methods that are used for bacterial culture have been described previously in detail (18,19). In this study, we optimized LukAB expression while using two different vectors: pET Duet-1 (Novagen), where LukA was cloned into the multiple cloning site 1 (MCS1) while using *Nco*I-*Hind*III, and LukB was cloned into MCS2 using *Nde*I-*Xho*I restriction sites within the same vector. In another system, LukA and LukB were expressed while using two different pET-24a(+) plasmids within the same *E. coli* cell. We compared LukAB WT expression, yield, and toxicity from these two systems. All of the mutants were expressed while using the latter system, where LukA and LukB were expressed using separate plasmids within the same *E. coli*. Towards this end, LukA (WT or mutants) was cloned into pET-24a(+) with a Kanamycin resistant marker within *NdeI-XhoI* sites. Similarly, LukB (WT/mutants) was inserted within *NdeI-XhoI* sites, but we replaced the inherent Kanamycin resistant cassette in pET-24a(+) with Ampicillin. The origin of replication of pET-24a(+) was replaced by p15a resulting in a recombinant pET24a(+)Amp^{R}p15a LukB vector to increase the plasmid compatibility. This LukB plasmid (WT/mutant) was transformed into BL21(DE3) (NEB) containing pET24a(+) LukA plasmid (WT/mutant) and the colonies were selected on LB plates with 50 ug/mL of Kanamycin (Kan$_{50}$) and 100 ug/mL of Ampicillin (Amp$_{100}$). The genes for LukAB (USA300_TCH1516) WT and mutants were codon optimized prior to transformation by GenScript®.

4.2. Growth Media and Bacterial Strains

Overnight cultures of *E. coli* that were grown at 37 °C in LB Kan$_{50}$Amp$_{100}$ were expanded to 0.5 L in a shaking incubator (225 rpm), until they reached an OD OD$_{600}$ of 0.5. The cultures were immediately chilled on ice for 10 min. with periodic shaking and then induced with 0.3 mM IPTG (Sigma, St.Louis, MO, USA) in a shaking incubator (225 rpm) overnight at 25 °C. The following day, the cells were harvested by centrifugation (14,000× *g*) and frozen at −80 °C. To lyse the cells, the pellet was resuspended in 3 mL of cell lysis buffer (20 mM Tris pH 8.0, 50 mM NaCl, 1 mM EDTA, 0.1% Triton X-100) per gram of cell paste and Hen egg lysozyme (Sigma, St. Louis, MO, USA) at 1 mg/mL final concentration prior to incubation at 37 °C for 30 min. The partially lysed cells were then cooled in a wet/dry ice ethanol bath, followed by sonication while using a microtip (10 × 10 s bursts with cooling between bursts, output 5, 50% duty). Lysis was confirmed by measuring the reduction of OD$_{600}$ absorbance. Post lysis, the NaCl concentration was adjusted to 0.5 M and nucleic acid precipitation was carried out by the dropwise addition of 0.3–0.5% polyethyleneimine (PEI) while maintaining constant mixing. The PEI pellet was separated by centrifugation at 12,000 rpm in a Sorvall SS34 rotor and the supernatant containing the toxoid was subjected to ammonium sulphate (AmSO$_4$) precipitation. Towards this end, 0.472 g/mL of AmSO$_4$ powder was added to the PEI supernatant and then placed on a rotating mixer for 20 min. at room temperature. The resulting pellet that was obtained by centrifugation at 12,000× *g* for 30 min. at 4 °C was frozen at −80 °C until purification.

The AmSO$_4$ pellet was resuspended and buffer exchanged into 20 mM NaPi pH 6.5, 25 mM NaCl, 5% glycerol using a GE Healthcare PD10 desalting column. The mixture was then clarified by filtration using 0.8/0.2 μm Supor® low protein binding syringe filter (Pall Life Sciences, Port Washington, NY, USA). The toxoid from the resulting solution was purified while using a two-column purification approach. The first purification was carried over a 10 mL Poros 50 HS column using a 40-column volume (C.V.) gradient from 25 to 1000 mM NaCl in the phosphate buffer. The peak fractions were analyzed by standard SDS-PAGE analysis and accordingly pooled together for dialysis into 20 mM NaPi pH 6.8, 50 mM NaCl, 5% glycerol, the equilibration buffer for the second column—a 10 mL ceramic Hydroxyapatite (HTP) (BioRad, Hercules, CA, USA; Type 1 40 μm). The toxoid was eluted using a 40 C.V. gradient from 50 to 1000 mM NaCl in the phosphate buffer. The appropriate fractions were pooled together upon SDS-PAGE evaluation and then dialyzed into the final storage buffer (20 mM NaPi pH 7.4, 150 mM NaCl, 5% glycerol). All purified LukAB proteins (MW = ~72 kDa) were concentrated while using Amicon 3K MWCO Ultra 15, filtered through 0.8/0.2 μm low protein binding membrane, and stored at −80 °C prior to use.

4.3. Growth Media and Bacterial Strains

Table S2 lists the bacterial strains that were used in this study. The SA strains were grown in brain heart infusion broth/agar (BHI) and tryptic soy broth/agar (TSB) media at 37 °C, whichever appropriate. The overnight bacterial culture supernatants were normalized based on culture OD$_{600}$ absorbance. The next day, culture supernatants were filtered through 0.2 μm filter to sterilize the supernatants. Sterility was confirmed by culturing 100 μL of the filtered supernatants on BHI or TSA agar plates overnight.

4.4. Cell Culture Maintenance and Induction

The HL-60 cells (ATCC, Manassas, VA, USA) were cultured in RPMI 1640 (Gibco, Gaithersburg, MD, USA) supplemented with 16.4% heat inactivated fetal bovine serum (FBS), 4 mM L-glutamine, 82 U/mL each of penicillin, and streptomycin. Cells were passaged twice a week at a concentration of $6–8 \times 10^5$ cells/mL. 1×10^7 cells were seeded in 30 mL and grown in culture media with 1.5% dimethyl sulfoxide (DMSO) for seven days to differentiate cells into neutrophils. CD11b expression using flow cytometry confirmed induction.

4.5. PMN-based Cytotoxicity Assay

The induced HL-60 cells were harvested by centrifugation at $420 \times g$ (Sorvall RT6000B rotor, ThermoFisher Scientific, Waltham, MA, USA) for 10 min. at 20 °C. Cells were washed and resuspended with phenol red-free RPMI 1640 (Gibco, Gaithersburg, MD, USA) supplemented with 2% FBS to a final concentration of 5×10^6 cells/mL. LukAB mutants (proteins) were serially diluted two-fold across 96-well plates (50 μL/well) and 100 μL of 5×10^6 cells/mL were added to each well. The plates were incubated at 37 °C, 5% CO_2, 95% humidity for 3 h. After 3 h, either XTT (Cell Signaling Technology) or CellTiter Glo (Promega) reagent was used to determine the cell viability and cytotoxicity [21,32]. When using XTT for readout, 50 μL of the activated XTT (50 μL electron coupling per 5 mL XTT) reagent was added to each well, and plate was returned to 37 °C, 5% CO_2, 95% humidity for 16–18 h. After incubation, the cells were pelleted by centrifugation at 3500 rpm $3565.9 \times g$ (Sorvall RT6000B rotor, ThermoFisher Scientific, Waltham, MA, USA) for 3 min. The supernatants were transferred to 96-well ELISA plates and absorbance was read at 470 nm while using Spectramax 190 plate reader (Molecular Devices, San Jose, CA, USA) and Softmax 5.4.5 software (Molecular Devices, Waltham, MA, USA). When using CellTiter Glo, 50 μL of the reconstituted CellTiter Glo reagent was added to each well. The plate was shaken on an orbital shaker for 10–15 min., followed by measurement of luminescence (emission at 560 nm) using Cytation 5 imaging reader (Biotek, Winooski, VT, USA) and Gen5 2.09 software to determine cell viability. For the kinetic cytotoxicity studies, the replicates were incubated for 6 and 19 h, in addition to a 3-h incubation period. After incubation, the CellTiter Glo reagent was used to determine the cell viability.

4.6. Rabbit Polyclonal Antibody Generation

Rabbit polyclonal antibody generation for anti-Hla$_{DM}$, LukS$_{mut1}$, LukF$_{mut9}$, and LukA$_{D39A}$B$_{R23E}$ toxoid as immunogens were generated by Genscript® (Piscataway, NJ, USA) using >98% pure proteins as immunogens. Briefly, four New Zealand white rabbits were immunized per toxoid on day 0, 14, and 21 with 0.2 mg protein per rabbit, along with Freund's Incomplete Adjuvant subcutaneously. The test bleeds and production bleeds were collected on day 21 and day 42. Hyperimmune sera were individually characterized for ELISA titer and TNA titers before pooling them together. The pooled serum was purified by Protein A affinity chromatography into total IgG and labeled, as follows: anti-Hla (IBT Cat: 1940-01 Rb pAb), anti-LukS- (IBT Cat: 1941-01 Rb pAb), anti-LukF-PV(IBT Cat: 1942-01 Rb pAb), and anti-LukAB (IBT Cat: 1944-02 Rb pAb). Full quality control (QC) were performed before use.

4.7. Toxin Neutralization Assay (TNA) in PMNs

The serum samples were serially diluted two-fold in RPMI across 96-well plates (25 μL/well). Twenty-five microliters of 2.5 nM LukAB toxin was added to each well. 100 μL of induced cells were prepared, as described above, at 5×10^6 cells/mL were added to each well. The plates were incubated at 37 °C, 5% CO_2, 95% humidity for 3 h, followed by XTT or CellTiter Glo readout to determine the cell viability and neutralization [21,32].

4.8. Reverse-TNA in PMNs

Select LukAB mutants were serially diluted from a starting concentration of 12 nM in RPMI semi-log across 96-well plates (25 µL/well). Polyclonal αLukAB at 100 µg/mL or RPMI was added to each well (12.5 µL/well), followed by incubation at RT for 10 min. Following incubation, 12.5 µL of 5 nM LukAB WT and 100 µL of 5×10^6 induced HL-60 cells/mL were added to each well. The plates were then incubated for 3 h at 37 °C, 5% CO_2, 95% humidity, followed by CellTiter Glo readout determine cell viability and neutralization.

4.9. Differential Scanning Fluorimetry (DSF)

The proteins in the storage buffer (20 mM NaPi pH 7.4, 150 mM NaCl, 5% glycerol) were mixed with 2X SYPRO orange dye (Invitrogen, Carlsbad, CA, USA) in a 96-well hard shell plate with clear bottom (BIO-RAD, Hercules, CA, USA) and then placed into a thermal cycler, wherein the temperature scan rate was fixed at 0.5 °C/min over a range of 30–99 °C (21). The fluorescence intensities were plotted against temperature to get a sigmoidal curve and the melting temperatures (Tm) were calculated while using the first derivative. BSA (Pierce) was used as a control, which recorded a melting temperature of 66 ± 0 °C in $1 \times$ PBS pH 7.4.

4.10. Cross-Linking with Glutaraldehyde

$LukAB_{WT}$ and mutants at 50 ug/mL concentration were incubated with 0.25% glutaraldehyde (Sigma, St. Louis, MO, USA) for 2 min. at 37 °C in a final volume of 100 µL in 20 mM HEPES pH 7.5, 50 mM NaCl. The reaction was stopped by adding 10 µL of 1M Tris, pH 8.0 and $4 \times$ LDS sample buffer, followed by SDS-PAGE and Western blot analysis while using αLukAB and rabbit pAb αLukB-specific (IBT Cat: 0313-001). αLukS-PV and αLukF-PV were used as the negative controls.

4.11. Reverse-Phase HPLC

LukAB (WT/Mutant; 100 µg) was injected into an AdvanceBio RP-mAb diphenyl column (Agilent 795975-944, Santa Clara, CA, USA, 4.6×100 mm, 3.5 micron) in a 1260 Infinity Quaternary instrument. A 30–90% gradient method was used consisting of 100% acetonitrile in Line A and 0.1% TFA (*v/v*) in Line B with a flow rate of 1.00 mL/min over 60 min. and a column temperature of 27 °C.

4.12. Animals and Immunizations

Female ICR (CD-1) mice, six weeks of age, were purchased from Envigo (US). The mice were maintained under pathogen-free conditions and fed laboratory chow and water ad libitum. All mouse work was conducted in accordance with protocols that were approved by institutional animal care and use committees (IACUC) of Integrated BioTherapeutics, where mouse studies were performed (approval date 28 February 2017; approval code: D17-00974). The mice were intramuscularly immunized (IM) three times two weeks apart with 10 µg of LukAB mutant in 50 µg of $Al(OH)_3$. For serological analysis, the mice were test bled via retro-orbital (RO) route prior to and 10 days after the third and final immunization.

4.13. Enzyme-Linked Immunosorbent Assay (ELISA) for Determination of Serum Titers

Blood samples from mice were centrifuged in serum separator tubes and the serum samples were stored at −80 °C until further use in ELISA. Briefly, 96-well plates were coated with 300 ng/well of LukAB WT overnight at 4 °C. The plates were blocked with Starting Block (SB) (Thermo Scientific, Waltham, MA, USA) for one hour at room temperature (RT). Serum samples were diluted in a semi-log manner starting from 1:100 to 1:316,228 in a 96-well plate, using starting block buffer as the diluent. The plates were washed three times and sample dilutions were applied in 100 µL volume/well. The plates were incubated for 1h at RT and washed three times before applying the conjugate, goat anti-mouse IgG (H+L)- Horseradish Peroxidase (HRP) in SB. Plates were incubated for 1 h at RT, washed, as described above, and incubated with TMB (3,3′,5,5′-tetramethylbenzidine) for 30 min. to detect HRP activity. Optical density at 650 nm was measured while using a Versamax™ plate reader (Molecular Devices, Waltham, CA, USA). Data analysis for full dilution curves was performed using Softmax program and graphed in GraphPad Prism.

4.14. Cytotoxicity Assay in Human Monocytes and THP-1 Cells

Purified human monocytes (CD14$^+$) up to 95.98% purity were purchased from BioIVT (San Carlos, CA, USA) and kept frozen in LiN$_2$ until use. Acute monocytic leukemia THP-1 cells were purchased from ATCC (ATCC® TIB-202™) and were cultured at 4E5 cells/mL every 2–3 days at 37 °C + 5% CO$_2$, as recommended in RPMI-1640, supplemented with 0.05 mM 2-mercaptoethanol and 10% FBS. To differentiate the THP-1 cells, 100 nM of Phorbol 12-myristate 13-acetate (PMA; Sigma, St. Louis, MO, USA) was added to THP-1 media and seeded at 4E5 cells/mL. The cells were allowed to differentiate for three days and analyzed for extent of cell adherence and surface marker for differentiation (i.e., CD11b) by flow cytometry. After three days, the media was replaced with non-PMA containing medium and allowed to rest for five days before use, feeding the cells with fresh THP-1 medium every two days.

To evaluate the cytotoxicity effects of purified LukAB (WT/mutants) or other toxins, such as α-toxin, LukSF, HlgAB, and HlgCB on differentiated THP-1 cells, the adherent cells were harvested by treatment with 1 × PBS + 0.05 mM EDTA at 37 °C for ~5 min. and immediately centrifuged at 420× g (Sorvall RT6000B rotor) for 10 min. The cells were then washed twice in 1% (*w/v*) RPMI+Cas (Bacto BD, Franklin Lakes, NJ, USA) and pelleted by centrifugation. Proteins (12.5 μL) were serially diluted two-fold or in semi-log fashion, as indicated in 1% RPMI+ Cas and incubated with 75 μL cells (seeded at 1E5 cells/well in a 96-well plate) in a final volume of 125 μL for 1 h at 37 °C, 5% CO$_2$, 95% humidity, followed by CellTiter Glo readout to determine the cell viability and % lysis. For experiments with supernatants from culture filtrates, the supernatants (as listed in Table S2) were normalized to OD$_{600}$ = 6 prior to serial dilution in RPMI+Cas. The incubation times were increased to 4 h at 37 °C, 5% CO$_2$, 95% humidity prior to CTG treatment and readout. For conditions with a polyclonal antibody, 12.5 μL of single or each antibody was added to the well, followed by incubation with purified protein or supernatant for 10 min. at RT prior to the addition of cells. Cells incubated with a final concentration of 0.2 or 0.4% Triton-X-100 in 1% RPMI+ Cas was used as the control for complete lysis.

4.15. Flow Cytometry Analysis

HL60, DMSO-treated HL-60, THP-1, and PMA-treated THP-1 cells were seeded at 1.5E5 cells/well on a clear 96-well plate. The cells were washed with 200 μL/well of PBS without Ca^{2+} and Mg^{2+}, supplemented with 2% FBS (FACS buffer). For HL60 cells, CD11b-FITC (BDPharmingen; Cat: 562,793 Clone: ICRF44) was added to each well at 1:40 dilution (50 μL) in FACS buffer and incubated at RT, being covered from light, for 15 min. For THP-1 cells, a combination of CD11b and CD14-APC/Cy7 stains (Cat: 301,820 Clone: M5E2) were used to monitor CD11b upregulation in addition to macrophage differentiation. The cells were washed twice with FACS buffer at 1400 rpm for 5 min. each and LIVE/DEAD Fixable Near-IR Dead Cell stain (ThermoFisher Scientific, Waltham, MA, USA) was added to each well at a dilution of 1:500 (50 μL) and incubated on ice, covered from light, for 15 min. The cells were washed again and resuspended in 200 μL FACS buffer. Fluorescence measurements were acquired using either Guava flow cytometer (EMD Millipore, Burlington, MA, USA) or they were acquired at a Symphony A3 (BD Biosciences, Franklin Lakes, NJ USA). Data was analyzed with FlowJo software V10. Induction was considered to be successful if CD11b expression in viable cells was found to be ~70% or higher.

5. Patents

An international patent application (WO2018232014A1) has been filed covering the composition of matter and method of used disclosed in this report.

Toxins **2019**, *11*, 339

Supplementary Materials: The following are available online at http://www.mdpi.com/2072-6651/11/6/339/s1, Table S1: Comparison of expression using pET Duet versus pET24a (+) dual plasmid system, Table S2: Bacterial Strains, Figure S1: Key residue mutations within LukAB, Figure S2: Expression of CD11b and/or CD14 on (**A**) PMN and (**B**) THP-1 cells.

Author Contributions: Conceptualization, S.K., M.J.A., R.P.A., T.K. (Thomas Kort) and F.W.H.; Data curation, S.K., T.K. (Thomas Kort), I.M., G.C.L., N.W., T.K. (Tulasikumari Kanipakala), N.G., A.V. and H.K.; Funding acquisition, M.J.A.; Investigation, S.K., T.K. (Thomas Kort), and I.M.; Methodology, S.K., R.P.A, T.K. (Thomas Kort); Supervision, M.J.A., R.P.A., H.K. and F.W.H.; Visualization, S.K. and M.J.A.; Writing—original draft, S.K., M.J.A. and R.P.A.; Writing—review & editing, All authors have read and approved the final manuscript.

Funding: This research was funded by National Institute of Allergy and Infectious Diseases (NIAID), grant number R01-AI111205, and CARB-X grant number 4500002691 to MJA.

Acknowledgments: The authors thank Dr. Binh Diep from UCSF, CA for providing SF8300 and isogenic knockout clones.

Conflicts of Interest: MJA has stocks and RPA, FWH, and HK have stock options in Integrated Biotherapeutics Inc. The funders had no role in the design of the study; in the collection, analyses, or interpretation of data; in the writing of the manuscript, or in the decision to publish the results.

References

1. Lowy, F.D. Staphylococcus aureus infections. *N. Engl. J. Med.* **1998**, *339*, 520–532. [CrossRef] [PubMed]
2. Hiramatsu, K. Vancomycin-resistant *Staphylococcus aureus*: A new model of antibiotic resistance. *Lancet Infect. Dis.* **2001**, *1*, 147–155. [CrossRef]
3. Klevens, R.M.; Morrison, M.A.; Nadle, J.; Petit, S.; Gershman, K.; Ray, S.; Harrison, L.H.; Lynfield, R.; Dumyati, G.; Townes, J.M.; et al. Invasive methicillin-resistant *Staphylococcus aureus* infections in the United States. *JAMA* **2007**, *298*, 1763–1771. [CrossRef] [PubMed]
4. Lee, B.Y.; Singh, A.; David, M.Z.; Bartsch, S.M.; Slayton, R.B.; Huang, S.S.; Zimmer, S.M.; Potter, M.A.; Macal, C.M.; Lauderdale, D.S.; et al. The economic burden of community-associated methicillin-resistant *Staphylococcus aureus* (CA-MRSA). *Clin. Microbiol. Infect.* **2013**, *19*, 528–536. [CrossRef] [PubMed]
5. Foster, T.J. Immune evasion by staphylococci. *Nat. Rev. Microbiol.* **2005**, *3*, 948–958. [CrossRef] [PubMed]
6. Aman, M.J.; Adhikari, R.P. Staphylococcal bicomponent pore-forming toxins: targets for prophylaxis and immunotherapy. *Toxins* **2014**, *6*, 950–972. [CrossRef] [PubMed]
7. Inoshima, N.; Wang, Y.; Bubeck Wardenburg, J. Genetic requirement for ADAM10 in severe *Staphylococcus aureus* skin infection. *J. Investig. Dermatol.* **2012**, *132*, 1513–1516. [CrossRef]
8. Spaan, A.N.; van Strijp, J.A.G.; Torres, V.J. Leukocidins: Staphylococcal bi-component pore-forming toxins find their receptors. *Nat. Rev. Microbiol.* **2017**, *15*, 435–447. [CrossRef]
9. Dumont, A.L.; Nygaard, T.K.; Watkins, R.L.; Smith, A.; Kozhaya, L.; Kreiswirth, B.N.; Shopsin, B.; Unutmaz, D.; Voyich, J.M.; Torres, V.J. Characterization of a new cytotoxin that contributes to *Staphylococcus aureus* pathogenesis. *Mol. Microbiol.* **2011**, *79*, 814–825. [CrossRef]
10. Menestrina, G.; Dalla Serra, M.; Comai, M.; Coraiola, M.; Viero, G.; Werner, S.; Colin, D.A.; Monteil, H.; Prevost, G. Ion channels and bacterial infection: The case of beta-barrel pore-forming protein toxins of *Staphylococcus aureus*. *FEBS Lett.* **2003**, *552*, 54–60. [CrossRef]
11. Badarau, A.; Rouha, H.; Malafa, S.; Logan, D.T.; Hakansson, M.; Stulik, L.; Dolezilkova, I.; Teubenbacher, A.; Gross, K.; Maierhofer, B.; et al. Structure-function analysis of heterodimer formation, oligomerization, and receptor binding of the *Staphylococcus aureus* bi-component toxin LukGH. *J. Biol. Chem.* **2015**, *290*, 142–156. [CrossRef] [PubMed]
12. Dumont, A.L.; Yoong, P.; Liu, X.; Day, C.J.; Chumbler, N.M.; James, D.B.; Alonzo, F., 3rd; Bode, N.J.; Lacy, D.B.; Jennings, M.P.; et al. Identification of a crucial residue required for *Staphylococcus aureus* LukAB cytotoxicity and receptor recognition. *Infect. Immun.* **2013**. [CrossRef] [PubMed]
13. Dumont, A.L.; Yoong, P.; Surewaard, B.G.; Benson, M.A.; Nijland, R.; van Strijp, J.A.; Torres, V.J. *Staphylococcus aureus* Elaborates Leukocidin AB To Mediate Escape from within Human Neutrophils. *Infect. Immun.* **2013**, *81*, 1830–1841. [CrossRef] [PubMed]

14. Melehani, J.H.; James, D.B.; DuMont, A.L.; Torres, V.J.; Duncan, J.A. *Staphylococcus aureus* Leukocidin A/B (LukAB) Kills Human Monocytes via Host NLRP3 and ASC when Extracellular, but Not Intracellular. *PLoS Pathog.* **2015**, *11*, e1004970. [CrossRef] [PubMed]

15. Badarau, A.; Rouha, H.; Malafa, S.; Battles, M.B.; Walker, L.; Nielson, N.; Dolezilkova, I.; Teubenbacher, A.; Banerjee, S.; Maierhofer, B.; et al. Context matters: The importance of dimerization-induced conformation of the LukGH leukocidin of *Staphylococcus aureus* for the generation of neutralizing antibodies. *mAbs* **2016**, *8*, 1347–1360. [CrossRef] [PubMed]

16. DuMont, A.L.; Yoong, P.; Day, C.J.; Alonzo, F., 3rd; McDonald, W.H.; Jennings, M.P.; Torres, V.J. *Staphylococcus aureus* LukAB cytotoxin kills human neutrophils by targeting the CD11b subunit of the integrin Mac-1. *Proc. Natl. Acad. Sci USA* **2013**, *110*, 10794–10799. [CrossRef] [PubMed]

17. Scherr, T.D.; Hanke, M.L.; Huang, O.; James, D.B.; Horswill, A.R.; Bayles, K.W.; Fey, P.D.; Torres, V.J.; Kielian, T. *Staphylococcus aureus* Biofilms Induce Macrophage Dysfunction Through Leukocidin AB and Alpha-Toxin. *mBio* **2015**, *6*, e01021-15. [CrossRef]

18. Thomsen, I.P.; Dumont, A.L.; James, D.B.; Yoong, P.; Saville, B.R.; Soper, N.; Torres, V.J.; Creech, C.B. Children with Invasive *Staphylococcus aureus* Disease Exhibit a Potently Neutralizing Antibody Response to the Cytotoxin LukAB. *Infect. Immun.* **2014**. [CrossRef]

19. Wood, J.B.; Jones, L.S.; Soper, N.R.; Nagarsheth, M.; Creech, C.B.; Thomsen, I.P. Commercial Intravenous Immunoglobulin Preparations Contain Functional Neutralizing Antibodies against the *Staphylococcus aureus* Leukocidin LukAB (LukGH). *Antimicrob. Agents Chemother.* **2017**, *61*. [CrossRef]

20. Adhikari, R.P.; Kort, T.; Shulenin, S.; Kanipakala, T.; Ganjbaksh, N.; Roghmann, M.C.; Holtsberg, F.W.; Aman, M.J. Antibodies to *S. aureus* LukS-PV Attenuated Subunit Vaccine Neutralize a Broad Spectrum of Canonical and Non-Canonical Bicomponent Leukotoxin Pairs. *PLoS ONE* **2015**, *10*, e0137874. [CrossRef]

21. Karauzum, H.; Adhikari, R.P.; Sarwar, J.; Devi, V.S.; Abaandou, L.; Haudenschild, C.; Mahmoudieh, M.; Boroun, A.R.; Vu, H.; Nguyen, T.; et al. Structurally Designed Attenuated Subunit Vaccines for S. aureus LukS-PV and LukF-PV Confer Protection in a Mouse Bacteremia Model. *PLoS ONE* **2013**, *8*, e65384. [CrossRef] [PubMed]

22. Jursch, R.; Hildebrand, A.; Hobom, G.; Tranum-Jensen, J.; Ward, R.; Kehoe, M.; Bhakdi, S. Histidine residues near the N terminus of staphylococcal alpha-toxin as reporters of regions that are critical for oligomerization and pore formation. *Infect. Immun.* **1994**, *62*, 2249–2256. [PubMed]

23. Guillet, V.; Roblin, P.; Werner, S.; Coraiola, M.; Menestrina, G.; Monteil, H.; Prevost, G.; Mourey, L. Crystal structure of leucotoxin S component: new insight into the Staphylococcal beta-barrel pore-forming toxins. *J. Biol. Chem.* **2004**, *279*, 41028–41037. [CrossRef] [PubMed]

24. Berube, B.J.; Bubeck Wardenburg, J. *Staphylococcus aureus* alpha-toxin: nearly a century of intrigue. *Toxins* **2013**, *5*, 1140–1166. [CrossRef] [PubMed]

25. Soong, G.; Chun, J.; Parker, D.; Prince, A. *Staphylococcus aureus* activation of caspase 1/calpain signaling mediates invasion through human keratinocytes. *J. Infect. Dis.* **2012**, *205*, 1571–1579. [CrossRef] [PubMed]

26. Fowler, V.G.; Allen, K.B.; Moreira, E.D.; Moustafa, M.; Isgro, F.; Boucher, H.W.; Corey, G.R.; Carmeli, Y.; Betts, R.; Hartzel, J.S.; et al. Effect of an investigational vaccine for preventing *Staphylococcus aureus* infections after cardiothoracic surgery: A randomized trial. *JAMA* **2013**, *309*, 1368–1378. [CrossRef]

27. Adhikari, R.P.; Ajao, A.O.; Aman, M.J.; Karauzum, H.; Sarwar, J.; Lydecker, A.D.; Johnson, J.K.; Nguyen, C.; Chen, W.H.; Roghmann, M.C. Lower Antibody Levels to *Staphylococcus aureus* Exotoxins Are Associated with Sepsis in Hospitalized Adults with Invasive *S. aureus* Infections. *J. Infect. Dis.* **2012**, *206*, 915–923. [CrossRef] [PubMed]

28. Jacobsson, G.; Colque-Navarro, P.; Gustafsson, E.; Andersson, R.; Mollby, R. Antibody responses in patients with invasive *Staphylococcus aureus* infections. *Eur. J. Clin. Microbiol. Infect. Dis.* **2010**, *29*, 715–725. [CrossRef]

29. Fritz, S.A.; Tiemann, K.M.; Hogan, P.G.; Epplin, E.K.; Rodriguez, M.; Al-Zubeidi, D.N.; Bubeck Wardenburg, J.; Hunstad, D.A. A Serologic Correlate of Protective Immunity against Community-Onset *Staphylococcus aureus* Infection. *Clin. Infect. Dis.* **2013**. [CrossRef]

30. Gouaux, E. α-Hemolysin from *Staphylococcus aureus*: an archetype of beta-barrel, channel-forming toxins. *J. Struct. Biol.* **1998**, *121*, 110–122. [CrossRef]

31. Berends, E.T.M.; Zheng, X.; Zwack, E.E.; Menager, M.M.; Cammer, M.; Shopsin, B.; Torres, V.J. *Staphylococcus aureus* Impairs the Function of and Kills Human Dendritic Cells via the LukAB Toxin. *mBio* **2019**, *10*. [CrossRef] [PubMed]

32. Rouha, H.; Badarau, A.; Visram, Z.C.; Battles, M.B.; Prinz, B.; Magyarics, Z.; Nagy, G.; Mirkina, I.; Stulik, L.; Zerbs, M.; et al. Five birds, one stone: neutralization of alpha-hemolysin and 4 bi-component leukocidins of *Staphylococcus aureus* with a single human monoclonal antibody. *mAbs* **2015**, *7*, 243–254. [CrossRef] [PubMed]

Article

A Derivative of Butyric Acid, the Fermentation Metabolite of *Staphylococcus epidermidis*, Inhibits the Growth of a *Staphylococcus aureus* Strain Isolated from Atopic Dermatitis Patients

Supitchaya Traisaeng [1], Deron Raymond Herr [2], Hsin-Jou Kao [3], Tsung-Hsien Chuang [4] and Chun-Ming Huang [3,5,*]

[1] Department of Life Sciences, National Central University, Taoyuan 32001, Taiwan; supit_trai@hotmail.com
[2] Department of Pharmacology, National University of Singapore, Singapore 117600, Singapore; phcdrh@nus.edu.sg
[3] Department of Biomedical Sciences and Engineering, National Central University, Taoyuan 32001, Taiwan; lulu21522@yahoo.com.tw
[4] Immunology Research Center, National Health Research Institutes, Zhunan, Miaoli 35053, Taiwan; thchuang@nhri.edu.tw
[5] Department of Dermatology, University of California, San Diego 3525 John Hopkins Court, Rm276, San Diego, CA 92121, USA
* Correspondence: chunming@ncu.edu.tw; Tel.: +886-3-422-7151 (ext. 36101); Fax: +886-3-425-3427

Received: 18 April 2019; Accepted: 24 May 2019; Published: 31 May 2019

Abstract: The microbiome is a rich source of metabolites for the development of novel drugs. Butyric acid, for example, is a short-chain fatty acid fermentation metabolite of the skin probiotic bacterium *Staphylococcus epidermidis* (*S. epidermidis*). Glycerol fermentation of *S. epidermidis* resulted in the production of butyric acid and effectively hindered the growth of a *Staphylococcus aureus* (*S. aureus*) strain isolated from skin lesions of patients with atopic dermatitis (AD) in vitro and in vivo. This approach, however, is unlikely to be therapeutically useful since butyric acid is malodorous and requires a high concentration in the mM range for growth suppression of AD *S. aureus*. A derivative of butyric acid, BA–NH–NH–BA, was synthesized by conjugation of two butyric acids to both ends of an –NH–O–NH– linker. BA–NH–NH–BA significantly lowered the concentration of butyric acid required to inhibit the growth of AD *S. aureus*. Like butyric acid, BA–NH–NH–BA functioned as a histone deacetylase (HDAC) inhibitor by inducing the acetylation of Histone H3 lysine 9 (AcH3K9) in human keratinocytes. Furthermore, BA–NH–NH–BA ameliorated AD *S. aureus*-induced production of pro-inflammatory interleukin (IL)-6 and remarkably reduced the colonization of AD *S. aureus* in mouse skin. These results describe a novel derivative of a skin microbiome fermentation metabolite that exhibits anti-inflammatory and *S. aureus* bactericidal activity.

Keywords: atopic dermatitis; butyric acid derivative; fermentation; microbiome; *S. aureus*

Key Contribution: An analog of butyric acid suppresses the growth of *Staphylococcus aureus* isolated from patients with atopic dermatitis.

---

1. Introduction

Skin dysbiosis has been defined as a state of microbial imbalance in the skin microbiome [1,2]. Increasing evidence indicates that probiotic bacteria in the skin microbiome can mediate fermentation [3] to rein in the overgrowth of opportunistic pathogens [4,5]. It has been well documented that the skin of patients with atopic dermatitis (AD) is more prone to the colonization and overgrowth of *Staphylococcus aureus* (*S. aureus*) [6]. AD is a chronic relapsing disease of pruritus and eczematous lesions that affects 15% to 20% of the childhood population [7]. It is a chronic, pruritic inflammatory skin disease of unknown origin that usually starts in early infancy but also affects a substantial number of adults [8,9].

S. aureus was found to co-exist with various skin commensal bacteria including *Propionibacterium acnes* (*P. acnes*) (now called *Cutibacterium acnes* (*C. acnes*) and *Staphylococcus epidermidis* (*S. epidermidis*) in an AD lesion [10]. *C. acnes* ferments carbohydrates into short-chain fatty acids (SCFAs) such as propionic acid. *S. epidermidis* ferments glycerol to butyric acid and acetic acid [11] that exert growth suppressive effects on USA300, a community-associated methicillin-resistant *S. aureus* (MRSA) [11]. In addition to the suppression of *S. aureus* growth, SCFAs, especially butyric acid, function as histone deacetylase (HDAC) inhibitors, thereby leading to increased acetylation of histones. Previous studies revealed that butyric acid can effectively inhibit HDAC in skin keratinocytes resulting in anti-inflammatory activity [12]. This is therapeutically relevant, since inflammatory processes are important drivers of AD. T helper type 2 (Th2) cytokines dominate in the early stage of AD, whereas a combination of Th1 and Th2 cytokines arises in chronic AD [13]. Many cytokines, such as interleukin (IL)-6, IL-5, IL-23, tumor necrosis factor alpha (TNFα), are detectable in AD lesions [13]. Functionally, previous studies have illustrated that alpha-toxin can provoke allergic skin diseases by activating mast cells and inducing both skin barrier disruption and AD-like skin inflammation. Furthermore, Toll-like receptor (TLR) 2 ligands provided by *S. aureus* promote AD through IL-4-mediated suppression of IL-10 [14]. SCFAs can regulate several immune cell functions including the production of cytokines (TNF-α, IL-2, IL-6, and IL-10). Butyric acid significantly attenuated lipopolysaccharides (LPS)-induced NFκB activation and nitric oxide production [15] and reduced IFNγ-induced IL-6 and TNF-α production in a macrophage cell line [16]. The ability of immune cells to migrate to the foci of infection can be regulated by SCFAs [17]. However, it is not clear yet how cytokines in a skin lesion of AD influence the growth of *S. aureus*. It has been known that IL-1β and IFN-γ, but not IL-6, induced a concentration-dependent increase of *S. aureus* growth [18]. Neutralization of IL-6 by monoclonal antibodies improved atopic dermatitis but was associated with bacterial superinfection [19].

Most SCFAs are malodorous and in general have short half-lives. Furthermore, a relatively high concentration (in the mM range) is necessary for a growth inhibitory effect of SCFAs toward pathogens [11]. Therapeutic levels of SCFAs in the mM range may be not feasible in vivo. In addition, high concentrations of SCFAs or their organic solvents may damage skin cells or the underlying tissues. These disadvantages present potentially insurmountable barriers that would prevent the use of native SCFAs as topical therapeutic agents. However, previous studies indicated that GW9508, an arylalkyl derivative of propionic acid, suppressed chemokine induction in keratinocytes and attenuated cutaneous inflammation at nanomolar to micromolar concentrations [20]. We have previously synthesized an esterified derivative of propionic acid which is not water soluble and has a minimum bactericidal concentration (MBC) value against USA300 of approximately 25 mM [21]. An analog of butyric acid, pivaloylomethyl butyrate (AN-9) [22], has been proposed as an anti-cancer prodrug that can produce effective concentrations of butyric acid. In the current study, with the aim to develop butyric acid analogs, a water-soluble derivative of butyric acid {N-[2-(2-Butyrylamino-ethoxy)-ethyl]-butyramide, BA–NH–NH–BA} was synthesized. The antimicrobial activity of BA–NH–NH–BA against an *S. aureus* strain that was isolated from the lesional skin of AD patients was examined.

2. Results

2.1. High Abundance of S. aureus in Lesional Skin of AD Patients

Tape strips were used to sample the skin microbiome from healthy skin and from non-lesional and lesional skin of AD patients. The bacteria on the tape strips were cultured on mannitol salt agar (MSA) plates for 3 d. As shown in Figure 1a, yellow and pink colonies formed in MSA plates. The yellow colonies were selected for 16S ribosomal RNA (rRNA) sequencing and identified as AD *S. aureus* (Figure S1). The pink colonies were recognized as non-*S. aureus* bacteria and were not sampled.

Approximately 40% of all culturable bacteria from healthy skin and non-lesional skin of AD patients produced yellow colonies. By contrast, the percentage of yellow colonies detected from the tape strips collected from lesional skin of AD patients was markedly higher (>80%) (Figure 1b), indicating that the ratio of *S. aureus* to other bacteria on the lesional skin of AD patients was higher than that on either healthy skin or non-lesional skin of AD patients. This result is in agreement with previous findings of *S. aureus* overabundance in the dysbiotic skin microbiome in AD patients [23]. A single yellow bacterial colony isolated from the lesional skin of AD patients ("AD *S. aureus*") was used for further experiments.

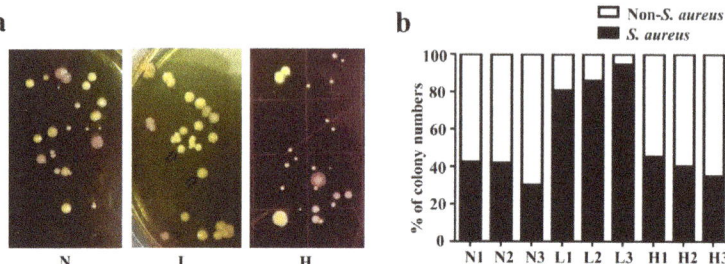

Figure 1. The colonization of *S. aureus* in healthy skin and non-lesional and lesional skin of atopic dermatitis (AD) patients. (**a**) Tape stripping was conducted on healthy skin (H), non-lesional AD skin (N), and lesional AD skin (L). After stripping, the pieces of tape strip were placed, adhesive side down, on mannitol salt agar (MSA) plates for bacterial growth. The colonies of *S. aureus* (arrows, yellow) and non-*S. aureus* (pink) were detected on the MSA plates. A representative picture of three independent experiments is shown. (**b**) The percentage of *S. aureus* and non-*S. aureus* colonies from three subjects with healthy skin (H1, H2, and H3) and AD patients with non-lesional (N1, N2, and N3) and lesional skin (L1, L2, and L3) are shown.

2.2. In Vitro Inhibition of AD S. aureus Growth by Glycerol Fermentation of S. epidermidis

Two experiments were conducted to determine whether the glycerol fermentation of *S. epidermidis*, a skin probiotic bacterium [24], hindered the growth of AD *S. aureus*. The first experiment, an overlay assay, was performed to examine the interference of *S. epidermidis* with AD *S. aureus* on agar plates with or without 2% glycerol (Figure 2a). *S. epidermidis* created a visible inhibitory zone against AD *S. aureus* in the presence of 2% glycerol. No inhibitory zone was detected when *S. epidermidis* and AD *S. aureus* were grown in the absence of glycerol. In the second experiment, *S. epidermidis* was co-cultured with AD *S. aureus* in media with or without glycerol. To establish an *S. aureus*-selective agar plate, the medium from the co-culture of *S. epidermidis* and AD *S. aureus* was spotted on tryptic soy broth (TSB) agar plates supplemented with 10 or 50 mM furazolidone. Furazolidone at 50 mM resulted in complete lethality of *S. epidermidis* (Figure S2) without affecting the growth of AD *S. aureus*. One day after the co-culture of *S. epidermidis* and AD *S. aureus* with or without glycerol, the medium was serially diluted and spotted on *S. aureus*-selective plates. As shown in Figure 2b,c, the concentration of AD *S. aureus* in the co-culture with glycerol ($3.25 \pm 0.48 \times 10^6$ colony-forming unit (CFU)/mL) was significantly lower than in the co-culture without glycerol ($5.25 \pm 0.48 \times 10^7$ CFU/mL). These results suggested that *S. epidermidis*-mediated glycerol fermentation interferes with the growth of AD *S. aureus* in vitro.

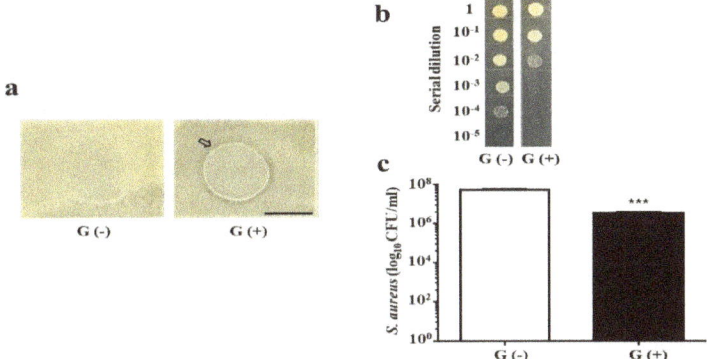

Figure 2. The interference of glycerol fermentation of *S. epidermidis* with the growth of AD *S. aureus* in vitro. (**a**) *S. epidermidis* (10^8 colony-forming unit, CFU) was spotted on top of a lawn of AD *S. aureus* in the absence (G−) or presence (G+) of 2% glycerol for 2 d. An inhibition/clear zone (arrow) was observed when *S. epidermidis* was spotted on the lawn of AD *S. aureus* in the presence of glycerol. Scale bar: 1 cm. (**b**) CFUs of AD *S. aureus* were enumerated by plating serial dilutions ($1:10^1–1:10^5$) of medium from co-cultures of *S. epidermidis* (10^8 CFU/mL) and AD *S. aureus* (10^5 CFU/mL) with (G+) or without (G−) 2% glycerol on a furazolidone (50 mM)-supplemented tryptic soy broth (TSB) plate for 24 h. (**c**) Quantification of AD *S. aureus* from the co-cultures. *** $p < 0.001$ (two-tailed *t*-test). Data shown represent the mean ± standard error (SE) of experiments performed in triplicate.

2.3. Inhibition of AD S. aureus Growth in Mouse Skin

To evaluate whether glycerol fermentation of *S. epidermidis* can hinder the growth of AD *S. aureus* in vivo, *S. epidermidis* and AD *S. aureus* with or without 2% glycerol were applied onto the wounded skin of Institute Cancer Research (ICR) mice for 3 d. The number of AD *S. aureus* in wounded skin applied with two bacteria in the presence of glycerol ($5.25 ± 0.48 × 10^5$ CFU/mL) was approximately one $\log_{10}$ lower than that in wounded skin applied with two bacterial species in the absence of glycerol ($2.25 ± 0.25 × 10^6$ CFU/mL) (Figure 3a,b). Furthermore, the level of IL-6 (Figure 3c) and the wound size (Figure S3) were markedly reduced when the wounded skin was treated with two bacterial species plus glycerol compared to the skin treated only with bacteria. The result indicated that *S. epidermidis* mediates glycerol fermentation to suppress skin colonization of AD *S. aureus* and the production of pro-inflammatory Il-6 cytokine.

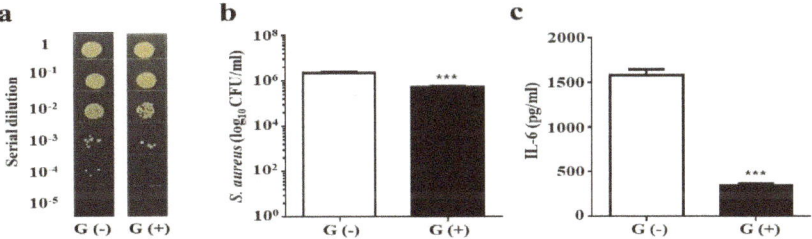

Figure 3. The effect of glycerol fermentation of *S. epidermidis* on the growth of AD *S. aureus* in vivo. (**a**) In the experiment, 10 μL of *S. epidermidis* (10^7 CFU) and AD *S. aureus* (10^7 CFU) with (G+) or without (G−) 2% glycerol was applied onto a skin wound on the dorsal skin of Institute Cancer Research (ICR) mice for 3 d. Bacterial CFUs in the skin wounds were enumerated by plating serial dilutions ($1:10^1–1:10^5$) of the skin homogenate on a furazolidone (50 mM)-supplemented TSB plate. The number ($\log_{10}$ CFU/mL) of AD *S. aureus* (**b**) and the level of pro-inflammatory IL-6 cytokine (**c**) were quantified. Data shown are the mean ± SE. *** $p < 0.001$ (two-tailed *t*-tests).

2.4. Anti-AD S. aureus Activities of Butyric Acid and Its Derivative

In our previous publication, we demonstrated that propionic acid suppressed the growth of USA300 [21]. Since butyric acid is the only SCFA fermentation metabolites of *S. epidermidis* [25], we determined the killing activity of butyric acid against AD *S. aureus*. Butyric acid (0–500 mM) was added into the culture of AD *S. aureus* (10^6 CFU/mL) overnight. The >1 $\log_{10}$ inhibition of butyric acid for AD *S. aureus* was 10 mM, and the concentration for complete inhibition was greater than 50 mM (Figure 4a,b). To circumvent the limitations of butyric acid's malodor and short half-life, BA–NH–NH–BA, a water-soluble derivative of butyric acid was synthesized by conjugating two butyric acids with a –NH–O–NH– linker (Figure 5a). The anti-*S. aureus* activity of BA–NH–NH–BA was examined by adding BA–NH–NH–BA (0–500 mM) into the culture of AD *S. aureus*. The >1 $\log_{10}$ inhibition of BA–NH–NH–BA for AD *S. aureus* was 0.02 mM, while concentrations greater than 250 mM completely inhibited the growth (Figure 5b,c). The >1 $\log_{10}$ inhibition of BA–NH–NH–BA for AD *S. aureus* was 500 times lower than that of butyric acid, indicating the higher potency of BA–NH–NH–BA as an anti-*S. aureus* agent.

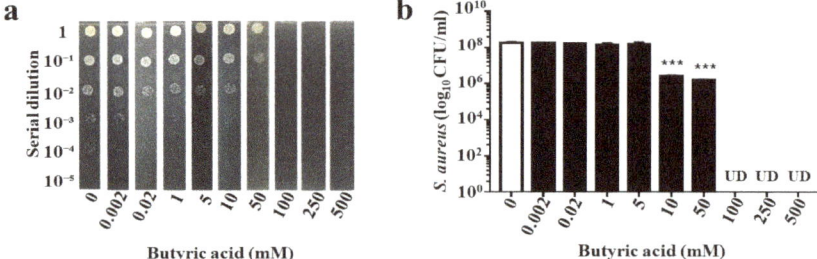

Figure 4. Suppression of AD *S. aureus* growth by butyric acid. AD *S. aureus* (10^6 CFU/mL) was incubated with butyric acid (**a,b**) (0.002–500 mM in PBS) for 24 h. Incubation of AD *S. aureus* with PBS served as a control (0). After incubation, AD *S. aureus* was diluted 1:10^0–1:10^5 with PBS, and 5 μL of the dilutions was spotted on an agar plate (**a**). The CFU counts are illustrated as the mean ± SE of three independent experiments (**b**). *** $p < 0.001$ (two-tailed *t*-test). UD, undetectable.

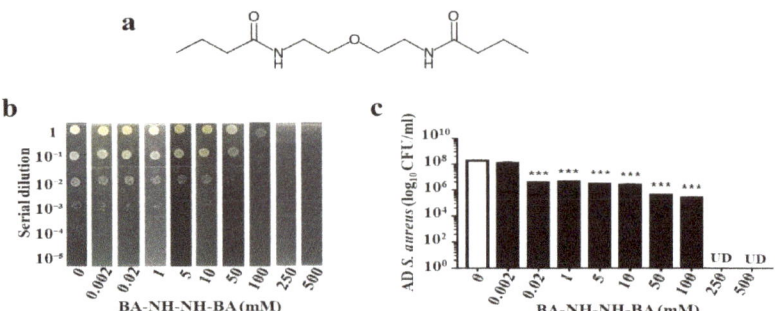

Figure 5. Suppression of AD *S. aureus* growth by BA–NH–NH–BA. (**a**) Chemical structure of BA–NH–NH–BA. (**b**) AD *S. aureus* (10^6 CFU/mL) was incubated with BA–NH–NH–BA (0–500 mM in PBS) for 24 h. After serial dilutions (1:10^0–1:10^5), colonies of AD *S. aureus* incubated with or without BA–NH–NH–BA were grown on agar plates, and the number ($\log_{10}$ CFU/mL) of bacteria was determined (**c**). Data are the mean ± SE of three individual experiments. ** $p < 0.01$, *** $p < 0.001$ (two-tailed *t*-tests). UD, undetectable.

2.5. Inhibition of HDAC and Suppression of AD S. aureus Growth In Vivo by BA–NH–NH–BA

SCFAs, especially butyric acid, act as HDAC inhibitors, thereby leading to increased histone acetylation and regulation of gene transcription. To test if butyric acid and BA–NH–NH–BA act as HDAC inhibitors in skin cells, human skin HaCaT keratinocytes were treated with butyric acid, BA–NH–NH–BA, or phosphate-buffered saline (PBS), as a control, for 8 h. As shown in Figure 6a, treatment with 4 mM butyric acid or BA–NH–NH–BA significantly increased the level of acetylated Histone H3 lysine 9 (AcH3K9). To assess the effects of BA–NH–NH–BA on AD *S. aureus* growth in mouse skin, skin wounds were treated with AD *S. aureus* and with BA–NH–NH–BA (100 μM, 0.4 mM, and 4 mM) or PBS. Three days after the application, the number of AD *S. aureus* was significantly reduced by 0.4 mM BA–NH–NH–BA. The concentration of BA–NH–NH–BA at 4 mM led to >1 $\log_{10}$ suppression of AD *S. aureus* growth in skin wounds (Figure 6b,c). It has been reported that HDAC inhibition by butyric acid modulated several leukocyte functions including the production of cytokines [26]. In the Figure 6d, we found that the level of IL-6 in the AD. *S. aureus*-colonized skin was dose-dependently reduced by BA–NH–NH–BA.

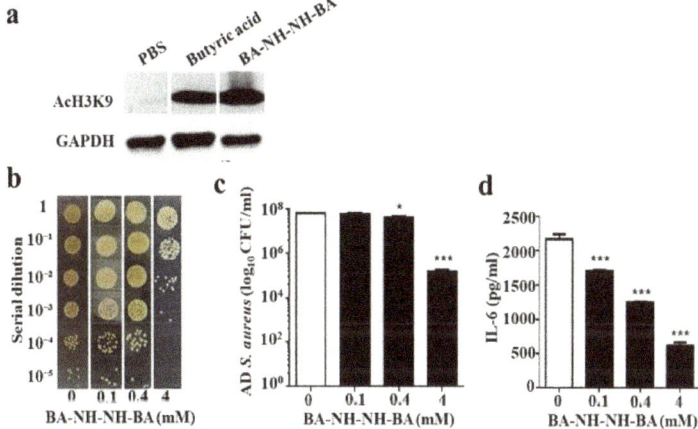

Figure 6. Induction of histone H3 lysine 9 acetylation (AcH3k9) and reduction of AD *S. aureus* growth and IL-6 production by BA–NH–NH–BA. (**a**) Human keratinocyte cells (HaCaT) were treated with 4 mM butyric acid or BA–NH–NH–BA in PBS for 8 h. Cells treated with PBS served as a control. The presence of AcH3K9 in cells was detected by Western blotting using an anti-AcH3K9 antibody. A representative result from three similar experiments is shown. (**b**) A 1 cm wound was made on the dorsal skin of ICR mice before applying AD *S. aureus* with PBS alone (0 mM) or BA–NH–NH–BA (0.1–4 mM in PBS) for 3 d. Bacterial CFUs in the skin wounds were enumerated by plating serial dilutions of the homogenate on a plate. The number ($\log_{10}$ CFU/mL) of AD *S. aureus* bacteria (**c**) and the level of IL-6 pro-inflammatory cytokine (**d**) were determined. Data are the mean ± SE of three separate experiments. * $p < 0.05$, *** $p < 0.001$ (two-tailed *t*-tests).

3. Discussion

S. aureus is a pathogen commonly found in patients with AD. It has been reported that *S. epidermidis* can produce the phenol-soluble modulins (PSMs) γ and δ to hamper the growth of *S. aureus* [27]. Our data in Figure 3 demonstrate that *S. epidermidis* can mediate glycerol fermentation to reduce skin colonization by AD *S. aureus*. Although a high dose is required, butyric acid, a metabolite of glycerol fermentation of *S. epidermidis*, can kill AD *S. aureus* (Figure 4). In addition to *S. epidermidis*, *Staphylococcus hominis* has also been recognized as a beneficial bacterium in AD skin. *S. hominis* can produce lantibiotics which are strain-specific, highly potent, *S. aureus*-selective bactericidal agents that synergize with human antimicrobial peptides such as LL-37 [28]. Clinical studies suggest that the topical application of commensal *S. hominis* reduces AD severity [29]. Several mouse models have been developed to mimic AD in humans. Application of 2,4-dinitrocholrlbenzene (DNCB) on mouse skin shows symptoms similar to those of human AD, including epidermal hyperplasia, dermal mast cell infiltration, and elevated serum IgE levels [30]. In another model, repeated exposure of NC/Nga mice to *Dermatophagoides farinae* (*D. farinae*) crude extract (DfE) induced AD-like lesions [31]. In this study, we applied an *S. aureus* strain isolated from an AD patient onto a skin wound in mice, thus simulating AD patients with a wound on a lesional skin after persistent scratching.

Because of its short half-life, a prohibitively large dose of butyric acid may be necessary for in vivo efficacy. One method to augment the efficacy of butyric acid is to convert it into a prodrug that would generate higher concentrations of intracellular butyric acid. Prodrugs of butyric acid such as AN-9 [32], isobutyramide [33], and tributyrin [34] are lipophilic and have faster rates of intracellular penetration and/or slower rates of metabolic degradation. Data in our previous publication have demonstrated that propionic acid killed USA300 by reducing its intracellular pH [21]. The 2-(2-propionyloxyethoxy) ethylester, a prodrug of propionic acid, and acetic acid 2-(2-acetyloxyethoxy) ethylester, a prodrug of acetic acid, effectively suppressed the growth of USA300 [21] and *Candida parapsilosis* [35], respectively. At least 10 mM butyric acid is required to cause greater than 1 $\log_{10}$ inhibition in the growth of AD *S. aureus* in vitro (Figure 4). Although the mechanism of action of BA–NH–NH–BA against AD *S. aureus* is not clear, the >1 $\log_{10}$ inhibition by BA–NH–NH–BA of AD *S. aureus* was 20 μM (Figure 5). Thus, BA–NH–NH–BA dramatically reduced the required dose of butyric acid for killing AD *S. aureus* in vitro. BA–NH–NH–BA was synthesized by conjugating two butyric acids to a non-cleavable –NH–O–NH– linker. It is a water-soluble compound with a molecular weight of 244.33. As shown in Figure S6, BA–NH–NH–BA did not cause significant cell death when it was applied onto mouse skin. Furthermore, results from gas chromatography (GC) analysis (Figure S7) indicated that a 4 mM BA–NH–NH–BA solution can be stored at 4 °C for six months without degradation. Like the antibacterial activity of water-soluble and hydrophilic chitosan [36], the anti-*S. aureus* activity of BA–NH–NH–BA may be the result of changes in the properties of plasma membrane permeability, which provoke internal osmotic imbalance, consequently inhibiting the growth of bacteria. It has been reported that AD *S. aureus* and other *S. aureus* strains have different characteristics including distinct activities of clumping factor B [37] and T cell responses [38]. Our future works will determine if BA–NH–NH–BA can selectively suppress the growth of AD *S. aureus* without affecting other skin commensal bacteria.

When skin wounds were treated with AD *S. aureus* and BA–NH–NH–BA for three days, both the number of AD *S. aureus* and the level of IL-6 in the skin were attenuated relative to wounds that received *S. aureus* alone (Figure 6b–d). To mimic the over-growth of *S. aureus* in AD lesions, future works will include the inoculation of AD *S. aureus* onto the skin for few days before topical application of BA–NH–NH–BA. To rule out the possibility of reduction of IL-6 due to the elimination of AD *S. aureus* by BA–NH–NH–BA, we applied heat-killed *S. aureus* with BA–NH–NH–BA to KERTr cells. As shown in Figure S4, BA–NH–NH–BA attenuated the level of IL-6 induced by heat-killed *S. aureus*. These data suggest that BA–NH–NH–BA may directly down-regulate the production of IL-6 via the inhibition of HDACs (Figure 6a). Butyric acid inhibits most HDACs except class III HDACs and class II HDAC-6 and -10 [39]. The study of the interaction of an HDAC-like enzyme with trichostatin A (TSA), a broad-spectrum HDAC inhibitor, revealed that the aliphatic chain of TSA occupies a hydrophobic cleft on the surface of HDAC-like enzymes [40]. It is possible that butyric acid and BA–NH–NH–BA also occupy this same hydrophobic pocket to inhibit HDAC activity.

HDAC inhibition may affect immune responses to bacterial infection. Suppression of TLR-induced cytokine production by HDAC inhibition may influence the quality of immune responses to pathogens. Reduction of the production of pro-inflammatory cytokines retards neutrophil recruitment to bacteria-infected sites [41]. It has been documented that phagocytosis of *S. aureus* by immune cells was reduced by HDAC inhibition [42], suggesting that the therapeutic application of HDAC inhibitors might significantly compromise the immune system and render patients more susceptible to *S. aureus* infections. It has been reported that inhibition of HDAC8 and HDAC9 by microbial SCFAs disrupted the tolerance of skin cells to TLR ligands [43], indicating that inhibition of HDACs by SCFAs may alter skin responses to damage-associated molecular patterns (DAMPs) and pathogen-associated molecular patterns (PAMPs). HDACs in mice were depleted by 3,3′-diindolylmethane (DIM) to examine if HDAC inhibition influenced skin colonization by AD *S. aureus*. As shown in Figure S5, depletion of HDACs lowered the level of IL-6 in skin but had no effect on skin colonization of AD *S. aureus*. This result indicates that a decrease in IL-6 production after HDAC inhibition by BA–NH–NH–BA is not sufficient to change skin colonization by AD *S. aureus* (Figure 6c). The reduction of skin colonization of AD *S. aureus* by BA–NH–NH–BA, therefore, is likely to be mainly due to its bactericidal activity. Cumulatively, our study introduces BA–NH–NH–BA as a derivative of butyric acid that, like butyric acid, exerts the activity of HDAC inhibition but has higher potency than butyric acid in terms of suppression of AD *S. aureus* growth.

4. Materials and Methods

4.1. Ethics Statement

This research was carried out in strict accordance with an approved Institutional Animal Care and Use Committee (IACUC) protocol at National Central University (NCU), Taiwan (NCU-106-015, December 19, 2017). The Institutional Review Board (IRB) at Landseed hospital in Taiwan approved the consent and bacterial sampling procedure under an approved protocol (No. 16018C0, June 15, 2018). Written consent was obtained from all participants before conducting bacterial sampling.

4.2. Bacterial Culture

S. aureus bacteria were isolated from AD patients of Landseed Hospital, Taiwan. The isolated *S. aureus* ("AD *S. aureus*") was validated by 16S rRNA sequencing using the 16S rRNA 27F and 534R primers [25] (Figure S1). AD *S. aureus* and *S. epidermidis* (ATCC 12228) were cultured in TSB (Sigma, St. Louis, MO, USA) overnight at 37 °C. The cultures were diluted 1:100 and cultured to an optical density 600 nm (OD_{600}) = 1.0. Bacteria were harvested by centrifugation at 5000× *g* for 10 min, washed with PBS, and suspended in PBS for further experiments.

4.3. Bacterial Sampling with Tape Strips

Bacteria on the skin area of each subject were collected via tape strips. A medical air-permeable tape with acrylic glue (2 × 3 cm, 3M, St. Paul, MN, USA) was sterilized by ultraviolet radiation and thoroughly applied to the skin of healthy subjects (*n* = 3), and to non-lesional (*n* = 3) and lesional (*n* = 3) skin of AD patients. The sterilized tapes were applied to each skin area of each subject for 1 min. The tapes were then peeled off from the skin with sterile forceps. The bacteria of the skin surface adhering on tapes were transferred to MSA (Merck, Billerica, MA, USA) plates and incubated for 72 h. Bacterial colonies appeared yellow or pink in color on MSA plates. Pink colonies were referred to as non-*S. aureus* bacteria. AD *S. aureus* bacteria appearing yellow in color were confirmed by 16S rRNA sequencing.

4.4. Co-culture of AD S. aureus and S. epidermidis In Vitro

An overlay assay was used for the detection of glycerol fermentation of *S. epidermidis* against AD *S. aureus*. Plates contained 1.5% molten (*w/v*) agar (Oxoid. Ltd., London, UK) with or without 2% glycerol in TSB. Bacteria (500 μL PBS containing AD *S. aureus* 10^7 CFU) were poured into the plates to produce a homogeneous lawn of AD *S. aureus*. Afterward, PBS (10 μL) containing *S. epidermidis* (10^8 CFU) was spotted on top of the lawn of AD *S. aureus*, and the bacteria were then cultured at 37 °C for 2 d. An inhibition/clear zone appeared on the surface of the agar when *S. epidermidis* interfered with the growth of AD *S. aureus*. For co-culture experiments, *S. epidermidis* (10^8 CFU/mL) and AD *S. aureus* (10^5 CFU/mL) were cultured in 5 mL of TSB with or without 2% glycerol at 37 °C for 24 h. CFUs of AD *S. aureus* were enumerated by plating serial dilutions ($1:10^0$–$1:10^5$) of the co-culture medium on a furazolidone (50 mM)-supplemented TSB plate. Plates were incubated overnight at 37 °C to count the colonies.

4.5. In Vivo Effects of S. epidermidis Glycerol Fermentation on Skin Colonization of AD S. aureus

ICR mice (8–12-month-old females; National Laboratory Animal Center, Taiwan) were anesthetized by isoflurane. A 1 cm wound was made on the dorsal skin following shaving with electrical clippers. Following skin wounding, 10 μL PBS of AD *S. aureus* (10^7 CFU) and *S. epidermidis* (10^7 CFU) in the presence or absence of 2% glycerol was applied onto the wounds for 3 d. To measure the extent of wound closure, transparent parafilm was placed over the wounded skin, and the area was marked by outlining the area of the wound. The lesion size (cm^2) was measured daily then calculated with ImageJ software [National Institutes of Health (NIH), Bethesda, MD, USA]. To determine bacterial counts, the skin wound was excised 3 d following bacterial application. The excised skin (20 mg) was homogenized in 200 μL of sterile PBS with a tissue grinder. Bacterial CFUs in the skin were enumerated by plating serial dilutions ($1:10^0$–$1:10^5$) of the homogenate on a furazolidone-supplemented TSB plate. The plates were incubated overnight at 37 °C to count the colonies. The bacterial number (CFUs/mL) in excised skin was calculated. The pro-inflammatory IL-6 cytokine was determined by sandwich enzyme-linked immunosorbent assay (ELISA) using IL-6 ELISA kits (R&D systems, Minneapolis, MN, USA).

4.6. Synthesis of BA–NH–NH–BA

Two butyric acids (Sigma-Aldrich, St. Louis, MO, USA) were conjugated to both ends of an –NH–O–NH– linker. Butyric acid (50 mmol) and –NH–O–NH– (20 mmol) in dichloromethane (100 mL) were added to N,N′-dicyclohexyl carbodimide (DCC) (60 mmol) portion-wise. The cloudy white suspensions were stirred at room temperature overnight, then filtered and washed with hexanes. The filtrate was concentrated under reduced pressure to yield pure (>97%) and colorless BA–NH–NH–BA which was further purified by silica gel chromatography eluting with 10% ethyl ethanoate (EtOAc)/hexanes. The conjugation of BA–NH–NH–BA was validated by ^{1}H nuclear magnetic resonance (NMR) (300 MHz) analysis (Bruker DPX-300, Billerica, MA, USA) using $CDCl_3$ as a solvent. To examine the anti-*S. aureus* activity, BA–NH–NH–BA was dissolved in PBS. Skin wounds of ICR mice were topically treated with 10 μL of AD *S. aureus* (10^8 CFU) along with BA–NH–NH–BA (0.1, 0.4 or 4 mM) for 3 d. Application of 10 μL of AD *S. aureus* (10^8 CFU) with PBS served as a control. The number of AD *S. aureus* and the level of IL-6 in skin wound were determined as described above.

4.7. Suppresion of Bacterial Growth

AD *S. aureus* (10^6 CFU/mL) was incubated with butyric acid or BA–NH–NH–BA at various concentrations in PBS as indicated, in media on a 96-well microtiter plate (100 μL per well) for 24 h. The bacteria were incubated with PBS alone as a control. After incubation, the bacteria were diluted $1:10^0$–$1:10^5$ with PBS. The number of bacteria was determined by spotting the dilution (5 μL) on an agar plate supplemented with medium for the counting of CFUs.

4.8. Cell Culture

Human HaCaT keratinocyte cells were cultured in Dulbecco's modified essential medium (Gibco-BRL, Grand Island, NY, USA) with 10% (*v/v*) fetal bovine serum (Irvine Scientific, Santa Ana, CA, USA), 100 units/mL penicillin, and 100 μg/mL streptomycin. The cells (5×10^4 cells/mL) were incubated for 3 d before treatment with PBS, 4 mM butyric acid, or BA–NH–NH–BA for 8 h. The cells were then harvested for detection of AcH3K9 levels by western blot.

Toxins **2019**, *11*, 311

4.9. Western Blotting

Lysates (30 μg) of HaCaT cells were separated by sodium dodecyl sulfate-polyacrylamide gel electrophoresis and then transferred onto a nitrocellulose membrane by use of a transfer cell (Bio-Rad, Hercules, CA, USA). Western blotting was carried out by sequential incubation in 5% non-fat milk blocking buffer at room temperature for 60 min, followed by incubation with primary antibodies against either AcH3K9 (Abcam, Cambridge, MA, USA) or glyceraldehyde 3-phosphate dehydrogenase (GAPDH) (Abcam, Cambridge, MA, USA) at 4 °C overnight, and finally horseradish peroxidase-conjugated anti-rabbit secondary antibodies (Abcam, Cambridge, MA, USA) at room temperature for 90 min. Immunoreactive bands were detected by reaction with the enhanced chemiluminescence (ECL) detection system reagent (Amersham, Arlington Heights, IL, USA).

4.10. Statistical Analysis

To determine significance between groups, comparisons were made using the two-tailed Student's *t*-test. For in vivo experiments, at least three mice per group per experiment were used. Data represent the mean ± SE from three independent experiments. For all statistical tests, *p*-values of < 0.05 (*), < 0.01 (**), and < 0.001 (***) were accepted for statistical significance.

Supplementary Materials: The following are available online at http://www.mdpi.com/2072-6651/11/6/311/s1, Figure S1: 16S ribosomal RNA (rRNA) sequence of AD *S. aureus* using the 16S rRNA 27F and 534R primers. Figure S2: Inhibition of *S. epidermidis* growth by furazolidone. *S. epidermidis* (10^7 CFU) was cultured with 0, 10, and 50 mM of furazolidone for 24 h. Figure S3: Effects of glycerol fermentation by *S. epidermidis* on wound healing in vivo. Figure S4: Levels of the pro-inflammatory IL-6 cytokine in KERTr cells treated with PBS, lysate of AD *S. aureus* (100 μg), BA–NH–NH–BA (0.1 mM), and lysate of AD *S. aureus* plus BA–NH–NH–BA for 24 h. Figure S5: Level of IL-6 and bacterial growth in AD *S. aureus*-colonized skin wounds of mice pretreated with DIM. HDACs in ICR mice were depleted by DIM as described in Materials and Methods. Figure S6: No significant cytotoxic effect of BA–NH–NH–BA. Figure S7: Stability of BA–NH–NH–BA by GC analysis.

Author Contributions: Conceptualization, C.-M.H. Investigation, C.-M.H., S.T. and H.-J.K. Resource, C.-M.H. Original Draft Preparation, S.T. Writing. C.-M.H., S.T., D.R.H., Review and Editing, C.-M.H. and D.R.H.

Funding: This work was mainly supported by an NHRI grant (NHRI-EX106-10607SI) and MOST grants (108-2622-8-008-003-TB1, 107-2622-B-008-002-CC1, 107-2622-B-008-001-CC1, 107-2314-B-008-001, and 1107-2923-B-008-001-MY3) and partially supported by an NIH STTR grant (1R41AR064046-01).

Acknowledgments: This work was mainly supported by an NIH STTR grant (1R41AR064046-01) and partially supported by an NHRI grant ((NHRI-EX106-10607SI) and MOST grants (105-2320-B-008-001,106-2314-B-008-002 and 106-2622-B-008-001-CC1). We thank Ya-Jen Chang at the Institute of biomedical sciences, Academia Sinica, Taiwan for providing KERTr cells.

Conflicts of Interest: The authors declare no conflict of interest.

References

1. Kaur, N.; Chen, C.-C.; Luther, J.; Kao, J.Y. Intestinal dysbiosis in inflammatory bowel disease. *Gut Microbes* **2011**, *2*, 211–216. [CrossRef]

2. Grice, E.A.; Segre, J.A. The human microbiome: Our second genome. *Annu. Rev. Genomics Human Genet.* **2012**, *13*, 151–170. [CrossRef]

3. Ren, T.; Glatt, D.U.; Nguyen, T.N.; Allen, E.K.; Early, S.V.; Sale, M.; Winther, B.; Wu, M. 16 S rRNA survey revealed complex bacterial communities and evidence of bacterial interference on human adenoids. *Environ. Microbiol.* **2013**, *15*, 535–547. [CrossRef]

4. Iwase, T.; Uehara, Y.; Shinji, H.; Tajima, A.; Seo, H.; Takada, K.; Agata, T.; Mizunoe, Y. Staphylococcus epidermidis Esp inhibits Staphylococcus aureus biofilm formation and nasal colonization. *Nature* **2010**, *465*, 346. [CrossRef]

5. Naik, S.; Bouladoux, N.; Wilhelm, C.; Molloy, M.J.; Salcedo, R.; Kastenmuller, W.; Deming, C.; Quinones, M.; Koo, L.; Conlan, S. Compartmentalized control of skin immunity by resident commensals. *Science* **2012**, *337*, 1115–1119. [CrossRef] [PubMed]

6. Leung, D.Y. New insights into atopic dermatitis: Role of skin barrier and immune dysregulation. *Allergol. Int.* **2013**, *62*, 151–161. [CrossRef] [PubMed]

7. Chung, C.S.; Yamini, S.; Trumbo, P.R. FDA's health claim review: Whey-protein partially hydrolyzed infant formula and atopic dermatitis. *Pediatrics* **2012**, *130*, e408–e414. [CrossRef] [PubMed]

8. Eichenfield, L. Consensus guidelines in diagnosis and treatment of atopic dermatitis. *Allergy* **2004**, *59*, 86–92. [CrossRef]

9. Tollefson, M.M.; Bruckner, A.L. Atopic dermatitis: Skin-directed management. *Pediatrics* **2014**, *134*, e1735–e1744. [CrossRef] [PubMed]

10. Kong, H.H.; Oh, J.; Deming, C.; Conlan, S.; Grice, E.A.; Beatson, M.A.; Nomicos, E.; Polley, E.C.; Komarow, H.D.; Murray, P.R. Temporal shifts in the skin microbiome associated with disease flares and treatment in children with atopic dermatitis. *Genome Res.* **2012**. [CrossRef]

11. Kao, M.S.; Huang, S.; Chang, W.L.; Hsieh, M.F.; Huang, C.J.; Gallo, R.L.; Huang, C.M. Microbiome precision editing: Using PEG as a selective fermentation initiator against methicillin-resistant Staphylococcus aureus. *Biotech. J.* **2017**, *12*. [CrossRef] [PubMed]

12. Meijer, K.; de Vos, P.; Priebe, M.G. Butyrate and other short-chain fatty acids as modulators of immunity: What relevance for health? *Curr. Opin. Clin. Nutr. Metab. Care* **2010**, *13*, 715–721. [CrossRef] [PubMed]

13. Mjösberg, J.; Eidsmo, L. Update on innate lymphoid cells in atopic and non-atopic inflammation in the airways and skin. *Clin. Exp. Allergy* **2014**, *44*, 1033–1043. [CrossRef] [PubMed]

14. Kaesler, S.; Volz, T.; Skabytska, Y.; Köberle, M.; Hein, U.; Chen, K.-M.; Guenova, E.; Wölbing, F.; Röcken, M.; Biedermann, T. Toll-like receptor 2 ligands promote chronic atopic dermatitis through IL-4–mediated suppression of IL-10. *J. Allergy Clin. Immunol.* **2014**, *134*, 92–99. [CrossRef]

15. Chakravortty, D.; Koide, N.; Kato, Y.; Sugiyama, T.; Mu, M.M.; Yoshida, T.; Yokochi, T. The inhibitory action of butyrate on lipopolysaccharide-induced nitric oxide production in RAW 264.7 murine macrophage cells. *J. Endotoxin Res.* **2000**, *6*, 243–247. [CrossRef]

16. Park, J.-S.; Lee, E.-J.; Lee, J.-C.; Kim, W.-K.; Kim, H.-S. Anti-inflammatory effects of short chain fatty acids in IFN-γ-stimulated RAW 264.7 murine macrophage cells: Involvement of NF-κB and ERK signaling pathways. *Int. Immunopharmacol.* **2007**, *7*, 70–77. [CrossRef] [PubMed]

17. Vinolo, M.A.; Rodrigues, H.G.; Nachbar, R.T.; Curi, R. Regulation of inflammation by short chain fatty acids. *Nutrients* **2011**, *3*, 858–876. [CrossRef] [PubMed]

18. Di Domenico, E.; Cavallo, I.; Bordignon, V.; Prignano, G.; Sperduti, I.; Gurtner, A.; Trento, E.; Toma, L.; Pimpinelli, F.; Capitanio, B. Inflammatory cytokines and biofilm production sustain Staphylococcus aureus outgrowth and persistence: A pivotal interplay in the pathogenesis of Atopic Dermatitis. *Sci. Rep.* **2018**, *8*, 9573. [CrossRef]

19. Navarini, A.A.; French, L.E.; Hofbauer, G.F. Interrupting IL-6–receptor signaling improves atopic dermatitis but associates with bacterial superinfection. *J. Allergy Clin. Immunol.* **2011**, *128*, 1128–1130. [CrossRef]

20. Fujita, T.; Matsuoka, T.; Honda, T.; Kabashima, K.; Hirata, T.; Narumiya, S. A GPR40 agonist GW9508 suppresses CCL5, CCL17, and CXCL10 induction in keratinocytes and attenuates cutaneous immune inflammation. *Invest. Dermatol.* **2011**, *131*, 1660–1667. [CrossRef] [PubMed]

21. Wang, Y.; Dai, A.; Huang, S.; Kuo, S.; Shu, M.; Tapia, C.; Yu, J.; Two, A.; Zhang, H.; Gallo, R. Propionic acid and its esterified derivative suppress the growth of methicillin-resistant Staphylococcus aureus USA300. *Benef. Microbes* **2014**, *5*, 161–168. [CrossRef]

22. Hobdy, E.; Murren, J. AN-9 (Titan). *Curr. Opin. Invest. Drugs* **2004**, *5*, 628–634.

23. Iwamoto, K.; Moriwaki, M.; Miyake, R.; Hide, M. Staphylococcus aureus in atopic dermatitis: Strain-specific cell wall proteins and skin immunity. *Allergol. Int.* **2019**. [CrossRef]

24. Yang, A.-J.; Marito, S.; Yang, J.-J.; Keshari, S.; Chew, C.-H.; Chen, C.-C.; Huang, C.-M. A Microtube Array Membrane (MTAM) Encapsulated Live Fermenting Staphylococcus epidermidis as a Skin Probiotic Patch against Cutibacterium acnes. *Int. J. Mol. Sci.* **2019**, *20*, 14. [CrossRef]

25. Wang, Y.; Kuo, S.; Shu, M.; Yu, J.; Huang, S.; Dai, A.; Gallo, R.L.; Huang, C.-M. Staphylococcus epidermidis in the human skin microbiome mediates fermentation to inhibit the growth of Propionibacterium acnes: Implications of probiotics in acne vulgaris. *Appl. Microbiol. Biotech.* **2014**, *98*, 411–424. [CrossRef] [PubMed]

26. Chriett, S.; Dąbek, A.; Wojtala, M.; Vidal, H.; Balcerczyk, A.; Pirola, L. Prominent action of butyrate over β-hydroxybutyrate as histone deacetylase inhibitor, transcriptional modulator and anti-inflammatory molecule. *Sci. Rep.* **2019**, *9*, 742. [CrossRef] [PubMed]

27. Cogen, A.L.; Yamasaki, K.; Sanchez, K.M.; Dorschner, R.A.; Lai, Y.; MacLeod, D.T.; Torpey, J.W.; Otto, M.; Nizet, V.; Kim, J.E. Selective antimicrobial action is provided by phenol-soluble modulins derived from Staphylococcus epidermidis, a normal resident of the skin. *J. Invest. Dermatol.* **2010**, *130*, 192–200. [CrossRef] [PubMed]

28. Nakatsuji, T.; Chen, T.H.; Narala, S.; Chun, K.A.; Two, A.M.; Yun, T.; Shafiq, F.; Kotol, P.F.; Bouslimani, A.; Melnik, A.V. Antimicrobials from human skin commensal bacteria protect against Staphylococcus aureus and are deficient in atopic dermatitis. *Sci. Transl. Med.* **2017**, *9*, eaah4680. [CrossRef]

29. Paller, A.S.; Kong, H.H.; Seed, P.; Naik, S.; Scharschmidt, T.C.; Gallo, R.L.; Luger, T.; Irvine, A.D. The microbiome in patients with atopic dermatitis. *J. Allergy Clin. Immunol.* **2019**, *143*, 26–35. [CrossRef] [PubMed]

30. Jin, W.; Huang, W.; Chen, L.; Jin, M.; Wang, Q.; Gao, Z.; Jin, Z. Topical Application of JAK1/JAK2 Inhibitor Momelotinib Exhibits Significant Anti-Inflammatory Responses in DNCB-Induced Atopic Dermatitis Model Mice. *Int. J. Mol. Sci.* **2018**, *19*, 3973. [CrossRef]

31. Maeda, N.; Yamada, C.; Takahashi, A.; Kuroki, K.; Maenaka, K. Therapeutic application of human leukocyte antigen-G1 improves atopic dermatitis-like skin lesions in mice. *Int. Immunopharmacol.* **2017**, *50*, 202–207. [CrossRef] [PubMed]

32. Rabizadeh, E.; Shaklai, M.; Nudelman, A.; Eisenbach, L.; Rephaeli, A. Rapid alteration of c-myc and c-jun expression in leukemic cells induced to differentiate by a butyric acid prodrug. *FEBS lett.* **1993**, *328*, 225–229. [CrossRef]

33. Perrine, S.P.; Dover, G.H.; Daftari, P.; Walsh, C.T.; Jin, Y.; Mays, A.; Faller, D.V. Isobutyramide, an orally bioavailable butyrate analogue, stimulates fetal globin gene expression in vitro and in vivo. *Br. J. Haematol.* **1994**, *88*, 555–561. [CrossRef] [PubMed]

34. Chen, Z.-X.; Breitman, T.R. Tributyrin: A prodrug of butyric acid for potential clinical application in differentiation therapy. *Cancer Res.* **1994**, *54*, 3494–3499. [PubMed]

35. Kao, M.-S.; Wang, Y.; Marito, S.; Huang, S.; Lin, W.-Z.; Gangoiti, J.A.; Barshop, B.A.; Hyun, C.; Lee, W.-R.; Sanford, J.A. The mPEG-PCL copolymer for selective fermentation of Staphylococcus lugdunensis against Candida parapsilosis in the human microbiome. *J. Microb. Biochem. Tech.* **2016**, *8*, 259.

36. Raafat, D.; Sahl, H.G. Chitosan and its antimicrobial potential—A critical literature survey. *Microb. Biotechnol.* **2009**, *2*, 186–201. [CrossRef]

37. Fleury, O.M.; McAleer, M.A.; Feuillie, C.; Formosa-Dague, C.; Sansevere, E.; Bennett, D.E.; Towell, A.M.; McLean, W.I.; Kezic, S.; Robinson, D.A. Clumping factor B promotes adherence of Staphylococcus aureus to corneocytes in atopic dermatitis. *Infection Immun.* **2017**, *85*, e00994-00916. [CrossRef] [PubMed]

38. Iwamoto, K.; Moriwaki, M.; Niitsu, Y.; Saino, M.; Takahagi, S.; Hisatsune, J.; Sugai, M.; Hide, M. Staphylococcus aureus from atopic dermatitis skin alters cytokine production triggered by monocyte-derived Langerhans cell. *J. Dermatol. Sci.* **2017**, *88*, 271–279. [CrossRef]

39. Davie, J.R. Inhibition of histone deacetylase activity by butyrate. *J. Nutr.* **2003**, *133*, 2485S–2493S. [CrossRef] [PubMed]

40. Finnin, M.S.; Donigian, J.R.; Cohen, A.; Richon, V.M.; Rifkind, R.A.; Marks, P.A.; Breslow, R.; Pavletich, N.P. Structures of a histone deacetylase homologue bound to the TSA and SAHA inhibitors. *Nature* **1999**, *401*, 188. [CrossRef]

41. Larsen, J.M. The immune response to Prevotella bacteria in chronic inflammatory disease. *Immunology* **2017**, *151*, 363–374. [CrossRef] [PubMed]

42. Mombelli, M.; Lugrin, J.; Rubino, I.; Chanson, A.-L.; Giddey, M.; Calandra, T.; Roger, T. Histone deacetylase inhibitors impair antibacterial defenses of macrophages. *J. Infect. Dis.* **2011**, *204*, 1367–1374. [CrossRef] [PubMed]

43. Sanford, J.A.; Zhang, L.-J.; Williams, M.R.; Gangoiti, J.A.; Huang, C.-M.; Gallo, R.L. Inhibition of HDAC8 and HDAC9 by microbial short-chain fatty acids breaks immune tolerance of the epidermis to TLR ligands. *Sci. Immunol.* **2016**, *1*, eaah4609. [CrossRef] [PubMed]

 *toxins*

Communication

Study on the Growth and Enterotoxin Production by *Staphylococcus aureus* in Canned Meat before Retorting

Luca Grispoldi [1,*]**, Paul Alexanderu Popescu** [2]**, Musafiri Karama** [3]**, Vito Gullo** [1]**, Giusi Poerio** [1]**, Elena Borgogni** [1]**, Paolo Torlai** [1]**, Giuseppina Chianese** [1]**, Anna Giovanna Fermani** [4]**, Paola Sechi** [1] **and Beniamino Cenci-Goga** [1]

[1] Medicina Veterinaria, Laboratorio di Ispezione degli Alimenti di Origine Animale, Università degli Studi di Perugia, 06126 Perugia, Italy; vito93.v@gmail.com (V.G.); giusi.poerio@gmail.com (G.P.); borgogni.ele@gmail.com (E.B.); si.to06@libero.it (P.T.); giusychianese@hotmail.com (G.C.); paola_sechi@outlook.it (P.S.); beniamino.cencigoga@unipg.it (B.C.-G.)

[2] Faculty of Biotechnology, University of Agronomical Science and Veterinary Medicine, 011464 Bucharest, Romania; paul.alex.popescu@gmail.com

[3] Faculty of Veterinary Science, Department of Paraclinical Sciences, University of Pretoria, Onderstepoort 0110, South Africa; Musafiri.Karama@up.ac.za

[4] Department of prevention, Azienda Unità Sanitaria Locale Latina, 04012 Lazio, Italy; angiferm@gmail.com

* Correspondence: grisluca@outlook.it; Tel.: +39-075-585-7973

Received: 16 April 2019; Accepted: 21 May 2019; Published: 23 May 2019

Abstract: Possible contamination by *Staphylococcus aureus* of the production environment and of the meat of a canned meat production factory was analysed. A total of 108 samples were taken from nine critical control points, 13 of them were positive for *S. aureus*. None of the isolates produced enterotoxins. To determine how much time can elapse between can seaming and sterilisation in the autoclave without any risk of enterotoxin production by *S. aureus*, the growth and enterotoxin production of three enterotoxin A producing strains of *S. aureus* (one ATCC strain and two field strains) in canned meat before sterilisation was investigated at three different temperatures (37, 20 and 10 °C). Two types of meat were used, one with and one without sodium nitrite. In the canned products, the spiked bacteria spread throughout the meat and reached high levels. Enterotoxin production was shown to start 10 hours after incubation at 37 °C and after 48 h after incubation at 20 °C; the production of enterotoxin was always detected in the transition between the exponential and the stationary growth phase. At 10 °C, the enterotoxin was never detected. The statistical analysis of the data showed that the difference between the two different types of meat was not statistically significant (p value > 0.05). Since it is well known that following heat treatment, staphylococcal enterotoxins, although still active (in in vivo assays), can be undetectable (loss of serological recognition) depending on the food matrix and pH, it is quite difficult to foresee the impact of heat treatment on enterotoxin activity. Therefore, although the bacteria are eliminated, the toxins may remain and cause food poisoning. The significance of the results of this study towards implementing good manufacturing practices and hazard analysis critical control points in a canned meat factory are discussed with reference to the management of pre-retorting steps after seaming.

Keywords: *Staphylococcus aureus*; canned meat; enterotoxin; HACCP

Key Contribution: This is a thorough study of CCPs in a large canned meat plant which includes a systematic sampling at nine CCPs along with a model to investigate the actual behaviour of an ATCC and two field strains of *S. aureus* in canned meat before sterilization. Although *S. aureus* enterotoxin production has been investigated extensively by several authors in the past; this is the first systematic study on the behaviour of *S. aureus* in canned meat matrix with and without added nitrites.

---

1. Introduction

Processed meats are protein-rich foods, which can serve as an excellent culture media for the growth of microorganisms [1]. Contamination by microorganisms can be exacerbated during the canning process, especially if the end product presents low acidity and is maintained in conditions of temperature abuse [2]. Contamination may occur not only during meat processing if manufacturing practices are poor, but also during the processes of transportation, storage and handling. There are several bacterial species, which are known to be able to contaminate canned meat, e.g., *Clostridium* spp., *Listeria* spp., *Bacillus* spp., *Escherichia coli* and *Staphylococcus aureus* [3–5].

Among those pathogens, staphylococci are of particular interest, as they are one of the most frequent microbial contaminants isolated from small- and medium-sized meat processing factories around the world [2,6]. *S. aureus* represents a serious hazard for the end consumer, as it is able to produce enterotoxins, which are stable at high temperatures (e.g., Crude enterotoxin A remains active at 100 °C for 2 h in broth and at 121 °C for 28 min in mushrooms) and can also resist under many environmental conditions (low pH, freezing, drying), in which *S. aureus* strains do not survive [7,8]. They are also resistant to human proteolytic enzymes and retain their activity in the digestive tract after ingestion [9]. The amount of enterotoxin required to cause the illness in susceptible subjects can be as little as 20–100 ng [10]. A fraction of the strains of *S. aureus* are also able to persist in the factory environment by forming biofilm [11].

To date, a total of 23 distinct SEs have been identified, based on their antigenicity (SEA to SElY) [12]. Among these, staphylococcal enterotoxin A (SEA) is the most frequently reported in cases of food poisoning [13,14].

The literature has reported several past cases of food poisoning caused by *S. aureus* contaminated canned meat [15–17]. In more recent years, despite the strict regulations on the processing and canning of food products, there are still cases of food poisoning related to the consumption of canned foods: e.g., the EU Rapid Alert System for Food and Feed (RASFF) reported the presence of *S. aureus* and staphylococcal enterotoxins in various meat products in Germany (2005) and the Netherlands (2007).

There is little information available in literature about the growth, survival and enterotoxin production by *S. aureus* in canned meat. Most of the available data is based on models consisting of liquid cultures, where the bacteria are in a planktonic state. Recent studies have shown that there are significant differences between those models and bacteria in a food matrix, where the association with surfaces and tissue and the communication with other bacteria by means of molecular signalling are prevalent [18,19].

In this study, we analysed the possible contamination by *S. aureus* of the production environment and of the meat in a large canned meat production factory. We investigated the behaviour of three enterotoxigenic *S. aureus* strains spiked in canned meat with and without sodium nitrite (a food additive used as a preservative and colour fixative in cured meat) at different temperatures, in order to implement good manufacturing practices and the hazard analysis and critical control point system at the factory, with specific attention to the management of pre-retorting steps after seaming.

2. Results

Table 1 shows the results of the microbiological analysis of the meat samples and swabs from the canned meat factory. The mesophilic flora ranged from a maximum of 3.91 ± 1.79 CFU g^{-1} in the meat on the belt after the metal detector to a minimum of 1.51 ± 0.88 CFU $(cm^2)^{-1}$ in the belt swabs. *Micrococcus* spp. reached a concentration of 2.84 ± 1.72 CFU g^{-1} in the meat on the belt after the metal detector. At the same point, *Staphylococcus* spp. reached a maximum concentration of 2.52 ± 1.62 CFU g^{-1}. No bacteria were found in the aspic samples taken from the doser. *S. aureus* isolates came from five out of the nine kinds of samples, for a total of 13 isolates. Six of those isolates came from samples of frozen cooked beef at reception. None of the *S. aureus* isolates produced enterotoxins.

Table 1. Results of the microbiological analysis.

Sampling Points	Number of Samples	Mesophilic (Mean ± SD)	*Staphylococcus* spp. (Mean ± SD)	*Micrococcus* spp. (Mean ± SD)	*Staphylococcus aureus* (Number of Isolated Strains)
Frozen cooked beef at reception	15	2.94 ± 1.57	1.92 ± 1.39	1.94 ± 1.76	6
Sliced beef at the slicer	12	2.06 ± 1.45	0.47 ± 0.85	1.24 ± 1.36	0
Belt swabs	9	1.51 ± 0.88	0.49 ± 0.78	0.34 ± 0.68	1
Piston swabs	9	1.52 ± 1.24	0.26 ± 0.51	0.19 ± 0.57	0
Frozen cooked beef at thawing	36	2.86 ± 0.83	1.49 ± 1.39	0.63 ± 1.12	4
Defrost water	6	2.02 ± 1.47	0.91 ± 0.74	0.00 ± 0.00	1
Cooked beef after thawing	6	3.12 ± 0.39	0.41 ± 1.01	0.33 ± 0.82	0
Meat on the belt after metal detector	12	3.91 ± 1.79	2.52 ± 1.62	2.84 ± 1.72	1
Aspic at the doser	3	0.00 ± 0.00	0.00 ± 0.00	0.00 ± 0.00	0

The initial concentration of *S. aureus* spiked ranged between 3.94 ± 0.16 CFU g^{-1} and 4.19 ± 0.11 CFU g^{-1} in meat with sodium nitrite and between 3.85 ± 0.27 CFU g^{-1} and 4.26 ± 0.17 CFU g^{-1} in meat without sodium nitrite.

At 37 °C, *S. aureus* counts reached the plateau phase after 24 h at a concentration between 8.79 ± 0.08 CFU g^{-1} and 8.98 ± 0.03 CFU g^{-1} in the meat with sodium nitrite and between 8.46 ± 0.14 CFU g^{-1} and 8.88 ± 0.1 CFU g^{-1} in the meat without nitrite, and maintained similar values throughout the experiment. The toxin was detected from the 10-hour at a concentration of *S. aureus* between 7.02 ± 0.15 CFU g^{-1} and 8.18 ± 0.11 CFU g^{-1} in the meat with sodium nitrite and at the same time at a concentration between 7.06 ± 0.09 CFU g^{-1} and 8.32 ± 0.02 CFU g^{-1} in the meat without nitrite (Figures 1 and 2).

At 20 °C, the plateau phase was reached 72 h after spiking at a concentration between 8.47 ± 0.16 CFU g^{-1} and 8.87 ± 0.11 CFU g^{-1} in meat with sodium nitrite and at the same time at a concentration between 8.38 ± 0.16 CFU g^{-1} and 8.88 ± 0.16 CFU g^{-1} in meat without sodium nitrite. The toxin was detected from the 48th hour with a *S. aureus* concentration between 7.81 ± 0.19 CFU g^{-1} and 8.34 ± 0.07 CFU g^{-1} in the meat with sodium nitrite and between 7.57 ± 0.19 CFU g^{-1} and 8.49 ± 0.15 CFU g^{-1} in the meat without nitrite (Figures 3 and 4).

After 28 days at 10 °C, the concentration of *S. aureus* was between 6.55 ± 0.27 CFU g^{-1} and 7.44 ± 0.14 CFU g^{-1} in the meat with sodium nitrite and between 6.93 ± 0.19 CFU g^{-1} and 7.52 ± 0.21 CFU g^{-1} in the meat without sodium nitrite. Enterotoxin was never detected throughout the incubation at 10 °C (Figures 5 and 6).

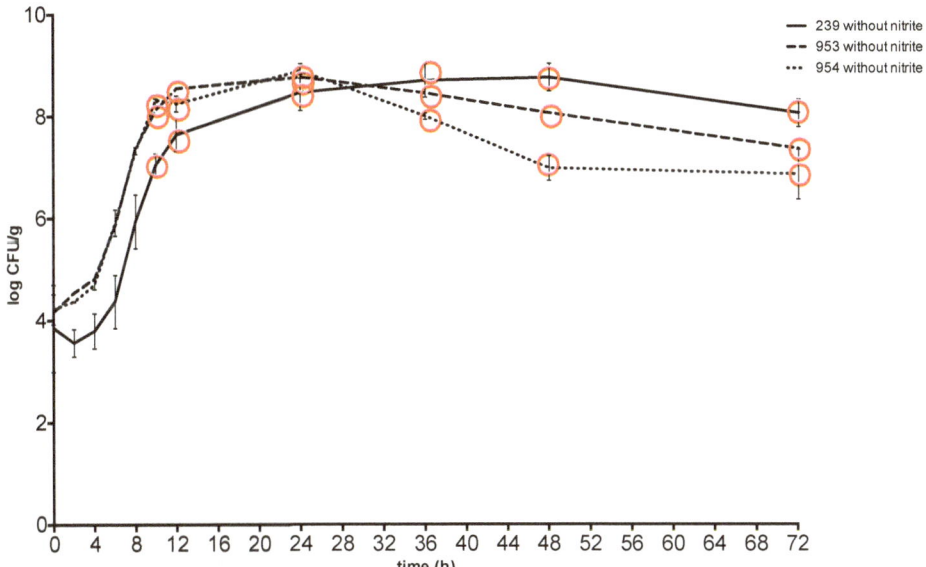

Figure 1. Growth and enterotoxin production of *S. aureus* at 37 °C in meat without sodium nitrite. Red circle: enterotoxin production. 239: *S. aureus* ATCC 29213; 953: *S. aureus* field strain; 954: *S. aureus* field strain.

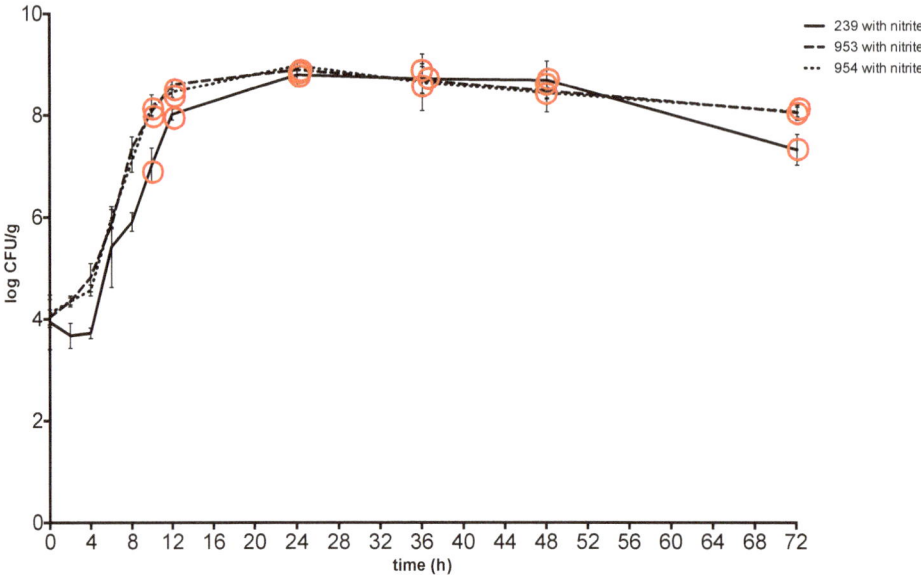

Figure 2. Growth and enterotoxin production of S. aureus at 37 °C in meat with sodium nitrite. Red Circle: enterotoxin production. 239: *S. aureus* ATCC 29213; 953: *S. aureus* field strain; 954: *S. aureus* field strain.

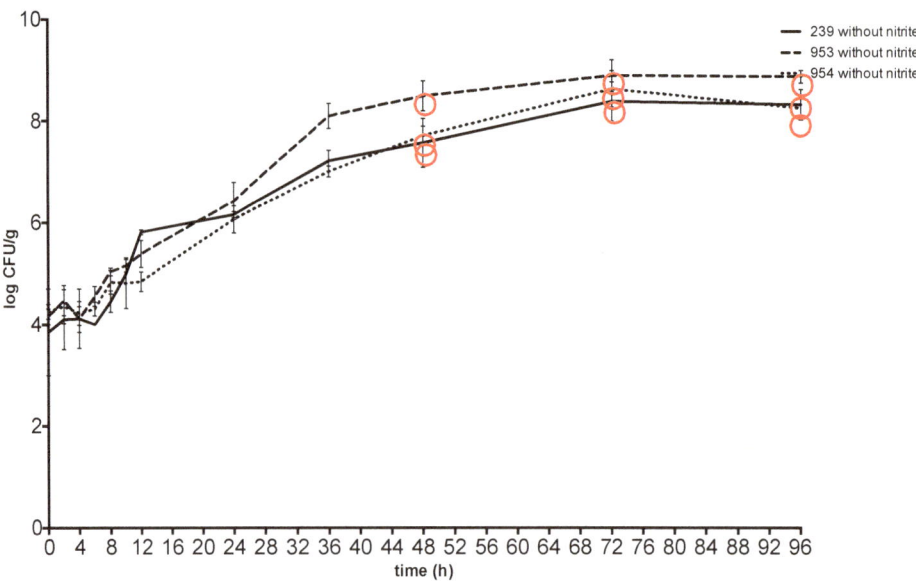

Figure 3. Growth and enterotoxin production of S. aureus at 20 °C in meat without sodium nitrite. Red circle: enterotoxin production. 239: *S. aureus* ATCC 29213; 953: *S. aureus* field strain; 954: *S. aureus* field strain.

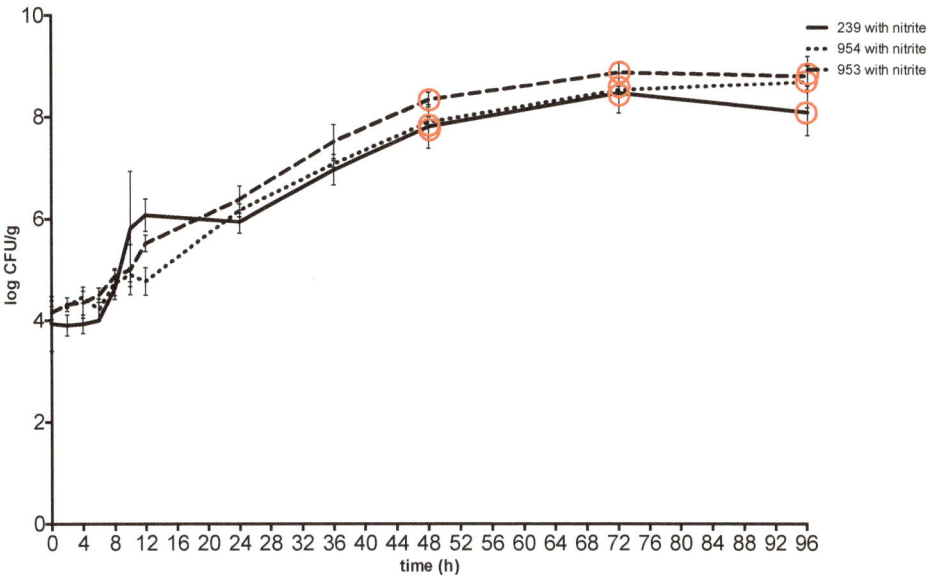

Figure 4. Growth and enterotoxin production of S. aureus at 20 °C in meat with sodium nitrite. Red circle: enterotoxin production. 239: *S. aureus* ATCC 29213; 953: *S. aureus* field strain; 954: *S. aureus* field strain.

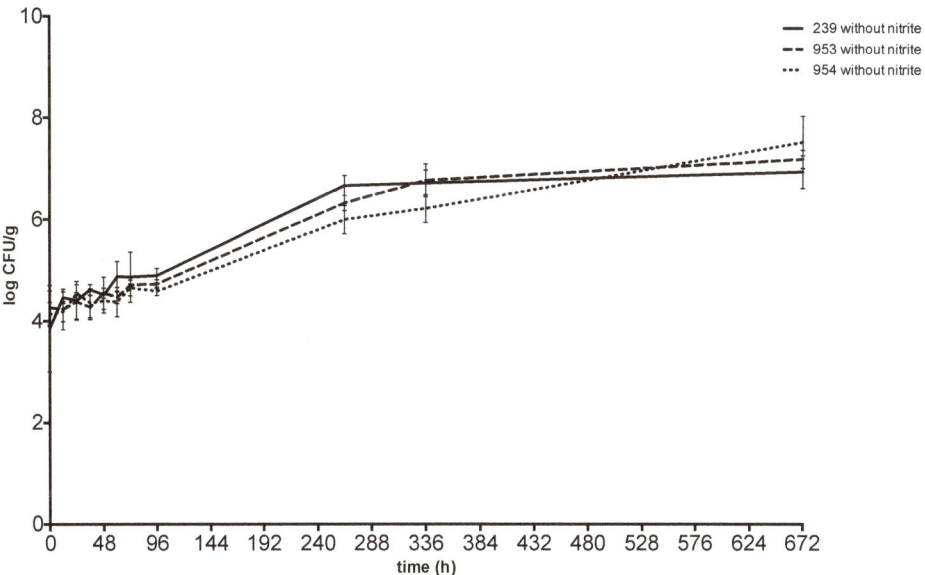

Figure 5. Growth and enterotoxin production of S. aureus at 10 °C in meat without sodium nitrite. 239: *S. aureus* ATCC 29213; 953: *S. aureus* field strain; 954: *S. aureus* field strain.

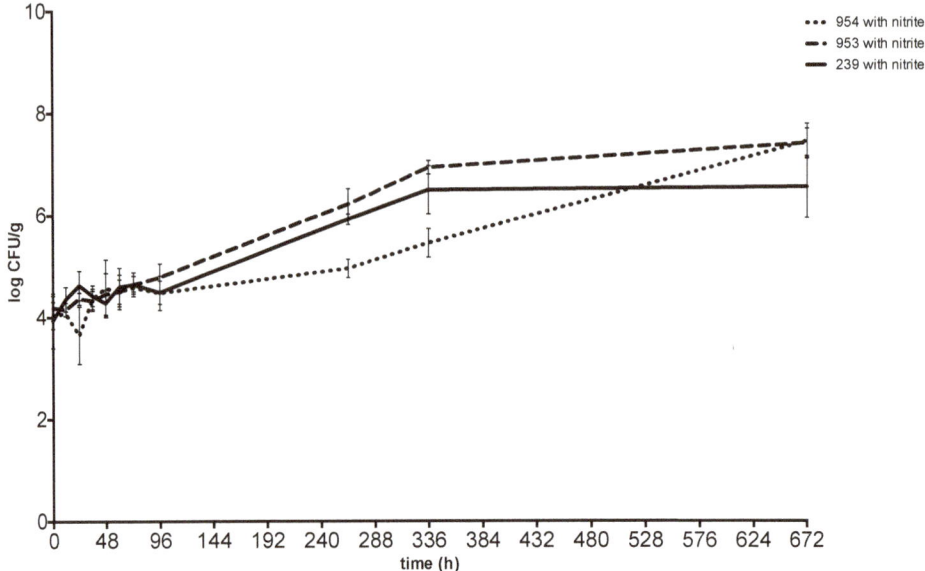

Figure 6. Growth and enterotoxin production of S. aureus at 10 °C in meat with sodium nitrite. 239: *S. aureus* ATCC 29213; 953: *S. aureus* field strain; 954: *S. aureus* field strain.

The statistical analysis of the data showed that the difference in the bacterial growth rate between the meat with sodium nitrite and the meat without sodium nitrite was not statistically significant, with a constant p value of >0.05.

3. Discussion

The data on the *S. aureus* contamination of the meat and the production environment is consistent with other studies: Koreňová et al. (2015) [6] reported 5 out of 144 samples positive for *S. aureus* in a small meat processing factory.

The growth profile and enterotoxin production of the three *S. aureus* strains tested in canned meat is similar to what is reported by other studies. Mansfield et al. (1983) [16] tested the growth and survival of *S. aureus* in canned meat at 15, 22, 30 and 37 °C and reported that the colony counts increased to a maximum of about 10^8 CFU g^{-1} at all temperatures, and the maximum concentration was reached first at the higher temperature. Enterotoxin production was detected when the colony count reached a concentration of >10^6 CFU g^{-1}. Wallin-Carlquist et al. (2010) [1] investigated the growth of *S. aureus* in different meat products stored at room temperature: starting from an initial concentration of approximately 10^4 CFU g^{-1}, the bacteria reached a concentration of 8.87 log CFU cm^{-2} in boiled ham and 8.52 log CFU cm^{-2} in smoked ham. The presence of SEA was detected after 24 h and for the entire incubation period. In our study, the bacteria reached similar concentrations after the same period of time and the production of enterotoxin was always detected in the transition between the exponential and the stationary growth phase.

Comparing our results to data based on a model consisting of liquid cultures, some differences can be observed. Tsutsuura et al. (2013) [20] incubated eleven SEA producer strains of *S. aureus* in BHI broth in temperatures ranging from 10 to 37 °C. SEA produced by these eleven strains were detected after 3 weeks of incubation at 10 °C at a concentration of 7.85 log CFU mL^{-1}, after 3–8 days at 15 °C at a concentration of 9.34 log CFU/mL^{-1}, after 30–58 h at 20 °C at a concentration of 6.96 log CFU/mL^{-1}, and after 6–8 h at 37 °C at a concentration of 7.42 log CFU mL^{-1} with an inoculum size of 10^2 CFU mL^{-1}.

In our study, the growth and enterotoxin production of *S. aureus* in canned meat was slower at 37 °C and 20 °C, whereas the enterotoxin was never detected at 10 °C.

Canned meat has different characteristics to a broth, e.g., nutrient availability, pH, salt content and water activity [21]: furthermore, whereas the bacteria in the broth are in a planktonic state, in the meat they grow in multi-cellular communities and can form biofilm. The enterotoxin A gene (*sea*) is carried by a family of temperate bacteriophages [22]. The bacteriophage is inserted in the bacterial chromosome as a prophage. Many studies have demonstrated that a stressful condition found in the food matrix can influence SEA production [23]. Tsutsuura et al. (2013) [20] demonstrated that SEA production by *S. aureus* is influenced not only by the intrinsic property of the strain, but also by the incubation temperature and inoculum size.

The statistical analysis of the data did not show any differences between the tests in meat with and without sodium nitrite. These results are consistent with previous studies, which demonstrated that there is no evidence of the inhibition of *S. aureus* growth in canned meat products by similar concentrations of nitrites [24].

4. Conclusions

The strict regulations on processing and canning of food products and the improvement of the HACCP system and of good manufacturing practices have strongly increased food safety. However, contamination of the meat before retorting by *S. aureus* is still possible and may represent a risk for the end consumer, due to its ability to produce highly thermostable toxins, which may not be inactivated during can sterilisation. Our study demonstrated that a SEA producer strain of *S. aureus* takes at least 10 h to produce detectable quantities of toxin at 37 °C and 48 h at 20 °C under the conditions tested, leaving quite a wide range of time to manage pre-retorting steps after seaming. Considering the fact that our study never detected SEA at 10 °C, even though the colony count reached a high level, it has demonstrated once again that the behaviour of *S. aureus* often differs in a complex food matrix than in liquid culture broth.

5. Materials and Methods

5.1. Microbiological Analysis of Samples from the Factory

In this study, the production flow chart of a canned meat production factory was analysed. Nine critical points for the contamination of meat by *S. aureus* (frozen cooked beef at reception, sliced beef at the slicer, belt, piston, frozen cooked beef at thawing, defrost water, cooked beef after thawing, meat on the belt after metal detector, aspic at the doser) were identified. A total of 108 samples were taken from these points, in order to determine the microbial flora and the possible presence of *S. aureus*. The samples were sent to the laboratory in a refrigerated container. To detect the quantitative and qualitative presence of the bacteria, swabs and food products were homogenised in a stomacher in 90 mL of peptone water (PW, Oxoid, Basingstoke, Hampshire, UK). 10-fold dilutions were made using sterile tubes with 9 mL of Maximum Recovery Diluent (MRD, Oxoid, Basingstoke, UK). Dilutions were inoculated in triplicate on Plate Count Agar (PCA, Oxoid) and Baird Parker Agar (BP, Oxoid), prepared with the addition of egg yolk tellurite emulsion (Liofilchem, Roseto degli Abruzzi, TE, Italy) using the spread plate technique, and incubated at 37 °C for 48 h. The colonies were then counted on all the plates, using a colony count viewer (Petri light, PBI, Milan, Italy) and colony counter pen (Colony Count, PBI, Milan, Italy). A shiny, greyish-black, convex colony, measuring from 1–1.5 mm up to 3 mm in diameter with a narrow, white, unbroken margin, surrounded by a 2–5 mm clear area, was identified as suspected *S. aureus* and confirmed by complementary biochemical tests with API 20 STAPH (BioMéhrieux, Marcy-l'Etoile, France).

5.2. Canned Meat Spiking

To determine how much time can elapse between can seaming and sterilisation in the autoclave without any risk of enterotoxin production by *S. aureus* and with the consequent risk of food poisoning for the consumer, the ATCC 29213 strain of *Staphylococcus aureus* (internal reference #239) and two field strains (internal reference #953 and #954), all producers of staphylococcal enterotoxin A (SEA), belonging to the collection of the Food Inspection laboratory of the Department of Veterinary Medicine, University of Perugia, were used to spike samples of the canned meat used throughout the experiments.

Two different types of canned meat, one with and one without sodium nitrite (20 ppm), were used to study the growth rate and enterotoxin production of *Staphylococcus aureus* at 10 °C, 20 °C and 37 °C to simulate the actua l temperatures that can be reached after aspic addition, seaming and during the layover before sterilisation. The canned meat samples presented a pH value of 5.83 and an a_w value of 0.971. Six replications were made for each trial (with and without sodium nitrite) and for each temperature on six different days. The strains were thawed and cultured in Brain Heart Infusion Broth (BHI, Oxoid). After a 48-h incubation at 37 °C, *S. aureus* reached a concentration of 10^8–10^9 CFU mL^{-1}. Ten-fold dilutions were made to obtain the correct concentration for spiking. The cans were opened in a sterile environment and 270 g of each canned meat sample (with and without sodium nitrite) were weighed in a sterile bag. Then, the meat samples were spiked with *S. aureus* to achieve a final concentration of 10^3–10^4 CFU mL^{-1} in the meat. The bags with the mixture of meat and *S. aureus* solution were homogenised for 1 min at 260 rpm in the Stomacher 400 (PBI International, Milan). 20 g of the sample were then placed in sterile glass jars (12 per sample). The jars were individually placed in vacuum bags in order to recreate can conditions and incubated at 37 °C for the first trial, at 20 °C for the second and at 10 °C for the third. Control samples with only canned meat were taken for each type of meat and incubated at the same temperature as the inoculated samples. The times of analysis for each trial were taken at 0, 2, 4, 6, 8, 10, 12, 24, 36, 48 and 72 h, plus 96 h for incubation at 20 °C and on day-11, day-14 and day-28 for incubation at 10 °C.

At each time of analysis, two different tests were conducted on the samples: the detection of the production of enterotoxin and a microbiological analysis to determine the growth of *S. aureus*.

5.3. Study of the Growth Rate

For the growth rate study, a 10 g portion of inoculated meat was transferred from each sample into sterile tubes containing 90 mL of Maximum Recovery Diluent (MRD, Oxoid) and 10-fold dilutions were prepared. Dilutions were inoculated in triplicate on Baird Parker Agar (BP, Oxoid) using the spread plate technique and incubated at 37 °C for 48 h. Colonies were then counted on all the plates, using a colony count viewer (Petri light, PBI, Milan) and colony counter pen (Colony Count, PBI, Milan). All values were converted into logs and the arithmetic mean was calculated for each sampling. All values were analysed with a Graph Pad In Stat, version 3.0b for Mac OS X (GraphPad Software, San Diego, CA, USA); the graphs and the statistical analysis (paired samples *t*-tests) were obtained with a Graph Pad Prism, version 6.0d for Mac OS X (GraphPad Software).

5.4. Enterotoxin Detection

For the study of the enterotoxin production, 10 g of each sample were weighed and then homogenised with 15 mL of PBS buffer. The prepared samples were shaken for 15 min and then centrifuged for 5 min at 3500 rpm at 10 °C. The supernatant was then collected and immediately analysed. A RIDASCREEN SET Total (R-Biopharm, Melegnano, Milan, IT), an enzyme immunoassay for the combined detection of *S. aureus* enterotoxins not only in fluid and solid foods, but also in bacterial cultures, was used to determine the production of enterotoxin. A photometrical interpretation of the results was made by measuring the absorbance at 450/630 ± 10 nm with a microwell plate photometer (SEAC, Radim Group, Freiburg, Germany). The cut-off value to evaluate the results as negative or positive of 0.17 was calculated by adding 0.15 to the value of the negative control (0.020).

A sample was considered positive when the test was valid and the absorbance of the sample was higher than, or equal to, the cut-off value.

Author Contributions: L.G., B.C.-G, P.T., G.C. and A.G.F conceived and designed the experiments; L.G., V.G., P.S. and G.P. performed the experiments; P.A.P. and M.K. analyzed the data; P.T., G.C. and E.B. contributed reagents/materials; L.G. and B.C.-G. wrote the paper."

Funding: This research was funded by Bolton Alimentari and NoNit srl, joint agreement 2017.

Acknowledgments: The authors wish to express sincere appreciation to members of Polyglot, Perugia for a careful reading and comments on the manuscript.

Conflicts of Interest: The authors declare no conflicts of interest.

References

1. Wallin-Carlquist, N.; Márta, D.; Borch, E.; Rådström, P. Prolonged expression and production of *Staphylococcus aureus* enterotoxin A in processed pork meat. *Int. J. Food Microbiol.* **2010**, *141*, S69–S74. [PubMed]

2. Nasser, L.A. Molecular identification of isolated fungi, microbial and heavy metal contamination of canned meat products sold in Riyadh, Saudi Arabia. *Saudi J. Biol. Sci.* **2015**, *22*, 513–520. [CrossRef]

3. Blake, P.A.; Horwitz, M.A.; Hopkins, L.; Lombard, G.L.; McCroan, J.E.; Prucha, J.C.; Merson, M.H. Type A botulism from commercially canned beef stew. *South. Med. J.* **1977**, *70*, 5–7.

4. Cragg, J.; Andrews, A.V. Observations on the microbiological flora of canned Parma ham. *J. Hyg.* **1973**, *71*, 417–422. [PubMed]

5. Mitrica, L.; Granum, P.E. The amylase-producing microflora of semi-preserved canned sausages: Identification of the bacteria and characterization of their amylases. *Z. Fur Lebensm.-Unters. Forsch.* **1979**, *169*, 4–8. [CrossRef]

6. Koreňová, J.; Rešková, Z.; Véghová, A.; Kuchta, T. Tracing *Staphylococcus aureus* in small and medium-sized food-processing factories on the basis of molecular sub-species typing. *Int. J. Environ. Health Res.* **2015**, *25*, 384–392. [CrossRef]

7. Cenci-Goga, B.T.; Karama, M.; Rossitto, P.V.; Morgante, R.A.; Cullor, J.S. Enterotoxin Production by *Staphylococcus aureus* Isolated from Mastitic Cows. *J. Food Prot.* **2003**, *66*, 1693–1696. [CrossRef] [PubMed]

8. Grispoldi, L.; Massetti, L.; Sechi, P.; Iulietto, M.F.; Ceccarelli, M.; Karama, M.; Popescu, P.A.; Pandolfi, F.; Cenci-Goga, B.T. Characterization of enterotoxin producing *Staphylococcus aureus* isolated from mastitic cows. *J. Dairy Sci.* **2019**, *102*, 1059–1065. [CrossRef] [PubMed]

9. Hennekinne, J.A.; De Buyser, M.L.; Dragacci, S. *Staphylococcus aureus* and its food poisoning toxins: Characterization and outbreak investigation. *FEMS Microbiol. Rev.* **2012**, *36*, 815–836. [CrossRef]

10. Asao, T.; Kumeda, Y.; Kawai, T.; Shibata, T.; Oda, H.; Haruki, K.; Nakazawa, H.; Kozaki, S. An extensive outbreak of staphylococcal food poisoning due to low-fat milk in Japan: Estimation of enterotoxin A in the incriminated milk and powdered skim milk. *Epidemiol. Infect.* **2003**, *130*, 33–40.

11. Notermans, S.; Dormans, J.A.M.A.; Mead, G.C. Contribution of surface attachment to the establishment of micro-organisms in food processing plants: A review. *Biofouling* **1991**, *5*, 21–36. [CrossRef]

12. Denayer, S.; Delbrassinne, L.; Nia, Y.; Botteldoorn, N. Food-Borne Outbreak Investigation and Molecular Typing: High Diversity of *Staphylococcus aureus* Strains and Importance of Toxin Detection. *Toxins* **2017**, *9*, 407. [CrossRef]

13. Cha, J.O.; Lee, J.K.; Jung, Y.H.; Yoo, J.I.; Park, Y.K.; Kim, B.S.; Lee, Y.S. Molecular analysis of *Staphylococcus aureus* isolates associated with staphylococcal food poisoning in South Korea. *J. Appl. Microbiol.* **2006**, *101*, 864–871. [CrossRef] [PubMed]

14. Kerouanton, A.; Hennekinne, J.A.; Letertre, C.; Petit, L.; Chesneau, O.; Brisabois, A.; De Buyser, M.L. Characterization of *Staphylococcus aureus* strains associated with food poisoning outbreaks in France. *Int. J. Food Microbiol.* **2007**, *115*, 369–375. [CrossRef] [PubMed]

15. Gilbert, R.J.; Kolvin, J.L.; Roberts, D. Canned foods - the problems of food poisoning and spoilage. *Health Hyg.* **1982**, *4*, 41–47.

16. Mansfield, J.M.; Farkas, G.; Wieneke, A.A.; Gilbert, R.J. Studies on the growth and survival of *Staphylococcus aureus* in corned beef. *J. Hyg.* **1983**, *91*, 467–478. [CrossRef]

17. Stersky, A.; Todd, E.; Pivnick, H. Food Poisoning Associated with Post-Process Leakage (PPL) in Canned Foods. *J. Food Protect.* **1980**, *43*, 465–476. [CrossRef]

18. Beenken, K.E.; Dunman, P.M.; McAleese, F.; Macapagal, D.; Murphy, E.; Projan, S.J.; Blevins, J.S.; Smeltzer, M.S. Global Gene Expression in *Staphylococcus aureus* biofilms. *J. Bacteriol.* **2004**, *186*, 4665–4684. [CrossRef]

19. Resch, A.; Rosenstein, R.; Nerz, C.; Götz, F. Differential Gene Expression Profiling of *Staphylococcus aureus* Cultivated under Biofilm and Planktonic Conditions. *Appl. Environ. Microbiol.* **2005**, *71*, 2663–2676. [CrossRef]

20. Tsutsuura, S.; Shimamura, Y.; Murata, M. Temperature dependence of the production of staphylococcal enterotoxin a by *Staphylococcus aureus*. *Biosci. Biotech. Biochem.* **2013**, *77*, 30–37. [CrossRef]

21. Valero, A.; Pérez-Rodríguez, F.; Carrasco, E.; Fuentes-Alventosa, J.M.; García-Gimeno, R.M.; Zurera, G. Modelling the growth boundaries of *Staphylococcus aureus*: Effect of temperature, pH and water activity. *Int. J. Food Microbiol.* **2009**, *133*, 186–194. [CrossRef]

22. Betley, M.J.; Mekalanos, J.J. Staphylococcal enterotoxin A is encoded by phage. *Science* **1985**, *229*, 185–187. [CrossRef]

23. Schelin, J.; Susilo, Y.B.; Johler, S. Expression of staphylococcal enterotoxins under stress encountered during food production and preservation. *Toxins* **2017**, *9*, 401. [CrossRef]

24. Scott, W.J. Factors in canned ham controlling Cl. botulinum and Staph. aureus. *Annales de l'Institut Pasteur de Lille* **1955**, *7*, 68–74.

Article

Staphylococcal Enterotoxin C Is an Important Virulence Factor for Mastitis

Rendong Fang [1,†], Jingchun Cui [2,†,*], Tengteng Cui [2], Haiyong Guo [2,3], Hisaya K. Ono [4,5], Chun-Ho Park [5], Masashi Okamura [5], Akio Nakane [4] and Dong-Liang Hu [1,4,5,*]

[1]   College of Animal Science and Technology, Southwest University, Congqing 400715, China; rdfang@swu.edu.cn
[2]   College of Life Science, Dalian Minzu University, Dalian 116600, China; tengteng.cui@pharmaron-bj.com (T.C.); guohaiyong78@jlnu.edu.cn (H.G.)
[3]   Department of Biological Science, School of Life Science, Jilin Normal University, Siping 136000, China
[4]   Department of Microbiology and Immunology, Hirosaki University Graduate School of Medicine, Hirosaki, Aomori 036-8562, Japan; hisaono@vmas.kitasato-u.ac.jp (H.K.O.); a27k03n0@hirosaki-u.ac.jp (A.N.)
[5]   Kitasato University School of Veterinary Medicine, Towada, Aomori 034-8628, Japan; baku@vmas.kitasato-u.ac.jp (C.-H.P.); okamuram@vmas.kitasato-u.ac.jp (M.O.)
*   Correspondence: cjc@dlnu.edu.cn (J.C.); hudl@vmas.kitasato-u.ac.jp (D.-L.H.); Tel.: +81-176-24-9451 (D.-L.H.)
†   These authors contributed equally to this work.

Received: 14 January 2019; Accepted: 26 February 2019; Published: 2 March 2019

Abstract: *Staphylococcus aureus* is an important bacterial pathogen causing bovine mastitis, but little is known about the virulence factor and the inflammatory responses in the mammary infection. Staphylococcal enterotoxin C (SEC) is the most frequent toxin produced by *S. aureus*, isolated from bovine mastitis. To investigate the pathogenic activity of SEC in the inflammation of the mammary gland and the immune responses in an animal model, mouse mammary glands were injected with SEC, and the clinical signs, inflammatory cell infiltration, and proinflammatory cytokine production in the mammary glands were assessed. SEC induced significant inflammatory reactions in the mammary gland, in a dose-dependent manner. SEC-injected mammary glands showed a severe inflammation with inflammatory cell infiltration and tissue damage. In addition, interleukin (IL)-1β and IL-6 production in the SEC-injected mammary glands were significantly higher than those in the PBS control glands. Furthermore, the SEC-induced inflammation and tissue damage in the mammary gland were specifically inhibited by anti-SEC antibody. These results indicated, for the first time, that SEC can directly cause inflammation, proinflammatory cytokine production, and tissue damage in mammary glands, suggesting that SEC might play an important role in the development of mastitis associated with *S. aureus* infection. This finding offers an opportunity to develop novel treatment strategies for reduction of mammary tissue damage in mastitis.

Keywords: staphylococcal enterotoxin; mastitis; superantigen; virulence factor

Key Contribution: This study, for the first time, demonstrates that SEC directly induces proinflammatory cytokine production in mammary glands and leads to mammary epithelial cell damage. This finding indicates that SEC is an important virulence factor for mastitis, and offers an opportunity to develop novel treatment strategies for reduction of mammary tissue damage in mastitis.

---

1. Introduction

Bovine mastitis is one of the most prevalent diseases which represents a huge economic loss for the dairy industry and is an increasing public health concern, worldwide [1–3]. *Staphylococcus aureus* is a common bacterial pathogen, isolated frequently from the milk produced by cows with mastitis, and is also an important bacterium causing bovine mastitis [4,5]. *S. aureus* can enter the mammary gland, adhere to epithelial cells, and start growing there, resulting in a mammary epithelial line damage that is initiated by the infiltration of inflammatory cells in the parenchyma, the alveoli, causing ulceration and occlusion of lactiferous ducts [6]. To treat acute inflammation of the udder tissue and improve animal health, cows have to be administrated with antibiotics, and the milk production becomes impaired for a long period and results in a huge economic loss. It is essential to reduce the prevalence of bovine mastitis infected with *S. aureus*, by different measures, such as vaccination and immunotherapy [4]. The clinical outcome of acute mastitis and progression to persistent mastitis might be related to the presence of virulence factors of *S. aureus* strains causing the infection. During the past decade, many studies have attempted to develop vaccines against some specific virulence factors that were considered to be responsible for the development and persistent infection of *S. aureus*-induced bovine mastitis. However, the vaccines have shown a limited protective efficacy in the dairy farm sites [7–9]. The lack of effective vaccines could be attributed, in part, to the incomplete understanding of the pathogenic mechanism of mastitis induced by *S. aureus* and the mechanisms of immune responses in the ruminant breasts [10].

Many *S. aureus* isolated from bovine mastitis harbor genes coding for superantigenic toxins, such as staphylococcal enterotoxins (SEs). SEs are members of the pyrogenic toxin family, including classic toxins, staphylococcal enterotoxin A (SEA) to SEE, and newly described toxins, SEG to SEI, SEK to SET, and staphylococcal enterotoxin like toxin J (SElJ), and SElU to SElY. The superantigenic toxins show strong T cell mitogenic activity by directly binding to major histocompatibility complex (MHC) class II molecules of antigen presenting cells and Vβ regions of T cell receptor (TCR), without the normal antigen presentation process [11–13]. Large amounts of activated cells release excessive inflammatory cytokines, such as interleukin-2 (IL-2), tumor necrosis factor alpha (TNF-α), and gamma interferon (IFN-γ), which are responsible for the development of inflammation, rashes, fever, multiorgan damage, and toxic shock syndrome, in humans and animals [14–16]. Previous studies have reported that *S. aureus* isolated from bovine mastitis produce superantigenic toxins, especially SEC [17,18]. SEC has also been detected in the *S. aureus* strains isolated from an outbreak case of skin and soft tissue infections, including mastitis in humans [19]. Several studies reported that the *sec* gene was frequently detected in *S. aureus* strains isolated from raw milk of goats and bovines with mastitis [20–22]. Although epidemiological studies showed that the presence of *sec* gene in the majority of strains of bovine mastitis and suggested a possible involvement of SEC in bovine mastitis pathogenesis [4,23–25], It is still unknown whether or not SEC is a direct virulence factor for bovine mastitis [25–27].

To understand the role of SEC in the development of bovine mastitis, we cloned and expressed the *sec* gene in the prokaryotic expression system, and the biological activities of purified SEC were analyzed in the present study. The pathogenicity of SEC and the immune responses of mammary gland induced by SEC were studied using the mouse model. Our results demonstrated that SEC can induce inflammatory reaction, cytokine production, and tissue damage in the mammary gland, indicating that SEC could be an important virulence factor for mastitis.

2. Results

2.1. Biological Activities of Purified Recombinant SEC

The biological and superantigenic activities of purified SEC were identified and assayed. Analysis of SEC by Coomassie-blue-stained SDS-PAGE revealed the presence of a purified protein band that was readily detectable (Figure 1). The level of endotoxin in the final purified SEC was not detected (less than 0.001 EU/mL). The superantigenic activity of SEC was evaluated by sandwich ELISA to determine the production of IL-2 and IFN-γ in mouse spleen cell culture supernatants. Compared with the BSA control group, the SEC-induced spleen cells produced large amounts of IL-2 and IFN-γ, in a dose dependent manner (Figure 1).

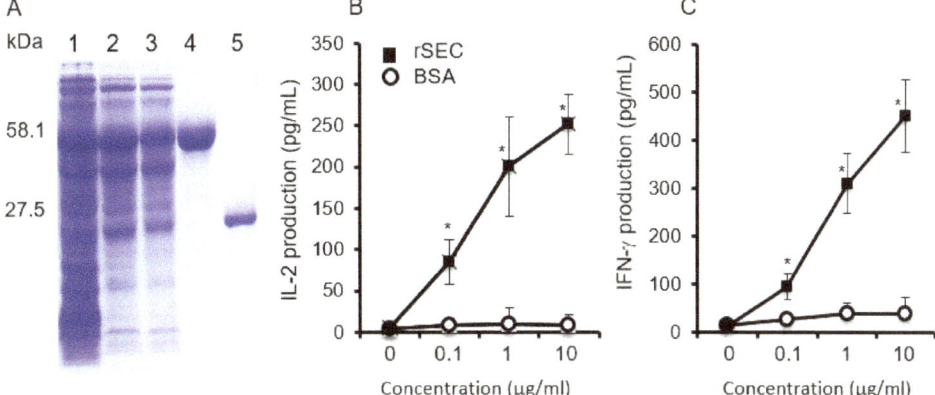

Figure 1. Biologic and superantigenic activities of purified staphylococcal enterotoxin C (SEC). (**A**) Purified SEC analyzed by SDS-PAGE. Lanes 1–3, supernatants of *E. coli* encoding glutathione S-transferase (GST)-SEC; lane 4, purified GST-SEC fusion protein; lane 5, purified SEC. (**B,C**) Production of cytokines in mouse spleen cell cultures stimulated with 0.1, 1, and 10 µg per mL of SEC (3.6, 36, and 360 nM) or BSA (1.5, 15, and 150 nM). Amounts of IL-2 (**B**) and IFN-γ (**C**) in the spleen cell cultures were measured by ELISA. Results are the mean ± SD, based on the samples obtained from five to six mice. Two experiments were performed independently. An asterisk indicates that the value was significantly different between the SEC and the BSA control group at $p < 0.05$.

2.2. SEC Exhibited Pathogenic Activity in the Murine Mammary Glands

To study whether SEC could induce mastitis in the murine mammary glands, mice were injected with PBS, SEC (5, 10, or 15 µg per gland), and non-treated normal control (Figure 2). Results showed that SEC significantly induced inflammatory responses and tissue damage in the mammary glands, 24 h after injection with the toxin (Figure 2). The pathogenic activity of SEC was found to be dose-dependent. The percentages of mice with clinical signs, including redness, swelling, congestion, bleeding, and discoloration of the mammary glands, were recorded and calculated. The mice injected with SEC at 5, 10, and 15 µg per gland showed 71.4% (10/14), 80.0% (12/16), and 91.7% (11/12) positive clinical signs, respectively. In contrast, the mice injected with PBS showed 7.7% (1/13) positive clinical signs (Figure 2B). The scores of clinical signs of mice for each group are shown in Figure 2C. The inflammation signs were still observed 72 h after SEC injection. In contrast, the PBS-treated and non-treated control mice showed no significant changes in the mammary glands.

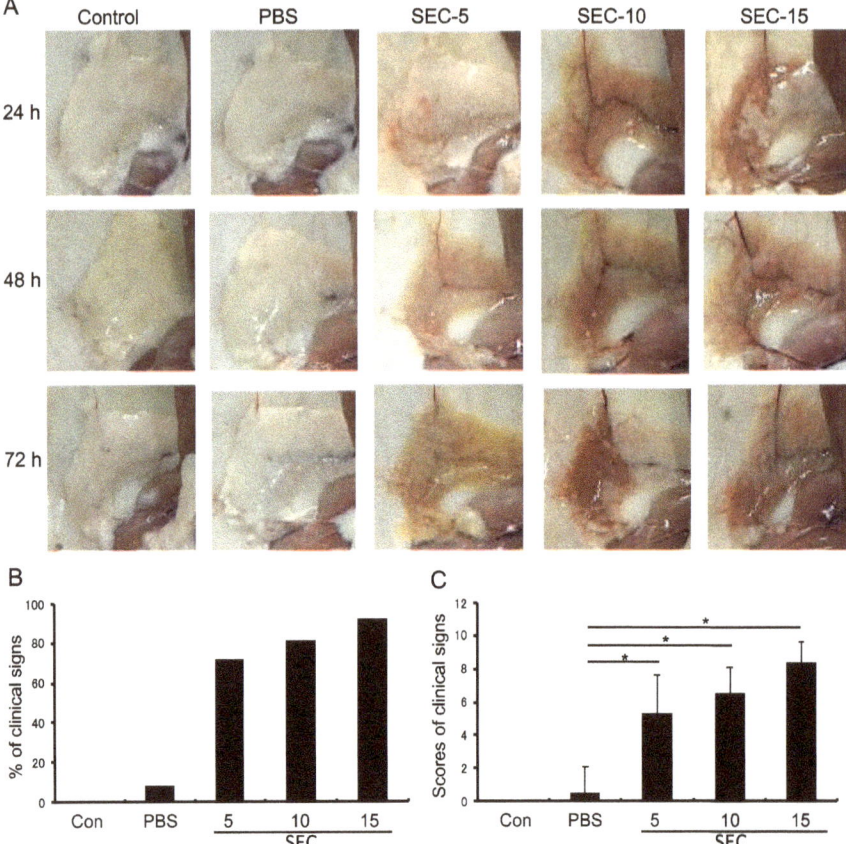

Figure 2. Appearance observation of mouse mammary glands injected with PBS, SEC (5, 10 or 15 μg per gland) or non-treated controls. (**A**) Mice were euthanized 24, 48, and 72 h after injection, and investigated for the extent of inflammation of the mammary gland area by observing the clinical signs, including redness, swelling, congestion, bleeding, and discoloration of the mammary glands. The results are representative of three independent experiments. (**B**) Percentages of mice with clinical symptoms for each group. (**C**) Scores of clinical signs of mice for each group. An asterisk indicates that the value was significantly different between the SEC and the PBS control group at $p < 0.05$.

2.3. Histopathological Changes of SEC-Injected Mammary Glands

Histopathological changes in the mammary glands injected with SEC or PBS were evaluated by hematoxylin and eosin (HE) staining (Figure 3). No significant abnormality was observed in the histopathological observation of the breast tissue of the non-treated group (data not shown). The PBS treated group with little histopathological change was almost identical to the untreated control group. In the SEC-treated group, there were inflammatory cells including neutrophils, macrophages, and monocytes infiltrating the mammary gland tissues. Quantification of the number of inflammatory cells, including neutrophils, macrophages, and monocytes, showed that the numbers of cells in mammary glands injected with PBS, 5 or 10 μg of SEC per gland, were 7 ± 4, 71 ± 12, and 80 ± 16 per field of view, respectively. The number of inflammatory cells in the mammary glands after injection with 5 and 10 μg SEC were significantly higher than that of the PBS control group ($p < 0.05$). In addition, interstitial edema, thickening of the gland alveolus wall and acutely damaged alveolar form were also observed (Figure 3).

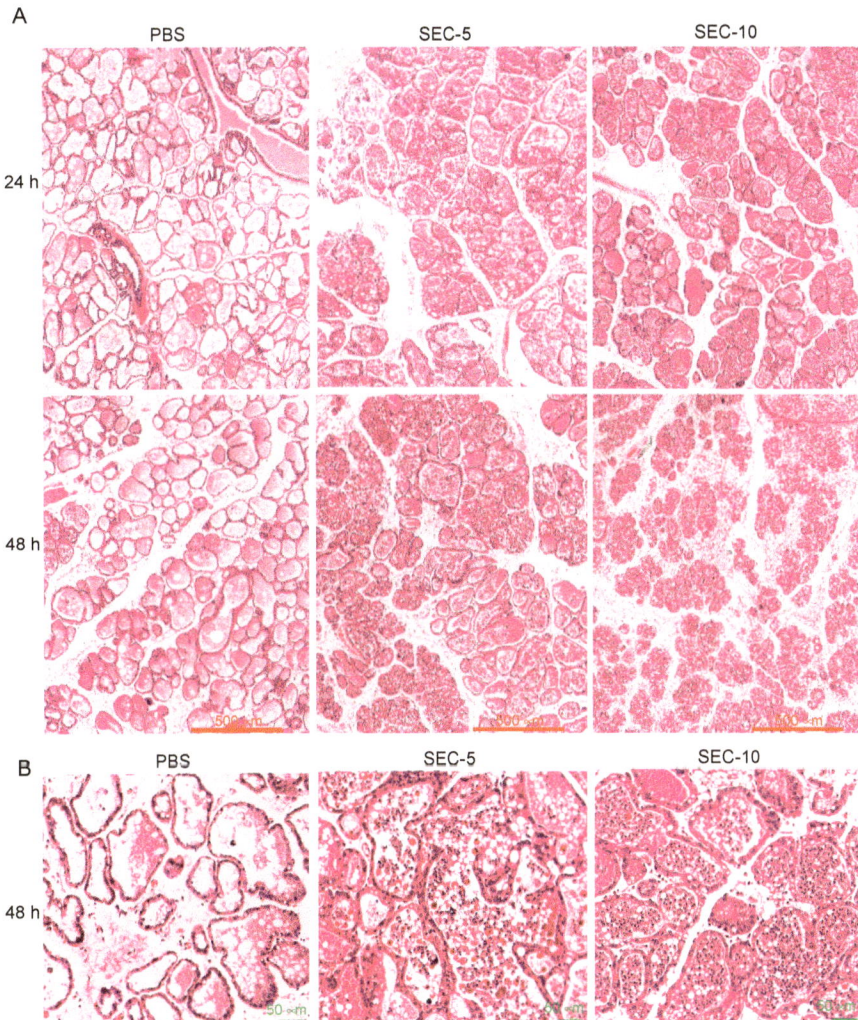

Figure 3. Histological changes of mammary glands of mice injected with PBS or SEC (5 or 10 μg per gland). The mammary glands were collected and the right glands were fixed in 10% neutral buffered formalin for 24 h, processed with automated tissue processor and embedded in paraffin wax. Sections were cut at three levels to a thickness of 4 μm and stained by HE staining. The results are representative of three independent experiments. (**A**) Low magnification photographs and the bars show 500 μm. (**B**) High magnification photographs and the bars show 50 μm.

2.4. SEC Induced the Production of Proinflammatory Cytokines

To evaluate the inflammatory and immune responses of the mammary glands injected with SEC, proinflammatory cytokines, IL-1β and IL-6, were measured in SEC-treated (10 μg/gland) or PBS-treated glands at 12, 24, and 48 h, after mammary injection (Figure 4). SEC-injection caused a significant increase of IL-1β and IL-6 in the mammary tissue of mice, compared with that of the control group at 12, 24, and 48 h, after mammary injection. Higher levels of cytokines were recorded in the SEC-treated glands which also showed neutrophils, macrophages, and monocytes infiltration during the histological observation. The results were consistent with the histological data and indicated the recruitment of immune cells in the alveolar lumen, at an earlier time point, in the mammary glands exposed to SEC.

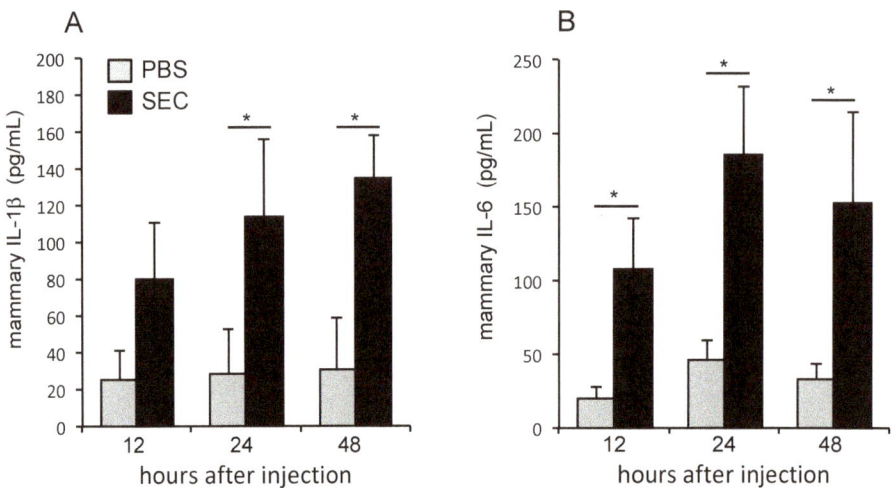

Figure 4. Production of proinflammatory cytokine in mammary glands of mice injected with PBS or SEC (10 µg per gland). Mammary tissues were taken out from each group, weighed and homogenized with PBS. Homogenate was removed into a centrifuge tube and centrifuged at 3,000 rpm, for 20 min, at 4 °C. The amounts of IL-1β (**A**) and IL-6 (**B**) in the supernatant of the homogenate were analyzed with the corresponding ELISA. The results are representative of three experiments, and the data are the mean ± SD for groups of three to five mice. An asterisk indicates that the value was significantly different between SEC-injected group and PBS control group at $p < 0.05$.

2.5. Anti-SEC Antibody Inhibits the Inflammation of Mammary Gland Induced by SEC

To further investigate whether the SEC-induced inflammation in mammary gland could be inhibited by anti-SEC antibody, SEC was preincubated with rabbit anti-SEC IgG, or with normal rabbit IgG, and then injected to the mammary glands of mice (Figure 5). The numbers and percentages of mice with clinical signs, including redness, swelling, congestion, bleeding, and discoloration of the mammary glands, were recorded and analyzed. The mice injected with SEC alone, SEC + anti-SEC IgG, or SEC + normal rabbit IgG showed an 87.5% (7/8), 20.0% (2/10), and 77.7% (7/9) positive clinical signs, respectively (Figure 5B). The results of the clinical signs for the mice of each group are shown in Figure 5C. The inflammatory reactions of the mammary glands were markedly inhibited by treatment with the anti-SEC antibody, compared with the SEC alone and SEC + normal rabbit IgG groups (Figure 5).

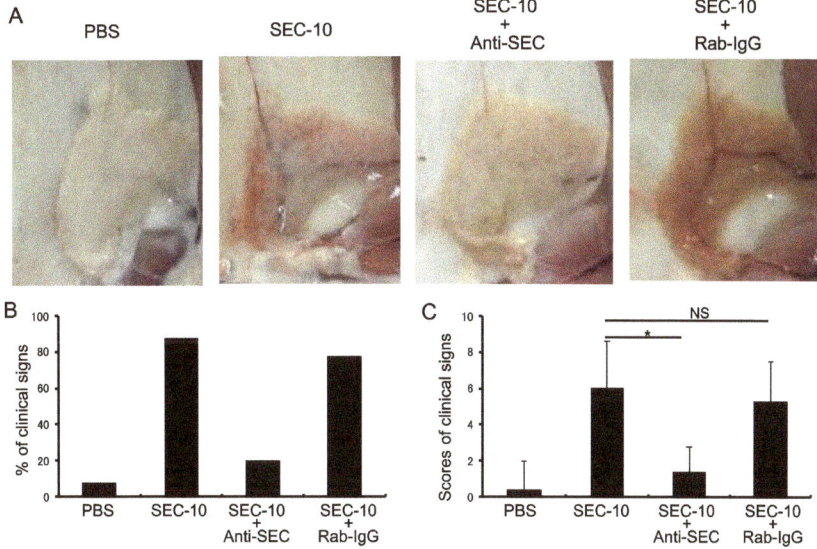

Figure 5. Neutralization ability of anti-SEC antibody against the inflammation of mammary gland induced by SEC in vivo. (**A**) Rabbit anti-SEC IgG or normal rabbit IgG, preincubated with SEC at 37 °C, for 1 h before SEC (10 μg per gland) were injected into the mammary glands of mice. The injected mice were then investigated for the extent of inflammation of the mammary gland at 48 h after injection, by observing clinical signs, swelling, redness, congestion, and bleeding of the mammary glands. The results are representative of two independent experiments, each with three to five mice per each group. (**B**) Percentages of mice with clinical symptoms for each group. (**C**) Results showing the clinical signs in mice for each group. An asterisk indicates that the value was significantly different between the SEC + anti-SEC IgG group and the SEC alone control group at $p < 0.05$; NS indicates no significant difference.

3. Discussion

S. aureus strains isolated from the bovine mastitis cases carried multiple genes encoding SEs or SEls, especially SEC, has been reported extensively. Previous studies have also reported that *S. aureus* pathogenicity island (SaPI), containing *sec* and *sel*, was detected in isolates of mastitis outbreak in cows, goats, and humans, indicating that *sec* and *sel* carrying *S. aureus* might be important and might be associated with mastitis in animals and humans [4,28–30]. However, little is known regarding if SEC is a virulence factor for bovine mastitis induced by *S. aureus*. In the present study, we analyzed the biological characteristics and potential pathogenic activity of SEC in mouse mastitis model that were directly injected with a purified SEC. Our results revealed that SEC, a superantigenic toxin, induced markedly inflammatory reactions and tissue damage in the mammary glands of the SEC-injected mice. These results indicated, for the first time, that SEC is an important virulence factor which contributes for occurrence and deterioration of mastitis.

Analysis of histopathological changes of mammary tissue is a widely used method to evaluate mammary tissue damage caused by bacterial invasion [6,31]. In the present study, SEC-induced acute mastitis was assessed, based on macroscopic clinical manifestations, histopathological changes, inflammatory cytokine production, and the level of mammary tissue damage. The inflammation with tissue degeneration and necrosis, and inflammatory cell infiltration in the intra-mammary tissue induced by SEC were observed. Moreover, SEC-induced inflammation and tissue damage were significantly inhibited by treatment of SEC with anti-SEC antibody, before SEC injection. Previous studies have demonstrated that *S. aureus*-induced bovine mastitis showed characteristic histopathological changes, including inflammatory cell infiltration, shrinkage of alveolar space, damage of secretory breast tissue, and necrosis of the mammary gland [32,33]. Inflammatory cells, such as neutrophil, macrophage, and monocyte play an important role in the host defenses against *S. aureus* invasion of the mammary gland [34]. On the other hand, however, excessive infiltration of inflammatory cells can also damage the mammary epithelia lines by respiratory burst and degranulation of the cells [35]. SEC-induced inflammatory reactions in mammary glands could be due to a large amount of inflammatory cell infiltration, excessive respiratory bursts, and degranulation and destruction of the mammary epithelial cells.

Inflammatory cytokines are vital mediators of the development and persistence of inflammation during mammary infections. Previous studies have reported that a large amount of IL-1 and TNF-α in milk and plasma of animals with mastitis might be important events for the inflammatory reactions in mammary glands [36–38]. The high level of IL-1 and TNF-α in the serum, could induce the activation and migration of neutrophils, and cause apoptosis of endothelial cells and mammary epithelial cells, in bovine and human suffering from acute clinical mastitis [39–42]. Inflammatory cytokines also increase a wide variety of functions of the inflammatory cells, including cell adhesion, expression of cell surface receptors, release of lysosomal constituents, and free radical production [34]. In the present study, the levels of proinflammatory cytokines, IL-1β and IL-6, were significantly higher in the mammary glands of mice injected with SEC than those of the control mice injected with PBS, indicating that the cytokines play an important role in the observed mammary tissue damage in mice. Although superantigen might generate varying effects, depending on the animal species [43,44], the effect of SEC-induced cytokines on breast tissue injury is likely to be mediated by the recruitment and activation of inflammatory cells in the mammary gland.

In conclusion, our results showed that mammary gland injected with SEC markedly caused inflammatory reactions, production of inflammatory cytokines, and tissue damage, in the mammary gland. To our knowledge, the present study is the first demonstration that SEC could directly induce mastitis in an animal model. Together with the molecular epidemiological data showing that the sec gene is frequently detected in isolates of bovine mastitis, our results suggest that SEC contributes to the pathogenesis through a superantigenic activity, which induces proinflammatory cytokine release, inflammation responses, and subsequently induces mammary tissue damage. This finding provides an opportunity to develop novel treatment strategies to alleviate tissue damage in the mammary glands infected by bacteria.

4. Materials and Methods

4.1. Expression and Purification of SEC

The *sec* gene was amplified from the genomic DNA of *S. aureus* 834 strain, which is a clinical isolate that produces SEC and induces mastitis, in the mouse model [45,46]. The PCR primers used were SEC/GST+ (5′-CCCCGGATTCGAGAGCCAACCAGACCCTACG), including a *Bam*HI site, and SEC/GST- (5′-CCCCGAATTCTTATCCATTCTTTGTTGTAAGGTGG), including an *Eco*RI site, and the predicted size of the PCR product was 669 bp. The PCR products were subcloned into a pGEX-6P-1 glutathione S-transferase (GST) fusion vector. The expression and purification of the recombinant SEC were performed, as described by Hu et al. [47]. Purified SEC protein was analyzed by sodium dodecyl sulfate polyacrylamide gel eletrophoresis (SDS-PAGE) and the Bradford method (Bio-Rad Laboratories, Richmond, CA, USA). Endotoxin contamination in the purified proteins was removed by ProteoSpin™ Total Protein Concentration, Detergent Clean-Up, and Endotoxin Removal Kit (Norgen Biotek Corp., Thorond, ON, Canada), and was detected by Limulus amebocyte lysate PYROTELL (CAPE COD, Falmouth, MA, USA).

4.2. Assay of Superantigenic Activity of SEC

To analyze the superantigenic activity of purified SEC, five six-week-old Balb/c mice were used for the spleen cell culture experiment. Analysis of the production of IL-2 and IFN-γ in the spleen cell cultures induced by the toxin were performed, as described previously [48]. The amount of IL-2 and IFN-γ in the supernatant of the cell cultures were determined by sandwich ELISAs, as described by our previous reports [47,48].

4.3. Intramammary Inoculation Model

Ten-week-old healthy Kunming mice were purchased from the Experimental Animal Center of Dalian Medical University. The mice were fed with a special diet for breeding, and drank freely. After one week of adaptive feeding, two female and one male mice were fed in the same cage to allow them to mate. The pregnant female mice were randomly divided into seven groups of 5 mice each, and used for the injection experiment. Injection of SEC into mammary glands was carried out using a modification to the procedure described previously [31,44]. Mice were anesthetized with ketamine (100 mg/kg) and xylazine (10 mg/kg) by intraperitoneal injection. With the mouse facing up, the mammary gland and its periphery were sterilized with 70% ethanol. A total of 20 μL of SEC was injected directly into the mammary duct, using a 31-gauge hypodermic needle to a depth of not more than 4 mm, without cutting or scraping the shallow layer of the mammary gland. The mice were observed to assess development of clinical signs of mastitis. For a control group, mice were injected with 20 μL of PBS, using the same procedure. The animal experiments described in this study were approved by the animal ethics committee of Daliang Minzu University (27 July 2015; DMU2015-002) and carried out in accordance with the guide for the management and use of experimental animals of Dalian Medical University.

4.4. Clinical Evaluation

The changes in mammary glands of mice injected with SEC were observed and evaluated for onset of mastitis. Clinical signs, swelling, redness, and discoloration of the breast were observed and recorded. Monitoring of mice for morbidity and mortality was carried out up to 72 h. Mice were euthanized at 24, 48, and 72 h after injection and investigated for the extent of inflammation of the mammary gland area, by observing redness, swelling, congestion, bleeding, and discoloration of the mammary gland. A total score of 10 could be obtained for the 5 symptoms, with 2 for each positive clinical sign and 0 for each negative clinical sign.

4.5. Histopathological Examination

To estimate the inflammation levels and histological changes, the mammary gland tissue of each group was collected and the right gland was fixed in 10% neutral buffered formalin, for 24 h, before being processed with an automatic tissue processor, and was embedded in paraffin. Sections were cut at three levels to a thickness of 4 μm and stained by the HE staining. The number of inflammatory cells, including neutrophils, macrophages, and monocytes, in the field of view were counted and quantified. Five fields of view of each sample were screened to obtain an average number of an animal, then the total mean and standard deviation of each group were calculated and analyzed.

4.6. Determination of Cytokines

To determinate cytokine production in the mammary gland, mammary tissues of each experiment group were taken out, weighed, and homogenized with PBS with 1:9 ratio. After the tissue was grinded, radically, the homogenate was centrifuged at 3,000 rpm, for 20 min, at 4 °C. The supernatant was stored at −30 °C. The contents of important inflammatory cytokines in the mammary gland, IL-1β and IL-6, were analyzed by the corresponding ELISA kits, according to the instructions of the manufacturers (BioLegend, Inc., Camino Santa Fe, San Diego, CA, USA).

4.7. Neutralization Assay of the Anti-SEC Antibody

To analyze the neutralizing activity of anti-SEC antibody, against the inflammation of mammary gland, induced by SEC, in vivo, 300 μL of rabbit anti-SEC IgG (1 mg/mL) or normal rabbit IgG (1 mg/mL), as a control antibody, were preincubated with 300 μL of SEC (1 mg/mL) at 37 °C, for 1 h, before the SEC was injected to the mammary gland. Twenty microliters of the preincubated solution, including 10 μg of antibody and 10 μg of SEC, was injected directly into the mammary duct, using a 31-gauge hypodermic needle. Both antibodies were prepared previously in our laboratory [45,46]. Mice were investigated for the extent of inflammation of the mammary gland area, 48 h after injection, by observing the clinical signs, swelling, redness, congestion, and bleeding of the mammary gland, as described above.

4.8. Statistical Analysis

Data in this study were showed as means ± standard deviations. The significance of differences in cytokine titers between the control and experimental groups were determined with the Mann-Whitney U test.

Author Contributions: Conceptualization, R.F., J.C., A.N., and D.-L.H.; Methodology, J.C., T.C., H.K.O., H.G., R.F., and C.-H.P.; Validation, R.F., T.C., and J.C.; Formal Analysis, R.F., J.C., and D.-L.H.; Investigation, R.F., J.C., H.G., H.K.O., and M.O.; Resources, J.C., R.F., and D.-L.H.; Data Curation, R.F., H.G., and J.C.; Writing-Original Draft Preparation, R.F. and J.C.; Writing-Review & Editing, D.-L.H., M.O., and A.N.; Visualization, R.F. and J.C.; Supervision, D.-L.H. and A.N.; Project Administration, R.F., J.C., and D.-L.H.

Funding: This research was funded and supported in part by the National Key Research and Development Program of China [2018YFD0500500 (R.F.)], the National Natural Science Foundation of China [31272584 (J.C.)] and the JSPS KAKENHI grant number [16H05030 (D.-L.H.)].

Acknowledgments: The authors gratefully acknowledge Guozhen Liu for reviewing our manuscript. The authors thank the National Key Research and Development Program of China, the National Natural Science Foundation of China, and the JSPS KAKENHI for supporting this research.

Conflicts of Interest: The authors declare no conflict of interest. The funders had no role in the design of the study; in the collection, analyses, or interpretation of data; in the writing of the manuscript, and in the decision to publish the results.

References

1. Sánchez, A.; Sierra, D.; Luengo, C.; Corrales, J.C.; de la Fe, C.; Morales, C.T.; Contreras, A.; Gonzalo, C. Evaluation of the MilkoScan FT 6000 milk analyzer for determining the freezing point of goat's milk under different analytical conditions. *J. Dairy Sci.* **2007**, *90*, 3153–3161. [CrossRef] [PubMed]
2. Gogoi-Tiwari, J.; Williams, V.; Waryah, C.B.; Costantino, P.; Al-Salami, H.; Mathavan, S.; Wells, K.; Tiwari, H.K.; Hegde, N.; Isloor, S.; et al. Mammary gland pathology subsequent to acute infection with strong versus weak biofilm forming *Staphylococcus aureus* bovine mastitis isolates: A pilot study using non-invasive mouse mastitis model. *PLoS ONE* **2017**, *27*, e0170668. [CrossRef] [PubMed]
3. White, D.G.; McDermott, P.F. Biocides, drug resistance and microbial evolution. *Curr. Opin. Microbiol.* **2001**, *4*, 313–317. [CrossRef]
4. Artursson, K.; Söderlund, R.; Liu, L.; Monecke, S.; Schelin, J. Genotyping of *Staphylococcus aureus* in bovine mastitis and correlation to phenotypic characteristics. *Vet. Microbiol.* **2016**, *25*, 156–161. [CrossRef] [PubMed]
5. Schabauer, A.; Pinior, B.; Gruber, C.M.; Firth, C.L.; Käsbohrer, A.; Wagner, M.; Rychli, K.; Obritzhauser, W. The relationship between clinical signs and microbiological species, spa type, and antimicrobial resistance in bovine mastitis cases in Austria. *Vet. Microbiol.* **2018**, *227*, 52–60. [CrossRef] [PubMed]
6. Zhao, X.; Lacasse, P. Mammary tissue damage during bovine mastitis: Causes and control. *J. Anim. Sci.* **2008**, *86* (Suppl. 13), 57–65. [CrossRef] [PubMed]
7. Tenhagen, B.A.; Edinger, D.; Baumgärtner, B.; Kalbe, P.; Klünder, G.; Heuwieser, W. Efficacy of a herd-specific vaccine against *Staphylococcus aureus* to prevent post-partum mastitis in dairy heifers. *J. Vet. Med. A Physiol. Pathol. Clin. Med.* **2001**, *48*, 601–607. [CrossRef] [PubMed]

8. Shkreta, L.; Talbot, B.G.; Diarra, M.S.; Lacasse, P. Immune responses to a DNA/protein vaccination strategy against *Staphylococcus aureus* induced mastitis in dairy cows. *Vaccine* **2004**, *15*, 114–126. [CrossRef] [PubMed]

9. Watson, D.L. Vaccination against experimental staphylococcal mastitis in dairy heifers. *Res. Vet. Sci.* **1992**, *53*, 346–353. [CrossRef]

10. Chang, B.S.; Moon, J.S.; Kang, H.M.; Kim, Y.I.; Lee, H.K.; Kim, J.D.; Lee, B.S.; Koo, H.C.; Park, Y.H. Protective effects of recombinant staphylococcal enterotoxin type C mutant vaccine against experimental bovine infection by a strain of *Staphylococcus aureus* isolated from subclinical mastitis in dairy cattle. *Vaccine* **2008**, *16*, 2081–2091. [CrossRef] [PubMed]

11. Wilson, G.J.; Tuffs, S.W.; Wee, B.A.; Seo, K.S.; Park, N.; Connelley, T.; Guinane, C.M.; Morrison, W.I.; Fitzgerald, J.R. Bovine *Staphylococcus aureus* superantigens stimulate the entire T cell repertoire of cattle. *Infect. Immun.* **2018**, *25*, e00505–e00518. [CrossRef] [PubMed]

12. Krakauer, T. Induction of CC chemokines in human peripheral blood mononuclear cells by staphylococcal exotoxins and its prevention by pentoxifylline. *J. Leukoc. Biol.* **1999**, *66*, 158–164. [CrossRef] [PubMed]

13. Fraser, J.D.; Proft, T. The bacterial superantigen and superantigen-like proteins. *Immunol. Rev.* **2008**, *225*, 226–243. [CrossRef] [PubMed]

14. Spaulding, A.R.; Salgado-Pabón, W.; Kohler, P.L.; Horswill, A.R.; Donald, Y.M.; Leung, D.Y.M.; Schlievert, P.M. Staphylococcal and streptococcal superantigen exotoxins. *Clin. Microbiol. Rev.* **2013**, *26*, 422–447. [CrossRef] [PubMed]

15. Uchiyama, T.; Saito, S.; Inoko, H.; Yan, X.J.; Imanishi, K.; Araake, M.; Igarashi, H. Relative activities of distinct isotypes of murine and human major histocompatibility complex class II molecules in binding toxic shock syndrome toxin 1 and determination of CD antigens expressed on T cells generated upon stimulation by the toxin. *Infect. Immun.* **1990**, *58*, 3877–3882. [PubMed]

16. Yarwood, J.M.; McCormick, J.K.; Schlievert, P.M. Identification of a novel two-component regulatory system that acts in global regulation of virulence factors of *Staphylococcus aureus*. *J. Bacteriol.* **2001**, *183*, 1113–1123. [CrossRef] [PubMed]

17. Fitzgerald, J.R.; Monday, S.R.; Foster, T.J.; Bohach, G.A.; Hartigan, P.J.; Meaney, W.J.; Smyth, C.J. Characterization of a putative pathogenicity island from bovine *Staphylococcus aureus* encoding multiple superantigens. *J. Bacteriol.* **2001**, *183*, 63–70. [CrossRef] [PubMed]

18. Ebling, T.L.; Fox, L.K.; Bayles, K.W.; Bohach, G.A.; Byrne, K.M.; Davis, W.C.; Ferens, W.A.; Hillers, J.K. Bovine mammary immune response to an experimental intramammary infection with a *Staphylococcus aureus* strain containing a gene for staphylococcal enterotoxin C1. *J. Dairy Sci.* **2001**, *84*, 2044–2050. [CrossRef]

19. Saiman, L.; O'Keefe, M.; Graham, P.L.; Wu, F.; Saïd-Salim, B.; Kreiswirth, B.; LaSala, A.; Schlievert, P.M.; Della-Latta, P. Hospital transmission of community-acquired methicillin-resistant *Staphylococcus aureus* among postpartum women. *Clin. Infect. Dis.* **2003**, *15*, 1313–1319. [CrossRef] [PubMed]

20. Zecconi, A.; Cesaris, L.; Liandris, E.; Daprà, V.; Piccinini, R. Role of several *Staphylococcus aureus* virulence factors on the inflammatory response in bovine mammary gland. *Microb. Pathog.* **2006**, *40*, 177–183. [CrossRef] [PubMed]

21. Günaydın, B.; Aslantaş, Ö.; Demir, C. Detection of superantigenic toxin genes in *Staphylococcus aureus* strains from subclinical bovine mastitis. *Trop. Anim. Health Prod.* **2011**, *43*, 1633–1637. [CrossRef] [PubMed]

22. Aydin, A.; Sudagidan, M.; Muratoglu, K. Prevalence of staphylococcal enterotoxins, toxin genes and genetic-relatedness of foodborne *Staphylococcus aureus* strains isolated in the Marmara Region of Turkey. *Int. J. Food Microbiol.* **2011**, *148*, 99–106. [CrossRef] [PubMed]

23. Morandi, S.; Brasca, M.; Lodi, R.; Cremonesi, P.; Castiglioni, B. Detection of classical enterotoxins and identification of enterotoxin genes in *Staphylococcus aureus* from milk and dairy products. *Vet. Microbiol.* **2007**, *124*, 66–72. [CrossRef] [PubMed]

24. Cui, J.C.; Zhang, B.J.; Lin, Y.C.; Wang, Q.K.; Qian, A.D.; Nakane, A.; Hu, D.L.; Tong, G.Z. Protective effect of glutathione S-transferase-fused mutant staphylococcal enterotoxin C against *Staphylococcus aureus*-induced bovine mastitis. *Vet. Immunol. Immunopathol.* **2010**, *135*, 64–70. [CrossRef] [PubMed]

25. Liu, Y.; Chen, W.; Ali, T.; Alkasir, R.; Yin, J.; Liu, G.; Han, B. Staphylococcal enterotoxin H induced apoptosis of bovine mammary epithelial cells in vitro. *Toxins (Basel)* **2014**, *6*, 3552–3567. [CrossRef] [PubMed]

26. Wang, W.; Lin, X.; Jiang, T.; Peng, Z.; Xu, J.; Yi, L.; Li, F.; Fanning, S.; Baloch, Z. Prevalence and characterization of *Staphylococcus aureus* cultured from raw milk taken from dairy cows with mastitis in Beijing, China. *Front. Microbiol.* **2018**, *9*, 1123. [CrossRef] [PubMed]

27. Piccinini, R.; Borromeo, V.; Zecconi, A. Relationship between *S. aureus* gene pattern and dairy herd mastitis prevalence. *Vet. Microbiol.* **2010**, *145*, 100–105. [CrossRef] [PubMed]

28. Orwin, P.M.; Fitzgerald, J.R.; Leung, D.Y.; Gutierrez, J.A.; Bohach, G.A.; Schlievert, P.M. Characterization of *Staphylococcus aureus* enterotoxin L. *Infect. Immun.* **2003**, *71*, 2916–2919. [CrossRef] [PubMed]

29. Kenny, K.; Reiser, R.F.; Bastida-Corcuera, F.D.; Norcross, N.L. Production of enterotoxins and toxic shock syndrome toxin by bovine mammary isolates of *Staphylococcus aureus*. *J. Clin. Microbiol.* **1993**, *31*, 706–707. [PubMed]

30. Franck, K.T.; Gumpert, H.; Olesen, B.; Larsen, A.R.; Petersen, A.; Bangsborg, J.; Albertsen, P.; Westh, H.; Bartels, M.D. *Staphylococcal aureus* enterotoxin C and enterotoxin-like L associated with post-partum mastitis. *Front. Microbiol.* **2017**, *8*, 173. [CrossRef] [PubMed]

31. Gogoi-Tiwari, J.; Williams, V.; Waryah, C.B.; Mathavan, S.; Tiwari, H.K.; Costantino, P.; Mukkur, T. Intramammary immunization of pregnant mice with staphylococcal protein A reduces the post-challenge mammary gland bacterial load but not pathology. *PLoS ONE* **2016**, *11*, e0148383. [CrossRef] [PubMed]

32. Heald, C.W. Morphometric study of experimentally induced *Staphylococcus bovis* mastitis in the cow. *Am. J. Vet. Res.* **1979**, *40*, 1294–1298. [PubMed]

33. Nickerson, S.C.; Heald, C.W. Histopathologic response of the bovine mammary gland to experimentally induced *Staphylococcus aureus* infection. *Am. J. Vet. Res.* **1981**, *42*, 1351–1355. [PubMed]

34. Paape, M.J.; Bannerman, D.D.; Zhao, X.; Lee, J.W. The bovine neutrophil: Structure and function in blood and milk. *Vet. Res.* **2003**, *34*, 597–627. [PubMed]

35. Bannerman, D.D.; Paape, M.J.; Lee, J.W.; Zhao, X.; Hope, J.C.; Rainard, P. *Escherichia coli* and *Staphylococcus aureus* elicit differential innate immune responses following intramammary infection. *Clin. Diagn. Lab. Immunol.* **2004**, *11*, 463–472. [CrossRef] [PubMed]

36. Riollet, C.; Rainard, P.; Poutrel, B. Cells and cytokines in inflammatory secretions of bovine mammary gland. *Adv. Exp. Med. Biol.* **2000**, *480*, 247–258. [PubMed]

37. Lee, J.W.; Bannerman, D.D.; Paape, M.J.; Huang, M.K.; Zhao, X. Characterization of cytokine expression in milk somatic cells during intramammary infections with *Escherichia coli* or *Staphylococcus aureus* by real-time PCR. *Vet. Res.* **2006**, *37*, 219–229. [CrossRef] [PubMed]

38. Kauf, A.C.; Vinyard, B.T.; Bannerman, D.D. Effect of intramammary infusion of bacterial lipopolysaccharide on experimentally induced *Staphylococcus aureus* intramammary infection. *Res. Vet. Sci.* **2007**, *82*, 39–46. [CrossRef] [PubMed]

39. Nakajima, Y.; Mikami, O.; Yoshioka, M.; Motoi, Y.; Ito, T.; Ishikawa, Y.; Fuse, M.; Nakano, K.; Yasukawa, K. Elevated levels of tumor necrosis factor-alpha (TNF-alpha) and interleukin-6 (IL-6) activities in the sera and milk of cows with naturally occurring coliform mastitis. *Res. Vet. Sci.* **1997**, *62*, 297–298. [CrossRef]

40. Messmer, U.K.; Briner, V.A.; Pfeilschifter, J. Tumor necrosis factor-alpha and lipopolysaccharide induce apoptotic cell death in bovine glomerular endothelial cells. *Kidney Int.* **1999**, *55*, 2322–2337. [CrossRef] [PubMed]

41. Bauer, D.; Redmon, N.; Mazzio, E.; Soliman, K.F. Apigenin inhibits TNFα/IL-1α-induced CCL2 release through IKBK-epsilon signaling in MDA-MB-231 human breast cancer cells. *PLoS ONE* **2017**, *12*, e0175558. [CrossRef] [PubMed]

42. Vieira, S.M.; Lemos, H.P.; Grespan, R.; Napimoga, M.H.; Dal-Secco, D.; Freitas, A.; Cunha, T.M.; Verri, W.A., Jr.; Souza-Junior, D.A.; Jamur, M.C.; et al. A crucial role for TNF-alpha in mediating neutrophil influx induced by endogenously generated or exogenous chemokines, KC/CXCL1 and LIX/CXCL5. *Br. J. Pharmacol.* **2009**, *158*, 779–789. [CrossRef] [PubMed]

43. Hu, D.L.; Omoe, K.; Sasaki, S.; Yokomizo, Y.; Sashinami, H.; Sakuraba, H.; Shinagawa, K.; Nakane, A. Vaccination with nontoxic TSST-1 protects *Staphylococcus aureus* infection. *J. Infect. Dis.* **2003**, *188*, 743–752. [CrossRef] [PubMed]

44. Hu, D.-L.; Omoe, K.; Sashinami, H.; Shinagawa, K.; Nakane, A. Immunization with nontoxic mutant staphylococcal enterotoxin A, D227A, protects against enterotoxin-induced emesis in house musk shrews. *J. Infect. Dis.* **2009**, *199*, 302–310. [CrossRef] [PubMed]

45. Omoe, K.; Hu, D.L.; Takahashi-Omoe, H.; Nakane, A.; Shinagawa, K. Comprehensive analysis of classical and newly described staphylococcal superantigenic toxin genes in *Staphylococcus aureus* isolates. *FEMS Microbiol. Lett.* **2005**, *246*, 191–198. [CrossRef] [PubMed]

46. Zhang, B.J.; Cui, J.C.; Zhang, X.; Qiang, A.D.; Tong, G.Z.; Hu, D.L.; Nakane, A.; Wang, Q.K. Development and evaluation of *Staphylococcus aureus*-induced mastitis model. *Chin. J. Prevent. Vet. Med.* **2009**, *31*, 365–369.

47. Hu, D.L.; Cui, J.C.; Omoe, K.; Sashinami, H.; Yokomizo, Y.; Shinagawa, K.; Nakane, A. A mutant of staphylococcal enterotoxin C devoid of bacterial superantigenic activity elicits a Th2 immune response for protection against *Staphylococcus aureus* infection. *Infect. Immun.* **2005**, *73*, 174–180. [CrossRef] [PubMed]

48. Brouillette, E.; Grondin, G.; Lefebvre, C.; Talbot, B.G.; Malouin, F. Mouse mastitis model of infection for antimicrobial compound efficacy studies against intracellular and extracellular forms of *Staphylococcus aureus*. *Vet. Microbiol.* **2004**, *101*, 253–262. [CrossRef] [PubMed]

 *toxins*

Article

Clinical *S. aureus* Isolates Vary in Their Virulence to Promote Adaptation to the Host

Lorena Tuchscherr [1,2,*], Christine Pöllath [1,2], Anke Siegmund [1], Stefanie Deinhardt-Emmer [1,2], Verena Hoerr [1,2], Carl-Magnus Svensson [3], Marc Thilo Figge [3,4], Stefan Monecke [5,6] and Bettina Löffler [1,2]

[1] Institute of Medical Microbiology, Jena University Hospital, 07747 Jena, Germany;
     Christine.Poellath@med.uni-jena.de (C.P.); Anke.Siegmund@med.uni-jena.de (A.S.);
     Stefanie.Deinhardt-Emmer@med.uni-jena.de (S.D.-E.); Verena.Hoerr@med.uni-jena.de (V.H.);
     Bettina.Loeffler@med.uni-jena.de (B.L.)
[2] Center for Sepsis Control and Care (CSCC), Jena University Hospital, 07747 Jena, Germany
[3] Applied Systems Biology, Leibniz-Institute for Natural Product Research and Infection Biology, 07745 Jena,
     Germany; Carl-Magnus.Svensson@hki-jena.de (C.-M.S.); Thilo.Figge@hki-jena.de (M.T.F.)
[4] Faculty of Biological Sciences, Friedrich Schiller University Jena, 07743 Jena, Germany
[5] Leibniz Institute of Photonic Technology (IPHT), 07745 Jena, Germany; stefan.monecke@leibniz-ipht.de
[6] Institute of Medical Microbiology and Hygiene, Medical Faculty Carl Gustav Carus, 01307 Dresden,
     Germany
*   Correspondence: lorena.tuchscherrdehauschopp@med.uni-jena.de

Received: 25 January 2019; Accepted: 21 February 2019; Published: 1 March 2019

Abstract: *Staphylococcus aureus* colonizes epithelial surfaces, but it can also cause severe infections. The aim of this work was to investigate whether bacterial virulence correlates with defined types of tissue infections. For this, we collected 10–12 clinical *S. aureus* strains each from nasal colonization, and from patients with endoprosthesis infection, hematogenous osteomyelitis, and sepsis. All strains were characterized by genotypic analysis, and by the expression of virulence factors. The host–pathogen interaction was studied through several functional assays in osteoblast cultures. Additionally, selected strains were tested in a murine sepsis/osteomyelitis model. We did not find characteristic bacterial features for the defined infection types; rather, a wide range in all strain collections regarding cytotoxicity and invasiveness was observed. Interestingly, all strains were able to persist and to form small colony variants (SCVs). However, the low-cytotoxicity strains survived in higher numbers, and were less efficiently cleared by the host than the highly cytotoxic strains. In summary, our results indicate that not only destructive, but also low-cytotoxicity strains are able to induce infections. The low-cytotoxicity strains can successfully survive, and are less efficiently cleared from the host than the highly cytotoxic strains, which represent a source for chronic infections. The understanding of this interplay/evolution between the host and the pathogen during infection, with specific attention towards low-cytotoxicity isolates, will help to optimize treatment strategies for invasive and therapy-refractory infection courses.

Keywords: *S. aureus*; low cytotoxic strains; chronic infection

Key Contribution: *S. aureus* low cytotoxic strains can successfully survive and are less efficiently cleared from the host than high cytotoxic strains, which represent a source for chronic and difficult-to-treat infections.

---

1. Introduction

Staphylococcus aureus is a very versatile pathogen that often colonizes the epithelial surfaces of healthy individuals [1,2]; but it is also the most common pathogen of the bloodstream, surgical side, and bone infections, which are often complicated, with several sites of metastatic foci, and the development of chronic infections [3]. In addition, many cases of staphylococcal bacteremia appear to have an endogenous source, since they originate from strains in the nasal mucosa [4]. The diversity of infections that are associated with *S. aureus* is due to the multiple virulence factors and its adaptation to different environments in the human host. This particular adaptation facilitates the bacteria's survival in the host, and their evasion of the host immune system [5].

Recently, we described that *S. aureus* passes different stages of infection through the use of a complicated gene regulatory network [6]. To establish an infection, *S. aureus* displays defined virulence factors, including adhesive surface proteins (adhesins) and toxic compounds that act in concert to destroy the host tissue and to resist the host defense system[2]. In particular, the quorum-sensing system accessory gene regulator (Agr) enhances the expression of toxins, e.g., α-hemolysin/α-toxin (hla), and other secreted cytotoxic factors, e.g., phenol-soluble modulins (PSMs), whereas the alternative sigma factor B SigB (σ^B) modulates stress responses and promotes *S. aureus* persistence [7,8]. During bacteremia or sepsis, the bacteria need to survive within the bloodstream, to defend against immune cells. Notably, secreted pore-forming toxins, such as *hla*, cause inflammation and contribute to sepsis development [9]. To commence an infection in the host tissue, e.g., bone tissue, bacteria need to adhere to the host structures, such as the extracellular matrix or the host cells. For this, *S. aureus* expresses various surface proteins with adhesive functions, such as fibronectin-binding proteins (FnBPs) [10]. After the infection is settled, the bacteria invade and adapt to the host tissue for persistence and escape from the host immune system, which is mainly mediated by the upregulation of SigB [6]. The intracellular location most likely represents a shelter against many antibiotics, and against the host immune defense system; this causes severe clinical problems in diagnosis and treatment. Chronic infections have been associated with an altered bacterial phenotype, the so-called "small colony variants" (SCVs). SCVs are adapted phenotypes with a reduced metabolism, which enables the bacteria to persist for long-lasting periods [11,12]. However, bacterial adaptation requires the fine-tuning of virulence factors [13]. Yet, the bacterial virulence and its relation to the type of infection are still under discussion. On the one hand, several studies have affirmed that highly virulent and toxin-producing strains cause severe infections [14,15]. On the other hand, a recent study did not find an association between elevated toxicity and the severity of infections [16]. Additionally, some authors have demonstrated that many staphylococcal isolates from invasive diseases have Agr dysfunctions that impair the production of toxins [17,18].

In this study, we investigated the role of staphylococcal strains with high and low capacities for inducing host cell death (cytotoxicity). *S. aureus* strains from nasal colonization, endoprosthesis infection, hematogenous osteomyelitis, and sepsis were collected and characterized by genotypic analysis, functional assays, and infection models. Knowing the functions and the interplay between high and low cytotoxic strains will help us to improve the therapeutic treatment and clearance of persistent staphylococcal infections.

2. Results

2.1. The Genetic Analysis of Staphylococcal Isolates Reveals Only Minor Differences in Bacterial Origin

To investigate the virulence of staphylococcal strains from different pathologies, 47 isolates were collected: 12 from nose swabs from healthy individuals (nasal colonization), 12 from orthopedic endoprosthesis infections, 13 from patients with hematogenous osteomyelitis, and 10 from sepsis patients. The main clinical characteristics of all of the patients are summarized in Table S1. The clonal complex affiliations are summarized in Table S2. All of the isolates (four MRSA and 43 MSSA) were analyzed with StaphyType DNA microarrays (Alere Technologies GmbH, Jena, Germany), facilitating the detection of specific genes, as well as their assignment to clonal complexes (CCs). Isolates belonged to a total of 13 different CCs. Although the small sample size excludes a statistically robust evaluation of the relationship between the virulence gene carriage and the clinical outcome, the carriage of some genes was higher, or not present, in some groups (Tables S2 and S3).

The rate of MRSA was not significantly elevated in one of the groups. The distribution of CCs in the defined groups was very diverse, as every group contained at least five different CCs. The most common CC for isolates with prosthetic origin was CC7 (33.3%) and CC5 (25%). Interestingly, CC7 was highly present in the prosthetic group, but not in other groups ($p = 0.007$, Table S2). Isolates from blood infections, such as hematogenous osteomyelitis and sepsis, showed similar distributions of CCs, where the CC22 was the most abundant ($p = 0.019$, Table S2). Moreover, CC45 was highly prevalent among sepsis isolates. In general, the clonal population structure of study isolates was similar to the one observed in another larger study in Germany [19] (Table S2).

The complete microarray hybridization data are provided as a Supplementary File (Table S4). The main characteristics of all isolates are summarized in Table S3. The presence of certain genes was significant in some groups, as outlined below. The enterotoxin gene cluster *egc* was present in more than 50% of the isolates of all groups other than prostheses infections (33.3%). Interestingly, the enterotoxin P gene (*entP/seP*) was significantly more common in the prosthetic group than in other groups (Table S3). The hemolysin genes *hla*, *hlb*, and *hld* were present in all isolates. The leukocidin genes *lukF/S-hlgA* were present in all strains, and only one strain from nasal colonization was positive for *lukF/S-PVL*. Yet, genes for *lukD/lukE* were significantly more common in isolates from prosthesis infections than in isolates from nasal colonization and sepsis (Table S3). Genes related to various enzymes and other virulence factors showed a heterogeneous distribution (Table S3). The aureolysin (*aur*), glutamyl endopeptidase (*sspA/B*), the staphylococcal complement inhibitor (*scn*) and staphylokinase (*sak*) genes were found in all isolates. The gene for chemotaxis inhibitory protein (*chip*) was less frequent among prosthesis isolates (25%) in comparison with the other groups (60–80%). Genes encoding serine proteases (*splA/B*) were more prevalent in the bone infection groups than in isolates from sepsis and nasal colonization. Additionally, serin protease E (*splE*) was rarely present in the sepsis group (Table S3). The gene encoding *ssl8* was more prevalent in the prosthetic infection group than in nasal colonization, hematogenous osteomyelitis, and sepsis (Table S3).

Genes encoding the capsule (*cap5/8*) and biofilm (*icaA-C*) were present in all isolates (Table S3). *cna* was less commonly present in the prosthetic infection group (Table S3). Interestingly, the presence of *ebh* was significantly reduced in isolates from hematogenous osteomyelitis (Table S3). All isolates harbored the genes encoding *ebh, ebps, eno, fib, fnbA, map, sdrC-, sdrD*, and *vwb*. Conversely, the presence of *fnbB* was reduced in the group of hematogenous osteomyelitis (Table S3).

Taken together, our results indicate that similar patterns can be found in hematogenous osteomyelitis and sepsis isolates. Despite some differences in the presence of certain genes between the analyzed groups of staphylococcal isolates, no clear genetic profile was associated with the focus of infection, and focus could be determined (Table S3).

2.2. High- as Well as Low-Cytotoxicity Strains are Equally Distributed among Isolates from Different Bacterial Origins

To determine whether the described genetic differences (Table S3) resulted in specific virulence behaviors, further functional assays were performed (Figure 1). The growth curves did not reveal any substantial differences between the isolated strains (Figure S1).

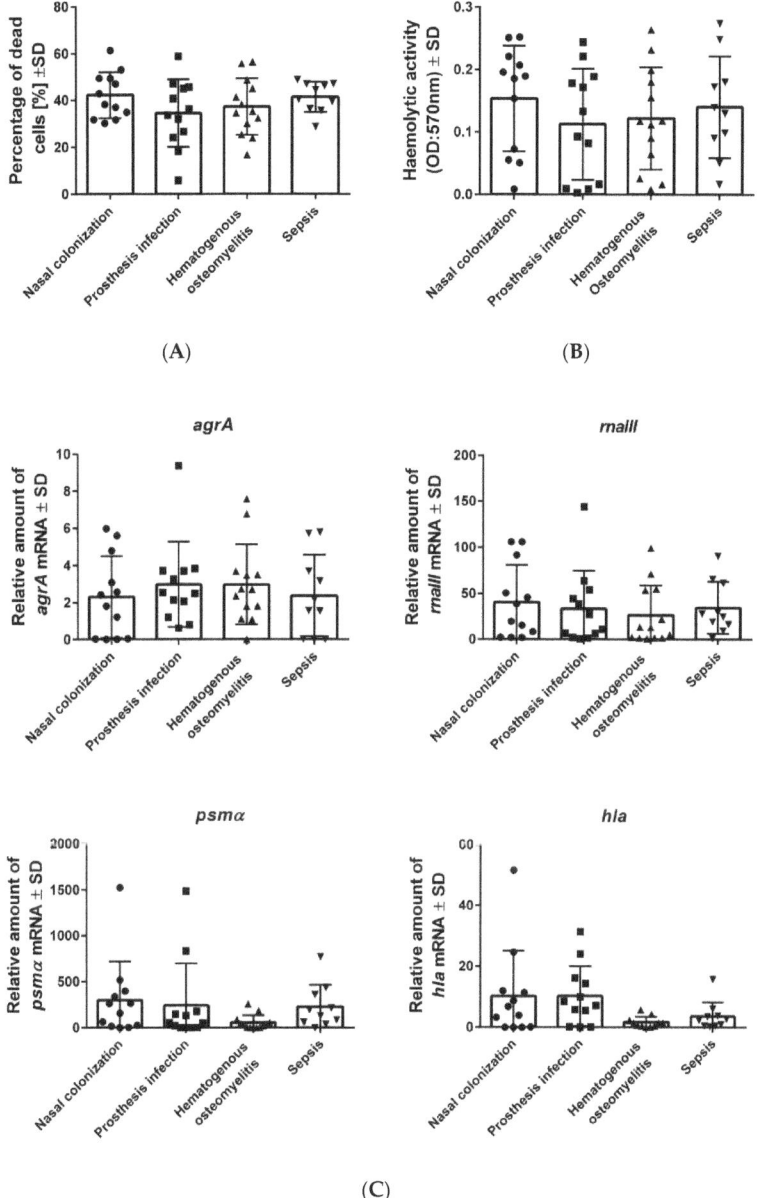

Figure 1. High- and low-cytotoxic strains are similarly distributed among *S. aureus* isolates from different origins. (**A**) Osteoblasts were infected with clinical strains of *S. aureus*, and the percentage of dead cells was measured after 24 h post-infection by staining with propidium iodide and quantification by flow cytometry. (**B**) Hemolysis was measured by detecting the release of hemoglobin from sheep blood erythrocytes (OD: 570 nm) post-infection with *S. aureus* strains. (**C**) Gene expression of *S. aureus* strains. All strains were cultivated in BHI for 4 h (*agrA*, *rnaIII*) and 6 h (*psmα*, *hla*) respectively. The RNA was isolated and transformed to complementary DNA (cDNA), in order to perform qPCR. The expression of the gene *gyrB* was used as a reference. The bars and whiskers represent the means ±SD of three independent experiments in duplicate. The differences between all of the groups of isolates were analyzed by a one-way ANOVA test, with Tukey's multiple comparisons test.

Initially we focused on the functional characteristics that are important for the spreading and settling of infections, such as cytotoxicity, hemolysis, and the expression of related virulence factors (Figure 1). During the first steps of infection, *S. aureus* induces host cell destruction through the expression of membrane disruption toxins, such as α-toxin (*hla*) and phenol-soluble modulins (PSMs) (*psmαβ*) to destroy host tissues, and to favor bacterial spreading [20]. To investigate the cytotoxic capacities of all staphylococcal strains, we analyzed cell death induction in osteoblasts (cytotoxicity), as well as the lysis of erythrocytes (hemolysis) (Figure 1A,B). A broad range of values was observed in each analyzed group in all our assays, but no significance differences based on isolate origins were found.

To analyze the toxins related to cell death, the expression of *agrA* (accessory gene regulator from the Agr quorum sensing system) and *rnaIII* (RNA effector molecule of this quorum system system), *psmα*, and *hla* were analyzed (see Mat. and Methods, Figure 1C). No differences were observed for *agr*, *rnaIII*, *and psmα*, regarding the bacterial origin. Only the *hla* expression was up-regulated in the prosthesis infection group, in comparison to strains that were isolated from hematogenous osteomyelitis. In general, a broad distribution of the high and low values was detected in all groups, independent of their bacterial origin.

After the acute phase when the bacteria induce cell destruction, *S. aureus* can invade host cells to escape from the host immune response. Thus, the invasion of *S. aureus* by osteoblasts was investigated (Figure 2). Similarly to the cytotoxic assays, high and low invasive isolates were found in all analyzed groups of isolates, which was unrelated to their bacterial origin.

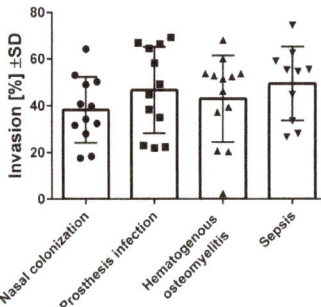

Figure 2. The internalization of *S. aureus* into osteoblasts was non-related to the origin of each isolate. Osteoblasts (hFOB 1.19) were infected with FITC-stained *S. aureus*, and the intracellular bacteria were quantified by flow cytometry. Bars and whiskers represent the means ±SD of at least three independent experiments in duplicate. The differences between all of the groups of isolates were analyzed by a one-way ANOVA test with Tukey's multiple comparisons test.

Next, we analyzed the pairwise correlation between all measurements, by calculating Spearman's correlation coefficient (Figure 3). A positive correlation was found between cytotoxicity and hemolysis ($p = 0.01$; $r = 0.36$) indicating an overlap of factors for both processes. Interestingly, a negative correlation was found between cytotoxicity/hemolysis and invasion, suggesting that both processes are inversely correlated ($p = 0.05$; $r = -0.3$) ($p = 0.004$; $r = -0.41$). As expected, a positive correlation was found between cytotoxicity and the expression of *hla* ($p = 0.02$; $r = 0.34$); *psmα* ($p < 0.001$; $r = 0.56$) and *rnaIII* ($p < 0.001$; $r = 0.51$). The induction of hemolysis is correlated with the expression of *hla* ($p = 0.004$; $r = 0.41$); *psmα* ($p < 0.001$; $r = 0.6$), *agrA* ($p = 0.02$; $r = 0.35$) and *rnaIII* ($p < 0.001$; $r = 0.51$). A positive correlation was found between the expression of *rnaIII* and *psmα* ($p < 0.001$; $r = .61$) and *agrA* ($p < 0.01$; $r = 0.39$) as previously described [21]. Even though the correlation *p*-values were significant, the moderate magnitude of the correlation coefficients, in a range between 0.1 and 0.6, reflected the large variation in the measurements between the strains.

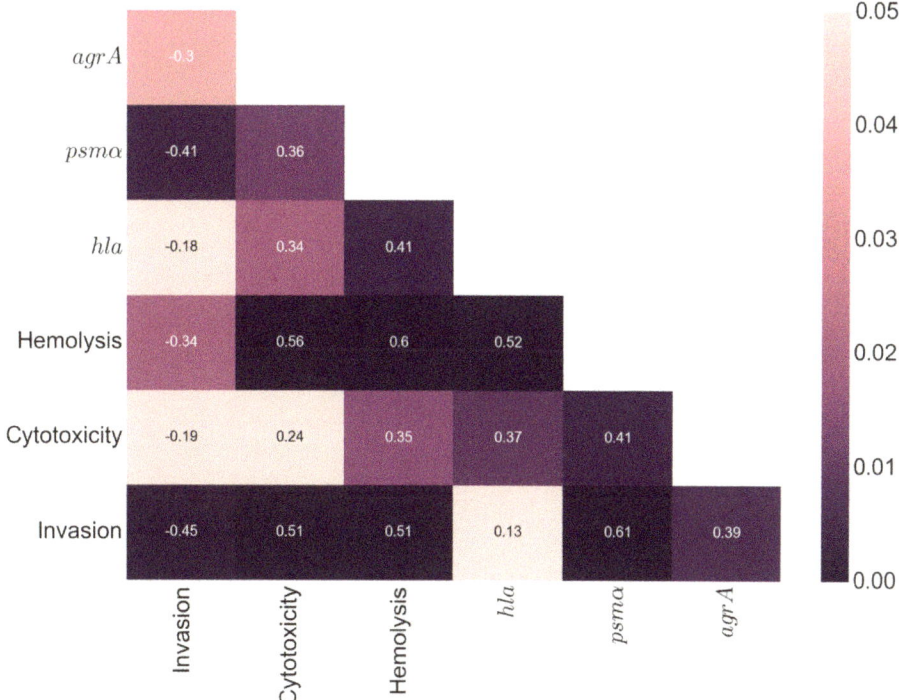

Figure 3. Correlation analysis. Spearman's correlation coefficients for all of the measured parameters. Significance testing of correlations was done by computing a two-sided *p*-value, assuming a t-distribution with two degrees of freedom. The values in the heat map are the pairwise correlation coefficients (*r*-values) and the colors indicate the significance level of the correlation coefficient (*p* values). Highly significant *p*-values are dark purple, and lighter colors indicate decreasing significance.

Cytotoxicity displayed a wide range of values, but it was not significantly different between the types of infection, and in each infection type, isolates from the entire spectrum of cytotoxicity were present (Figure S2). The correlation analysis suggests that highly cytotoxic strains display a strong induction of cell death in general (cytotoxic and hemolysis), a low invasion capacity, and a high expression of *hla*, *psmα*, *rnaIII*, and *agrA*, which enabled the bacteria to spread to different tissues by killing host cells during the acute phase. In contrast, strains with low cytotoxic levels showed reduced rates of cell death, but they achieved a higher degree of host cell internalization compared to highly cytotoxic strains. This internalization is necessary for their escape from the host immune response, and antimicrobial treatment.

2.3. Highly Invasive but Low-Cytotoxicity Strains Persist at High Numbers within Host Cells

S. aureus can invade its host cells, and it is able to persist intracellularly for long time periods, while down-regulating its metabolism and toxin expression [22]. To investigate the association between virulence and persistence, four highly cytotoxic and three low-cytotoxicity strains were selected according the cytotoxic and hemolysis values (Table S5). The persistence was analyzed by infecting osteoblasts for up to seven days (Figure 4A). We found that both high- and low-cytotoxicity strains were able to survive within the host cells (Figure 4A). Strikingly, the highly cytotoxic strains had reduced intracellular survival, compared to the low-cytotoxicity strains after seven days (Figure 4B). Moreover, the low-cytotoxicity strains survived in higher numbers, and induced a higher rate of

SCV formation (Figure 4C). Next, the inflammatory response in osteoblasts was investigated by the determination of CCL5 (RANTES) by ELISA after 24 h of infection (Figure 4D). All highly cytotoxic strains induced more inflammation than the low-cytotoxicity strains. Taken together, our results show that staphylococcal low-cytotoxicity strains promote long-term intracellular persistence by inducing a reduced inflammatory response and cell death, which contributes to their survival within host cells in higher numbers.

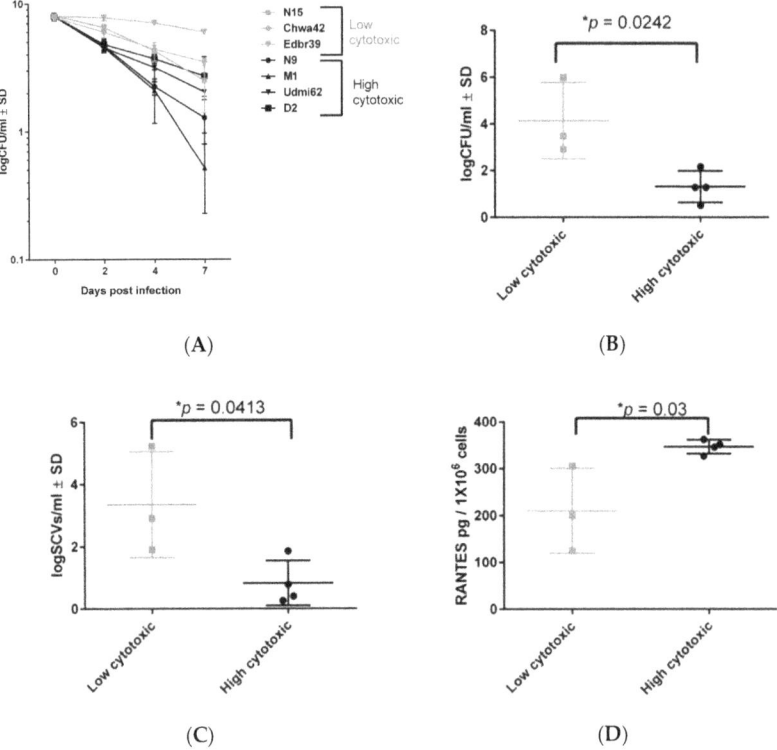

(A)   (B)

(C)   (D)

Figure 4. Long-term persistence of low- and high-cytotoxicity strains in osteoblasts. (**A**) Intracellular bacteria recovered from infected osteoblasts for four high cytotoxic strains (black lines) and three low cytotoxic strains (gray lines) as a function of time post-infection. (**B**) Intracellular bacteria recovered from infected osteoblasts with high- and low-cytotoxicity strains at day 7 post-infection. (**C**) SCVs recovered at day 7 post-infection. (**D**) RANTES levels were measured in the cell culture supernatant of infected osteoblasts after 24 h post-infection, using the ELISA test. All results represent the mean ±SD of at least three independent experiments. The difference between the low- and high-cytotoxicity strains was analyzed by the Unpaired *t*-test. (* $p < 0.05$; ** $p < 0.01$; *** $p < 0.001$ and **** $p < 0.0001$)

2.4. Low-Cytotoxicity Strains can Persist in High Numbers in a Murine Sepsis Model

To test whether bacterial virulence patterns affect the development of infection, we performed our hematogenous osteomyelitis model [23]. Mice were infected with a low-cytotoxicity (Chwa42; Table S5) or a high-cytotoxicity *S. aureus* strain (D2; Table S5) and we analyzed the bacterial loads in the bones after six weeks post-infection. No difference in murine survival was observed after infection between both strains (Figure 5A). The bacterial loads in bones after six weeks post-infection revealed a higher persistence and less efficient clearance of the low-cytotoxicity strain Chwa42, in comparison to the highly cytotoxic strain D2 (Figure 5B). Furthermore, low inflammatory levels of CCL5 (RANTES) was measured in the serum, from mice infected with the low-cytotoxicity strain Chwa42, whereas the highly cytotoxic strain D2 enhanced the release of the indicated chemokine (Figure 5C). Taken together, our results suggest that low-cytotoxicity strains are able to maintain infection by evading host clearance.

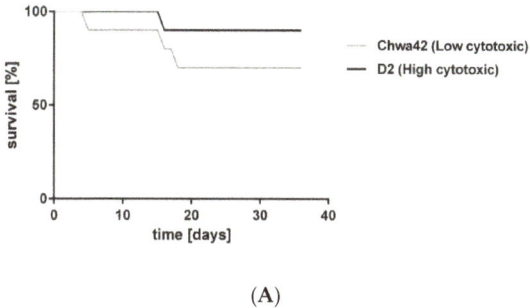

(A)

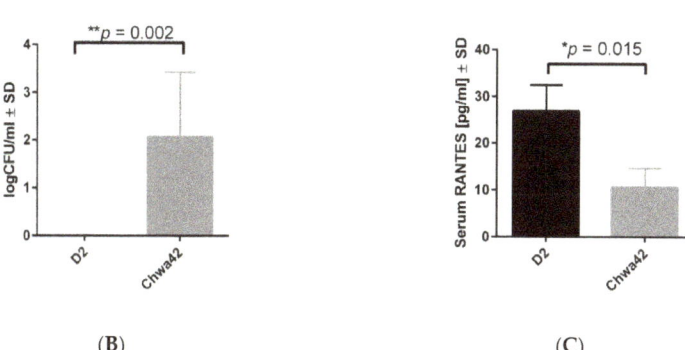

(B)   (C)

Figure 5. A highly cytotoxic strain induced high inflammation and fast clearance by the host in the mouse sepsis model. C57/BL6 mice were infected with Chwa42 (low cytotoxic) or D2 (high cytotoxic) strains for six weeks. (**A**) Survival curves of the infected mice with Chwa42 (n = 8) and D2 (n = 10). No differences were observed according to the long-rank (Mantel–Cox) test. (**B**) The bacterial loads within the tibiae were analyzed after six weeks post-infection. The bar and whiskers represent the mean ± SD. Statistical analysis was performed with an unpaired *t*-test comparing the bacterial load in the tibiae (** $p = 0.0021$). (**C**) RANTES levels in the serum were measured after three days post-infection. A Statistical analysis was performed, using the unpaired *t*-test, comparing the abundance of RANTES in the serum (* $p = 0.0015$).

3. Discussion

S. aureus is a versatile microorganism that causes a diverse array of infections. For different pathologies connected to *S. aureus* a controversial link between virulence and clinical outcome has been discussed [16,24,25]. Bacterial virulence is defined as the capacity for a pathogen to cause disease in the host [26] The virulence is influenced by a diversity of ecological factors, such as the surrounding environment during the course of infection, which either enhances or reduces the expression of specific virulence factors [5]. Thus, we assumed as a first hypothesis of our work, that different environments/origins of infection may determine the virulences of the staphylococcal isolates.

To investigate the associations between genes for virulence and infection types, we collected staphylococcal strains from nasal colonization, and from patients suffering from prosthesis infection, hematogenous osteomyelitis, and sepsis. By genotypic analysis, we found only a few genes with different prevalence in the respective groups. Moreover, some differences were related to clonal complex affiliations, and not to the type of infection.

To analyze whether the genetic characteristics affected the interactions between *S. aureus* and the host, we performed several functional assays related to the course of infection. During the first few steps of infection, *S. aureus* expressed several toxins and exoenzymes to induce tissue destruction, and to facilitate bacterial dissemination [27]. Thus, we investigated cell death induced by staphylococcal isolates from all groups on osteoblasts (cytotoxicity), and their capacity to lyse erythrocytes (hemolysis). Furthermore, the expression of main toxins, such as *hla* and *psmα*, as well as the expression of virulence global regulators such as *agr* and *rnaIII* were measured. We found a wide range in the cytotoxic capacity between the different strains, but we could not correlate cytotoxicity or hemolysis with the different infection types. By studying our bacterial collections from different origins, we demonstrated that low-cytotoxicity strains were present in all of the analyzed groups, including the healthy nasal carriers, indicating that low- and not only highly destructive strains are part of the pathogenic process of *S. aureus*. In line with these results, recent studies have demonstrated a higher propensity for low cytotoxic isolates to cause bacteremia, which can be explained by evolutionary trade-offs and bacterial fitness [16]. Toxins must not only be considered as tissue-destructive agents, but they are also involved in many functions, such as activating or manipulating the host immune system, and biofilm and colonization processes. Supporting the multiple functions of toxins besides tissue destruction, we also found a positive correlation between cytotoxicity and toxin expression [28]. Consequently, the role of cytotoxicity for the long-term survival of bacteria within the host is highly complex, and it needs to be tightly regulated to favor bacterial persistence.

Following tissue destruction and bacterial spreading, *S. aureus* invades the host cell to settle an infection. Thus, the invasiveness of staphylococcal isolates was analyzed by infecting osteoblasts and measuring the number of intracellular bacteria. Again, no differences were found to be related with the bacterial origins, but here also, a wide distribution between the highly and low-invasive strains was found in all of the analyzed groups.

Furthermore, positive and significant correlations were found between cytotoxicity as well as hemolysis, and the expression of *hla, psmα, agr,* and *rnaIII*, as these genes are largely responsible for the cytotoxic capacity of *S. aureus* [27]. By contrast, most of these parameters showed a significant but negative correlation with invasiveness. According to our functional assays, we defined highly cytotoxic strains that trigger high cell death, as having a high expression of *hla, psmα, agr,* and *rnaIII*, but a low degree of invasiveness. In contrast, the low-cytotoxicity strains were characterized by high invasiveness and a low expression of toxins.

To analyze the roles of high- and low-cytotoxicity strains during the course of infection, we investigated the ability for the long-term persistence of selected strains from each group. Furthermore, the appearance of SCVs was analyzed as a sign of adaptation and persistence. It is well-known that the SCVs can be induced/formed, or selected under stress conditions, such as intracellular locations [11,22]. Interestingly, high- and low-virulence strains were able to survive within host cells for up to seven days and form SCVs, indicating that the persistence mechanisms and SCV formation are general features of all *S. aureus* strains. Nevertheless, the low-cytotoxicity strains persisted in higher numbers, and presented more SCVs than the highly cytotoxic strains. These results indicate that low-cytotoxicity strains are better adapted for long-term persistence and survival within the host. In particular, in our murine model, we found that highly cytotoxic strains induced greater immune responses than low-cytotoxicity strains, which contributed to clearing the infecting bacteria. Consequently, the host is often not able to fully clear infections by low-cytotoxicity strains, which can turn into chronic and therapy-refractory disease courses.

These results are in line with recently published work demonstrating that genetic and phenotypic diversity favor infection development and the survival of the bacteria within the host [16,24,29,30]. Bacterial diversity must not be neglected as a virulence strategy. On one hand, highly virulent bacteria are able to attack and spread within the host, and in a population of hosts. On the other hand, less virulent bacteria can easier escape from the host immune system and survive within the host organism, and/or spread to other hosts (Figure 6). Apparently, different invasive infections (such as bone, foreign material infections, and sepsis) follow similar pathogenic courses, as we could not detect significant differences between strains isolated from different foci, nor in their degree of virulence. Moreover, the evolutionary success of *S. aureus* may be caused by heterogeneity within bacterial populations, and thus by its ability to vary, to evolve and to adapt, allowing it to induce a wide range of different types of infection in a range of different host species. It is already known that diversity within a bacterial population is an advantage for survival under environmental changes, and this has already been demonstrated for other bacterial species [29,31,32]. Furthermore, the low-cytotoxicity strains were present, not only in groups of different infections, but also in healthy nasal colonization. Thus, low-cytotoxicity strains are already positively selected during colonization, because they enhance bacterial fitness, resulting in a large remainder population that can turn into SCVs. The SCVs, the dormant phenotype, are able to persist under non-favorable conditions, such as low nutrition, high oxidative stress, and antibiotic treatment.

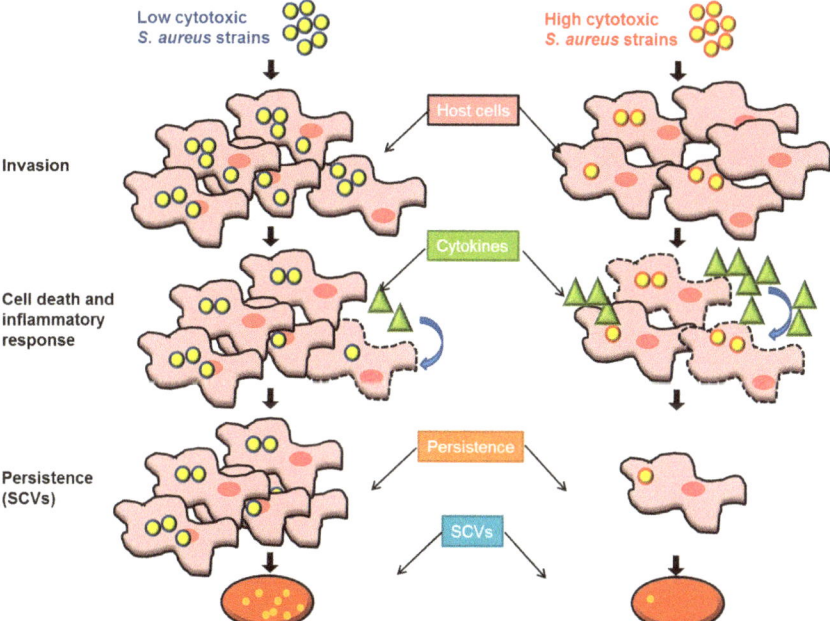

Figure 6. Low-cytotoxicity strains are able to persist in higher numbers, and they represent a bacterial reservoir for further infections. Low-cytotoxicity and highly cytotoxic staphylococcal strains are present in every staphylococcal infection, but they promote different mechanisms of infection. Highly cytotoxic strains display low invasion capacities and a high expression of toxins such as *hla*, *psmα* and others regulated by *rnaIII* and *agrA*, which promote the release of high amounts of cytokines. In this scenario, the bacteria are able to kill the host cells. Thus, only a few cells contain bacteria which can switch to SCVs. This pathway allows the bacteria to disseminate into other tissues, and this takes place during the acute phase. On the contrary, low-cytotoxicity strains tend to cause lower rates of cell death, but they achieve a higher level of host cell internalization compared to highly cytotoxic strains. Due to their low expression of virulence factors, the low–cytotoxicity strains are able to maintain the integrity of the host cells, and higher numbers of bacteria can switch to SCVs, favoring the silent persistence of *S. aureus*.

In summary, our results emphasize the importance of low-cytotoxicity strains in different pathologies induced by *S. aureus*. The low-cytotoxicity strains can adapt to the host, are poorly eliminated, and remain in higher numbers to select/form SCVs as a reservoir for chronic infection courses. These characteristics enhance the bacterial transmissibility, fitness, and antimicrobial tolerance. The understanding of this interplay/evolution between the host pathogen during infection, with specific attention on low cytotoxic isolates, will help us to optimize the treatment strategies for invasive and therapy-refractory infection courses.

4. Material and Methods

4.1. Collection of S. aureus Isolates and Definitions

The bacterial isolates were obtained from the Jena University Hospital. Identification procedures were performed in our lab by Vitek (Biomerieux, Germany). The corresponding patients resided in Thuringia and the adjacent federal states (Saxony-Anhalt, Bavaria and Hessen). The isolates were grouped into different categories: sepsis, hematogenous osteomyelitis, prosthesis (orthopedic patients with endoprosthesis), and nasal isolates from healthy persons. The main clinical characteristics are summarized in Table S1.

Sepsis was defined according to a definition on the basis of the quick Sepsis-related Organ Failure Assessment score (qSOFA) [33]. Patients belonging to the group of hematogenous osteomyelitis were characterized by a bone infection and an *S. aureus*-positive blood culture. The prosthesis groups were isolates from patients with prosthesis infections and a negative score for the blood culture. The nasal isolates were taken from a larger collection from our institute, and have previously been investigated [34].

4.2. Genotypic Characterization

The *S. aureus* isolates were genotyped with the Alere StaphyType DNA microarray (Alere Technologies GmbH, Jena, Germany). This array allowed for their assignment to clonal complexes, and the detection of 170 genes with their allelic variants.

4.3. Growth Curve and Generation Time

For growth curves, overnight cultures from *S. aureus* strains in BHI were diluted to $OD_{578nm} = 0.05$, and incubated at 37 °C with shacking (160 rpm). Turbidity (OD_{578nm}) was measured every 15 min for 17 hours (hr). The growth rate (μ, growth speed) and the generation time (g) for each strain used in this study (Figure S1) were calculated according the standard formula by Madigan et al. [35].

4.4. Measurement of Haemolysis

The hemolysis assay was performed by using the protocol described previously [6].

4.5. Cell Death and Invasion Assay in Osteoblasts

The human osteoblast cell line hFOB 1.19 (ATCC CRL-11372) was cultured according to the manufacturer's protocol. The invasion assay was performed by flow cytometry (BD Accuri™ C6), as described previously [36], but only raw values were used. The rate of cell death in the hFOB 1.19 cells was determined by measuring the uptake of propidium iodide (PI), as described previously [37].

4.6. Long-Term Persistence

The human osteoblasts cell line hFOB 1.19 (ATCC CRL-11372) was cultivated, following the manufacturer's indications, and it was infected with different *S. aureus* strains (Table S5). Briefly, osteoblasts cells were infected with a MOI (multiplicity of infection) of 50. After 1.5 hr, cells were washed with PBS, and lysostaphin (20 µg/mL) was added for 30 min, to lyse all extracellular or adherent staphylococci, and then fresh culture medium with Penicillin and streptomycin were added to the cells. The washing, the lysostaphin, and medium-exchange steps were repeated every two days, to remove all of the extracellular staphylococci. To detect live intracellular bacteria at different time points post-infection (p.i.), host cells were lysed in H_2O, and the number of colony-forming units (CFU) was determined by serial dilutions on blood agar. The colony phenotypes were analyzed by a Colony Counter Shuett (Biosys, Karben, Germany). SCVs were colonies with a diameter < 0.6 mm.

4.7. RNA Isolation and Real-Time PCR

To determine gene expression in the early stationary phase, overnight bacterial cultures were diluted to an OD_{578nm} of 0.05, and incubated at 37 °C with rotation (160 rpm). The time points for the higher expression of each factor were estimated by analyzing the expression of each gene in the different staphylococcal strains used in our lab (SH1000, LS1, 6850, USA300, and selected clinical strains). After 4 h (*agrA*, *rnaIII*) and 6 h (*psma*, *hla*) of incubation, 1 mL of bacterial suspension was mixed with RNAprotect Bacteria Reagent (Qiagen), centrifuged, and stored at −20 °C. For isolation of RNA, the pellet was mixed with RNApro™ Solution (MP Biomedicals, Eschwege, Hessen, Germany). The mixture was transferred to a Lysing Matrix B tube (MP Biomedical) and homogenized with a FastPrep® (MP Biomedicals) homogenizer. After subsequent centrifugation, the supernatant was used for RNA isolation with the peqGOLD Total RNA Kit (VWR) following the manufacturer's instructions. DNA was digested with TURBO™ DNase (ThermoFisher Scientific, Darmstadt, Germany). The concentration of the RNA was determined by spectrophotometric analysis with a NanoDrop (ThermoFisher Scientific, Darmstadt, Germany) before reverse transcription to complementary DNA (cDNA) (qScript cDNA SuperMix, Quantabio, Beverly, MA, USA). The cDNA was analyzed with QuantiNova SYBR Green PCR Kit (QIAGEN, Hilden, Germany) in a Rotor-Gene Q (QIAGEN) thermocycler. The reaction mixtures were incubated for 15 min at 95 °C, followed by 40 cycles of 15 s at 95 °C, 30 s at 55 °C, and 30 s at 72 °C. The primers used are described in Table S6. Fold-changes in expression were calculated by the Pfaffl equation [38]. The *gyrB* gene was used as a reference.

4.8. Murine Sepsis Model

Our murine sepsis model was performed as described before [23]. C57BL/6 10-week-old female mice were obtained from the central laboratory animal facility of the Jena University hospital. The animals were maintained according to institutional guidelines in individually ventilated cages, and were given food and water ad libitum. Mice were inoculated with 1×10^6 CFU of *S. aureus* in 200 µL of PBS, via a lateral tail vein, and sacrificed by CO_2 asphyxiation after three days (acute) and six weeks (chronic) post-infection. For the enumeration of bacteria in the tibiae of infected mice, homogenates were prepared in PBS and plated in 10-fold serial dilutions on blood agar.

4.9. Release of the Chemokine RANTES

The cell supernatants from the long-term persistence experiment were analyzed with RANTES human Instant ELISA™ (ThermoFisher Scientific).

Serum samples from infected mice were taken after three days post-infection, and analyzed by RANTES Mouse Instant ELISA™ (ThermoFisher Scientific).

4.10. Ethical Permissions

The ethical approval for the bacterial strains and the patients' data was approved by the local ethical committee of the University of Jena 4874-07/16 and 4449-06/15 (Thüringen, Jena). The murine sepsis model infection model was conducted in accordance with the recommendation and guidelines of the German regulations of the Society for Laboratory Animal Science 22-2684-04-02-006/15 and 22-2684-04-02-046/16 (Thüringen, Jena).

4.11. Statistics

The distribution of clonal complexes and the frequency of selected genes among all groups were analyzed by Fisher's exact test. The functional assays and statistical analyses of gene expression were performed by using GraphPad Prism version 4.00 (Graphpad, La Jolla, CA, USA). The normality of the distribution was analyzed with the D'Agostino & Pearson omnibus, and the Shapiro–Wilk normality test. Additionally, all of the results were tested for outliers with the ROUT test. An unpaired t-test was used when two groups were compared. Multiple groups were compared by and ordinary one-way ANOVA test, followed by Tukey's multiple comparisons test. According to the p-values, the differences were: either not significant (ns, $p > 0.05$); or significant (* $p < 0.05$; ** $p < 0.01$; *** $p < 0.001$ and **** $p < 0.0001$). The pairwise correlations between all of the measured parameters were performed by calculating Spearman's correlation coefficient and a two-sided p-value was used. A t-distribution with two degrees of freedom for testing the significance of the coefficients was assumed. Coefficients and p-values were both calculated using the Python library SciPy [39]. The difference between both survival mice curves was analyzed by the long-rank (Mantel Cox) test.

Supplementary Materials: The following are available online at http://www.mdpi.com/2072-6651/11/3/135/s1, Figure S1: Growth dynamics of *S. aureus* isolates, Figure S2: Cytotoxicity of *S. aureus* isolates, Table S1: Clinical characteristics of all patients, Table S2: Distribution of clonal complexes amongst staphylococcal isolates from nasal colonization, prosthesis infection, hematogenous osteomyelitis, and sepsis, Table S3: Frequencies of selected genes in the defined isolate groups, Table S4: Alere complete results, Table S5: Main characteristics of selected strains for long-term cell culture and the mouse sepsis model, Table S6: Primers used in this study.

Author Contributions: Conceptualization, L.T. and B.L.; Formal analysis, L.T.; Funding acquisition, B.L.; Methodology, C.P., A.S., S.D.-E., V.H. and S.M.; Supervision, L.T.; Validation, C.-M.S., M.T.F. and S.M.; Writing—original draft, L.T., C.-M.S., M.T.F., S.M. and B.L.

Funding: This work was funded by the German Federal Ministry of Education and Research (BMBF, CSCC Staphbone, FKZ 01EO1502) and by a Strategy and Innovation Grant from the Free State of Thuringia (41-5507-2016) in association to the Leibniz Science Campus InfectoOptics (SAS-2015-HKI-LWC).

Acknowledgments: We thank Sindy Wendler, Yvonne Ozegowski and Jennifer Geraci for excellent technical assistance and Thomas Lehmann from the institute of Institute for Medical Statistics, Computer Science and Data Science (Jena University Hospital) for his contribution with the statistical analysis.

Conflicts of Interest: The authors declare no conflict of interest.

References

1. Kluytmans, J.; van Belkum, A.; Verbrugh, H. Nasal carriage of *Staphylococcus aureus*: Epidemiology, underlying mechanisms, and associated risks. *Clin. Microbiol. Rev.* **1997**, *10*, 505–520. [CrossRef] [PubMed]

2. Abu-Qatouseh, L.F.; Chinni, S.V.; Seggewiss, J.; Proctor, R.A.; Brosius, J.; Rozhdestvensky, T.S.; Peters, G.; von Eiff, C.; Becker, K. Identification of differentially expressed small non-protein-coding RNAs in *Staphylococcus aureus* displaying both the normal and the small-colony variant phenotype. *J. Mol. Med.* **2010**, *88*, 565–575. [CrossRef] [PubMed]

3. Lowy, F.D. Medical progress—*Staphylococcus aureus* infections. *N. Engl. J. Med.* **1998**, *339*, 520–532. [CrossRef] [PubMed]

4. Von Eiff, C.; Becker, K.; Machka, K.; Stammer, H.; Peters, G. Nasal carriage as a source of *Staphylococcus aureus* bacteremia. Study Group. *N. Engl. J. Med.* **2001**, *344*, 11–16. [CrossRef] [PubMed]

5. Brown, S.P.; Cornforth, D.M.; Mideo, N. Evolution of virulence in opportunistic pathogens: Generalism, plasticity, and control. *Trends Microbiol.* **2012**, *20*, 336–342. [CrossRef] [PubMed]

6. Tuchscherr, L.; Bischoff, M.; Lattar, S.M.; Noto Llana, M.; Pfortner, H.; Niemann, S.; Geraci, J.; Van de Vyver, H.; Fraunholz, M.J.; Cheung, A.L.; et al. Sigma Factor SigB Is Crucial to Mediate *Staphylococcus aureus* Adaptation during Chronic Infections. *PLoS Pathog.* **2015**, *11*, e1004870. [CrossRef] [PubMed]

7. Bischoff, M.; Dunman, P.; Kormanec, J.; Macapagal, D.; Murphy, E.; Mounts, W.; Berger-Bachi, B.; Projan, S. Microarray-based analysis of the *Staphylococcus aureus* sigma(B) regulon. *J. Bacteriol.* **2004**, *186*, 4085–4099. [CrossRef] [PubMed]

8. Novick, R.P.; Geisinger, E. Quorum Sensing in Staphylococci. *Annu. Rev. Genet.* **2008**, *42*, 541–564. [CrossRef] [PubMed]

9.   Sonnen, A.F.P.; Henneke, P. Role of Pore-Forming Toxins in Neonatal Sepsis. *Clin. Dev. Immunol.* **2013**. [CrossRef] [PubMed]

10. Foster, T.J.; Geoghegan, J.A.; Ganesh, V.K.; Hook, M. Adhesion, invasion and evasion: The many functions of the surface proteins of *Staphylococcus aureus*. *Nat. Rev. Microbiol.* **2014**, *12*, 49–62. [CrossRef] [PubMed]

11. Proctor, R.A.; von Eiff, C.; Kahl, B.C.; Becker, K.; McNamara, P.; Herrmann, M.; Peters, G. Small colony variants: A pathogenic form of bacteria that facilitates persistent and recurrent infections. *Nat. Rev. Microbiol.* **2006**, *4*, 295–305. [CrossRef] [PubMed]

12. Tuchscherr, L.; Heitmann, V.; Hussain, M.; Viemann, D.; Roth, J.; von Eiff, C.; Peters, G.; Becker, K.; Loffler, B. Staphylococcus aureus Small-Colony Variants Are Adapted Phenotypes for Intracellular Persistence. *J. Infect. Dis.* **2010**, *202*, 1031–1040. [CrossRef] [PubMed]

13. Tuchscherr, L.; Geraci, J.; Loffler, B. Staphylococcus aureus Regulator Sigma B is Important to Develop Chronic Infections in Hematogenous Murine Osteomyelitis Model. *Pathogens* **2017**, *6*, 31. [CrossRef] [PubMed]

14. Cheung, A.L.; Eberhardt, K.J.; Chung, E.; Yeaman, M.R.; Sullam, P.M.; Ramos, M.; Bayer, A.S. Diminished virulence of a sar-/agr- mutant of *Staphylococcus aureus* in the rabbit model of endocarditis. *J. Clin. Investig.* **1994**, *94*, 1815–1822. [CrossRef] [PubMed]

15. Seidl, K.; Bayer, A.S.; McKinnell, J.A.; Ellison, S.; Filler, S.G.; Xiong, Y.Q. In vitro endothelial cell damage is positively correlated with enhanced virulence and poor vancomycin responsiveness in experimental endocarditis due to methicillin-resistant *Staphylococcus aureus*. *Cell. Microbiol.* **2011**, *13*, 1530–1541. [CrossRef] [PubMed]

16. Laabei, M.; Uhlemann, A.C.; Lowy, F.D.; Austin, E.D.; Yokoyama, M.; Ouadi, K.; Feil, E.; Thorpe, H.A.; Williams, B.; Perkins, M.; et al. Evolutionary Trade-Offs Underlie the Multi-faceted Virulence of *Staphylococcus aureus*. *PLoS Biol.* **2015**, *13*, e1002229. [CrossRef] [PubMed]

17. Suligoy, C.M.; Lattar, S.M.; Noto Llana, M.; Gonzalez, C.D.; Alvarez, L.P.; Robinson, D.A.; Gomez, M.I.; Buzzola, F.R.; Sordelli, D.O. Mutation of Agr Is Associated with the Adaptation of *Staphylococcus aureus* to the Host during Chronic Osteomyelitis. *Front. Cell. Infect. Microbiol.* **2018**, *8*, 18. [CrossRef] [PubMed]

18. Joo, E.J.; Choi, J.Y.; Chung, D.R.; Song, J.H.; Ko, K.S. Characteristics of the community-genotype sequence type 72 methicillin-resistant *Staphylococcus aureus* isolates that underlie their persistence in hospitals. *J. Microbiol.* **2016**, *54*, 445–450. [CrossRef] [PubMed]

19. Holtfreter, S.; Grumann, D.; Balau, V.; Barwich, A.; Kolata, J.; Goehler, A.; Weiss, S.; Holtfreter, B.; Bauerfeind, S.S.; Doring, P.; et al. Molecular Epidemiology of *Staphylococcus aureus* in the General Population in Northeast Germany: Results of the Study of Health in Pomerania (SHIP-TREND-0). *J. Clin. Microbiol.* **2016**, *54*, 2774–2785. [CrossRef] [PubMed]

20. Otto, M. Staphylococcus aureus toxins. *Curr. Opin. Microbiol.* **2014**, *17*, 32–37. [CrossRef] [PubMed]

21. Cheung, G.Y.; Joo, H.S.; Chatterjee, S.S.; Otto, M. Phenol-soluble modulins–critical determinants of staphylococcal virulence. *FEMS Microbiol. Rev.* **2014**, *38*, 698–719. [CrossRef] [PubMed]

22. Tuchscherr, L.; Medina, E.; Hussain, M.; Volker, W.; Heitmann, V.; Niemann, S.; Holzinger, D.; Roth, J.; Proctor, R.A.; Becker, K.; et al. Staphylococcus aureus phenotype switching: An effective bacterial strategy to escape host immune response and establish a chronic infection. *EMBO Mol. Med.* **2011**, *3*, 129–141. [CrossRef] [PubMed]

23. Horst, S.A.; Hoerr, V.; Beineke, A.; Kreis, C.; Tuchscherr, L.; Kalinka, J.; Lehne, S.; Schleicher, I.; Kohler, G.; Fuchs, T.; et al. A novel mouse model of *Staphylococcus aureus* chronic osteomyelitis that closely mimics the human infection: An integrated view of disease pathogenesis. *Am. J. Pathol.* **2012**, *181*, 1206–1214. [CrossRef] [PubMed]

24. Recker, M.; Laabei, M.; Toleman, M.S.; Reuter, S.; Saunderson, R.B.; Blane, B.; Torok, M.E.; Ouadi, K.; Stevens, E.; Yokoyama, M.; et al. Clonal differences in *Staphylococcus aureus* bacteraemia-associated mortality. *Nat. Microbiol.* **2017**, *2*, 1381–1388. [CrossRef] [PubMed]

25. Sandulescu, O.; Bleotu, C.; Matei, L.; Streinu-Cercel, A.; Oprea, M.; Dragulescu, E.C.; Chifiriuc, M.C.; Rafila, A.; Pirici, D.; Talapan, D.; et al. Comparative evaluation of aggressiveness traits in staphylococcal strains from severe infections versus nasopharyngeal carriage. *Microb. Pathog.* **2017**, *102*, 45–53. [CrossRef] [PubMed]

26. Casadevall, A.; Pirofski, L.A. Host-pathogen interactions: Redefining the basic concepts of virulence and pathogenicity. *Infect. Immun.* **1999**, *67*, 3703–3713. [PubMed]

27. Oliveira, D.; Borges, A.; Simoes, M. Staphylococcus aureus Toxins and Their Molecular Activity in Infectious Diseases. *Toxins* **2018**, *10*, 252. [CrossRef] [PubMed]

28. Rudkin, J.K.; McLoughlin, R.M.; Preston, A.; Massey, R.C. Bacterial toxins: Offensive, defensive, or something else altogether? *PLoS Pathog.* **2017**, *13*, e1006452. [CrossRef] [PubMed]

29. Davis, K.M.; Isberg, R.R. One for All, but Not All for One: Social Behavior during Bacterial Diseases. *Trends Microbiol.* **2018**. [CrossRef] [PubMed]

30. Priest, N.K.; Rudkin, J.K.; Feil, E.J.; van den Elsen, J.M.; Cheung, A.; Peacock, S.J.; Laabei, M.; Lucks, D.A.; Recker, M.; Massey, R.C. From genotype to phenotype: Can systems biology be used to predict *Staphylococcus aureus* virulence? *Nat. Rev. Microbiol.* **2012**, *10*, 791–797. [CrossRef] [PubMed]

31. Davis, K.M.; Isberg, R.R. Defining heterogeneity within bacterial populations via single cell approaches. *Bioessays News Rev. Mol. Cell. Dev. Biol.* **2016**, *38*, 782–790. [CrossRef] [PubMed]

32. Avery, S.V. Microbial cell individuality and the underlying sources of heterogeneity. *Nat. Rev. Microbiol.* **2006**, *4*, 577–587. [CrossRef] [PubMed]

33. Abajy, M.Y.; Kopec, J.; Schiwon, K.; Burzynski, M.; Doring, M.; Bohn, C.; Grohmann, E. A type IV-secretion-like system is required for conjugative DNA transport of broad-host-range plasmid pIP501 in gram-positive bacteria. *J. Bacteriol.* **2007**, *189*, 2487–2496. [CrossRef] [PubMed]

34. Deinhardt-Emmer, S.; Sachse, S.; Geraci, J.; Fischer, C.; Kwetkat, A.; Dawczynski, K.; Tuchscherr, L.; Loffler, B. Virulence patterns of *Staphylococcus aureus* strains from nasopharyngeal colonization. *J. Hosp. Infect.* **2018**, *100*, 309–315. [CrossRef] [PubMed]

35. Schijffelen, M.J.; Boel, C.E.; van Strijp, J.A.; Fluit, A.C. Whole genome analysis of a livestock-associated methicillin-resistant *Staphylococcus aureus* ST398 isolate from a case of human endocarditis. *BMC Genom.* **2010**, *11*, 376. [CrossRef] [PubMed]

36. Tuchscherr, L.; Korpos, E.; van de Vyver, H.; Findeisen, C.; Kherkheulidze, S.; Siegmund, A.; Deinhardt-Emmer, S.; Bach, O.; Rindert, M.; Mellmann, A.; et al. Staphylococcus aureus requires less virulence to establish an infection in diabetic hosts. *Int. J. Med. Microbiol.* **2018**, *308*, 761–769. [CrossRef] [PubMed]

37. Haslinger, B.; Strangfeld, K.; Peters, G.; Schulze-Osthoff, K.; Sinha, B. Staphylococcus aureus alpha-toxin induces apoptosis in peripheral blood mononuclear cells: Role of endogenous tumour necrosis factor-alpha and the mitochondrial death pathway. *Cell. Microbiol.* **2003**, *5*, 729–741. [CrossRef] [PubMed]

38. Pfaffl, M.W. A new mathematical model for relative quantification in real-time RT-PCR. *Nucleic Acids Res.* **2001**, *29*, e45. [CrossRef] [PubMed]

39. Oliphant, T.E. Python for scientific computing. *Comput. Sci. Eng.* **2007**, *9*, 10–20. [CrossRef]

Article

Sphingomyelin Depletion from Plasma Membranes of Human Airway Epithelial Cells Completely Abrogates the Deleterious Actions of *S. aureus* Alpha-Toxin

Sabine Ziesemer [1], Nils Möller [1], Andreas Nitsch [1], Christian Müller [1], Achim G. Beule [2] and Jan-Peter Hildebrandt [1,*]

[1] University of Greifswald, Animal Physiology and Biochemistry, Felix Hausdorff-Straße 1, D-17489 Greifswald, Germany; sabine.ziesemer@uni-greifswald.de (S.Z.); nils.moeller@uni-greifswald.de (N.M.); an124100@uni-greifswald.de (A.N.); christian.mueller@uni-greifswald.de (C.M.)

[2] Department of Otorhinolaryngology, University Hospital, Münster, Germany and Department of Otorhinolaryngology, Head and Neck Surgery, Greifswald University Hospital, D-17489 Greifswald, Germany; achimgeorg.beule@ukmuenster.de

* Correspondence: jph@uni-greifswald.de; Tel.: +49-3834-420-4295

Received: 21 December 2018; Accepted: 15 February 2019; Published: 20 February 2019

Abstract: Interaction of *Staphylococcus aureus* alpha-toxin (hemolysin A, Hla) with eukaryotic cell membranes is mediated by proteinaceous receptors and certain lipid domains in host cell plasma membranes. Hla is secreted as a 33 kDa monomer that forms heptameric transmembrane pores whose action compromises maintenance of cell shape and epithelial tightness. It is not exactly known whether certain membrane lipid domains of host cells facilitate adhesion of Ha monomers, oligomerization, or pore formation. We used sphingomyelinase (hemolysin B, Hlb) expressed by some strains of staphylococci to pre-treat airway epithelial model cells in order to specifically decrease the sphingomyelin (SM) abundance in their plasma membranes. Such a pre-incubation exclusively removed SM from the plasma membrane lipid fraction. It abrogated the formation of heptamers and prevented the formation of functional transmembrane pores. Hla exposure of rHlb pre-treated cells did not result in increases in $[Ca^{2+}]_i$, did not induce any microscopically visible changes in cell shape or formation of paracellular gaps, and did not induce hypo-phosphorylation of the actin depolymerizing factor cofilin as usual. Removal of sphingomyelin from the plasma membranes of human airway epithelial cells completely abrogates the deleterious actions of *Staphylococcus aureus* alpha-toxin.

Keywords: sphingomyelin; airway epithelial cells; cell physiology; *Staphylococcus aureus*; alpha-toxin

Key Contribution: Alpha-toxin is an important soluble virulence factor of *Staphylococcus aureus*, whose expression has a major impact on the pathogenicity of this bacterium. The result of this study that removal of sphingomyelin from the plasma membranes of host cells abrogates the deleterious actions of alpha-toxin on their cell physiology and cell integrity may open avenues to take preventive measures in patients at risk for staphylococcal infections by interfering with the interaction between alpha-toxin and plasma membrane sphingolipids.

1. Introduction

Airway epithelia form major barriers between inhaled air and the internal space of the body [1]. In vivo, respiratory epithelia are covered by a thick mucus layer. Inhaled microorganisms and other particles stick to that mucus layer and are removed from the airways by the ciliary activity in the periciliary liquid (mucociliary clearance) [2]. Therefore, it is unlikely that inhaled bacteria like the human commensal and opportunistic pathogen *Staphylococcus aureus* (*S. aureus*) readily come into direct contact with the apical surfaces of epithelial cells. However, when the mucociliary clearance is attenuated (as in bedridden or immune deprived patients, or patients with virus infections or cystic fibrosis) bacteria may reach critical densities in the mucus layer and start to secrete soluble virulence factors.

Virulence factors play a central role in the pathogenicity of *S. aureus* [3]. Secreted soluble virulence factors like alpha-toxin (hemolysin A, Hla) may diffuse through the mucus layer and reach the apical surfaces of the epithelial cells [4]. The assumption that Hla may play a role in the onset of *S. aureus* lung infection is supported by the findings of pneumonia patients having generated antibodies against Hla [5,6] and by animals being protected from developing *S. aureus*-mediated pneumonia when vaccinated against Hla [7].

Alpha-toxin is a pore-forming bacterial toxin [8]. It is lytic to red blood cells [9] and toxic to a wide range of mammalian cells [10]. Hla pores increase the membrane permeability for ATP [11,12] and cations like calcium, potassium, and sodium [13–17]. Cation-entry depolarizes the membrane potential in airway epithelial cells and enhances phosphorylation of p38 MAP kinase [17]. Additionally, Hla induces alterations in cell shape by remodeling the actin cytoskeleton and disrupts cell-matrix adhesions in human airway epithelial cells [18,19]. Actin remodeling seems to be mediated by hypo-phosphorylation and activation of cofilin [19], an actin depolymerizing factor [20].

Hla is secreted by the bacteria as a water-soluble 33 kDa monomer that binds to plasma membranes (PM) of eukaryotic host cells. Seven Hla monomers form a non-lytic heptameric pre-pore [21]. Subsequently, all seven subunits simultaneously unfold their pre-stem domains, which are then inserted into the lipid bilayer and form a cylindrical transmembrane pore [4,22,23].

Binding of Hla monomers to the plasma membranes of host cells may be facilitated by protein-receptors. Potential Hla receptors are the metalloproteinase ADAM10 (a disintegrin and metalloproteinase 10) [24,25], which is also expressed in the plasma membranes of airway epithelial model cells [26], alpha5beta1-integrin [27,28], rabbit erythrocyte band 3, or caveolin [29,30]. Specific toxin binding through Hla receptors appears to be important at low concentrations of Hla [8,31]. At higher concentrations, Hla is able to bind directly to PM lipids [31], probably by interacting preferentially with phosphocholine headgroups of sphingomyelin (SM) and phosphatidylcholine (PC) [32–35].

SM and PC form clusters with cholesterol in so-called lipid rafts [36]. It is conceivable that such microdomains act as concentration platforms for membrane-associated proteins, and hence may mediate quick oligomerization of pore forming toxins [35,37,38]. It has actually been shown that pore formation of *S. aureus* Hla is highly effective in biological membranes, which have a high proportion of SM [39]. However, it was not clear whether the lipid composition affects the binding of the monomers, the assembly of membrane-bound monomers to heptamers, or the final step of pore formation, namely the coordinated unfolding of the stem loops of each of the assembled monomers to form the transmembrane portion of the pore.

To answer these questions, we used the recombinant form of another toxin of *S. aureus*, hemolysin B (beta-hemolysin, Hlb), which has enzymatic activity and functions as a neutral sphingomyelinase [40,41] cleaving SM to phosphorylcholine and ceramide [42,43]. Immortalized human airway epithelial cells (S9, 16HBE14o-), as well as freshly isolated human nasal epithelium, were pre-treated with recombinant Hlb to deplete sphingomyelin from the PM and subsequently exposed to recombinant Hla. Cells were tested for the well-known effects of Hla on cell signaling like calcium influx [17], changes in cell morphology, and cell layer integrity [18], or actin remodeling induced by hypo-phosphorylation of cofilin at Ser3 [19].

2. Results

2.1. Pre-Treatment of Airway Cells with rHlb Allows rHla Monomer Binding to the PM, but Prevents Formation of Heptamers

To investigate whether SM is necessary for binding of Hla monomers to the host cell plasma membrane or assembly of heptameric transmembrane pores in these membranes, we pre-treated confluent cell layers (16HBE14o- and S9) for 1 h with 5000 ng/mL rHlb (sphingomyelinase) followed by a 0–4 h incubation with 2000 ng/mL rHla. Semi-quantitative Western blot analysis in whole cell protein extracts showed no changes in abundance of Hla monomers, whether cells had been pre-treated with rHlb or not (Figure 1A,B,D,E). This indicates that binding of rHla monomers to airway cell plasma membranes was obviously not affected by the removal of SM from the plasma membranes in the two cell types (c.f. Figure S1). Experiments using freshly prepared human nasal tissue showed similar results confirming the above conclusion (Figure 1G,H). In contrast, the abundance of rHla heptamers was lower in cell or tissue samples that were pre-treated with sphingomyelinase (rHlb) (Figure 1C,F,I). In 16HBE14o- as well as S9 cells that had not been pre-treated with rHlb, heptamer assembly started immediately after the addition of rHla at 0 h and increased steadily with the duration of exposure up to 4 h (Figure 1C,F). However, when cells had been pre-treated with rHlb, heptamer abundance was significantly lower and did not show any increases over the time of exposure (Figure 1C,F). Similar results were obtained when freshly prepared human nasal tissue was used in the experiments (Figure 1H,I). These observations indicate that the presence of SM in the plasma membranes of human airway epithelial cells is essential for *S. aureus* Hla to form multimeric complexes.

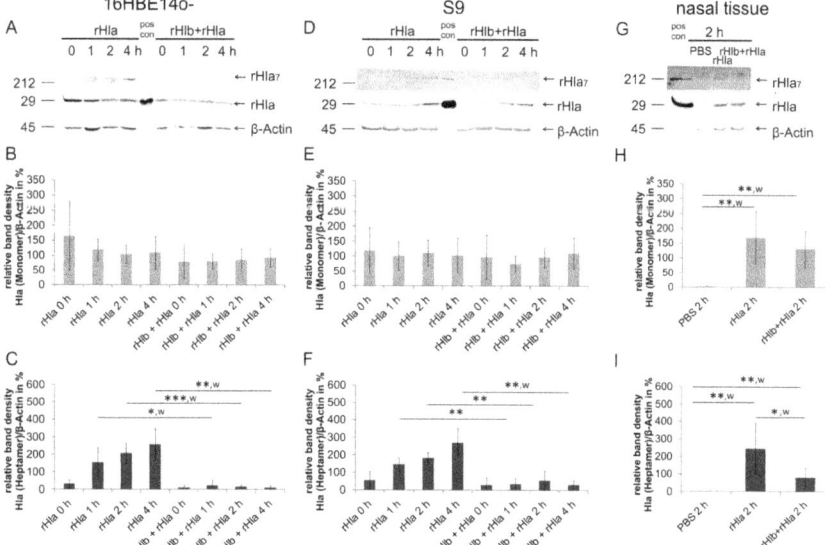

Figure 1. Pre-incubation of cells with sphingomyelinase (rHlb) prevented formation of rHla heptamers (rHla7), but not plasma membrane binding of rHla monomers in airway epithelial cells and nasal tissue. Confluent layers of immortalized airway epithelial cells (16HBE14o- (**A–C**) and S9 (**D–F**)) were treated with 2000 ng/mL rHla after pre-treatment of cells in the presence or absence of 5000 ng/mL rHlb (sphingomyelinase) for 0–4 h. Cells treated with rHla showed binding of Hla monomers (33 kDa, rHla) and Hla heptamers (231 kDa, rHla7). The rHla monomer abundances were independent of the incubation time with rHla und independent of the pre-treatment regime with sphingomyelinase (**A,B,D,E**). Formation of Hla heptamers, however, was significantly reduced in 16HBE14o- or S9 cells which had been pre-treated with sphingomyelinase (rHlb) compared with control cells without sphingomyelinase

pre-treatment (**A,C,F**). Experiments using freshly prepared human nasal tissue showed similar results (**G–I**). Representative example Western blot signals of Hla heptamers (rHla$_7$), Hla monomers (rHla), and β-actin are shown (**A,D,G**). Recombinant Hla (approximately 40 ng/lane) was used to indicate the position of Hla monomers (pos con), and in some cases, heptamers that form spontaneously when aqueous solutions of rHla are left at room temperature for 10 min. The positions of molecular mass standards (in kDa) are indicated. Mean values ± S.D. of densitometry signals of Western blot analyses normalized to the densities of the respective β-actin bands (n = 5, each) were assembled in histograms. Individual means were tested for significant differences using Student's *t*-test or Welch's *t*-test (w): * $p < 0.05$, ** $p < 0.01$, or *** $p < 0.001$.

2.2. Effects of Sphingomyelinase Pre-Treatment of Airway Epithelial Cells on rHla-Mediated Changes in $[Ca^{2+}]_i$

As previously shown in human airway epithelial cells, treatment with rHla induced elevations in the cytosolic calcium concentration ($[Ca^{2+}]_i$) [15,17]. As observed previously, $[Ca^{2+}]_i$ started to increase with a lag phase of approximately 5–10 min after the addition of rHla and reached levels significantly different ($p < 0.05$) from the controls at 20–22 min recording time. These results were confirmed in this study as treatments of 16HBE14o- (Figure 2A), as well as S9 cells (Figure 2B), with 2000 ng/mL rHla resulted in significant increases in $[Ca^{2+}]_i$ (traces PBS + rHla). Pre-treatment of 16HBE14o- (Figure 2A) or S9 cells (Figure 2B) with 5000 ng/mL rHlb (sphingomyelinase) and subsequent exposure to 2000 ng/mL rHla (traces rHlb + rHla), however, did not result in any significant increases in $[Ca^{2+}]_i$. These traces were not significantly different from those that were obtained using cells that had been pre-treated with PBS (instead of rHlb) and treated with PBS instead of rHla during the experiment (Figure 2, traces PBS + PBS). Treatments of 16HBE14o- or S9 cells with 5000 ng/mL rHlb (traces rHlb + PBS), per se, did not induce any changes in $[Ca^{2+}]_i$ when compared to untreated control cells. These results indicate that pre-treatment of airway epithelial model cells with sphingomyelinase (rHlb) prevented rHla-mediated increases in $[Ca^{2+}]_i$. Because acute addition of sphingomyelinase (rHlb) to airway epithelial cells did not elicit any sustained changes in $[Ca^{2+}]_i$ (Figure S2A) and rHlb-pre-treated cells showed strong calcium influx upon addition of calcium ionophores (Figure S2B), it can be concluded that the suppression of rHla-mediated calcium signaling in rHlb-pretreated airway epithelial cells is not a consequence of indirect effects (like rHlb-mediated emptying of calcium stores before the addition of rHla or ceramide-mediated internalization of rHla-containing plasma membrane).

2.3. Effects of Sphingomyelinase Pre-Treatment of Airway Epithelial Cells on rHla-Mediated Formation of Paracellular Gaps

Previous investigations had demonstrated alterations in cell shape, loss of cell-cell contacts, and the formation of paracellular gaps in initially confluent cell layers of airway epithelial model cells upon exposure to rHla [18], with 16HBE14o- cells' reaction being much more pronounced than S9 cells. Thus, we tested in this study whether these effects of rHla could be moderated or abrogated by pre-treatment of cells with sphingomyelinase (rHlb). As shown in the still pictures taken from time lapse movies shown in Figure 3 (third row, each), confluent cell layers of 16HBE14o-, as well as S9 cells that had been pre-treated with 5000 ng/mL rHlb, did not develop microscopically visible gaps or other rHla-typical cellular changes upon treating the cells with 2000 ng/mL rHla (added at 0 h). Cell growth, division, and shape were comparable with control cells treated with PBS in both cell lines (Figure 3, first row, each), but clearly different from those cell cultures that had not been pre-treated with sphingomyelinase and exposed to rHla (Figure 3, second row, each).

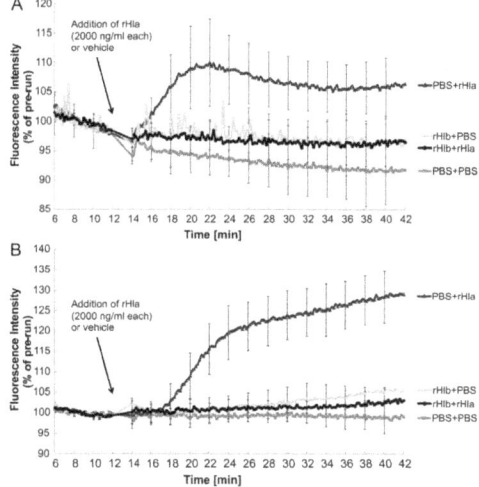

Figure 2. Pre-incubation of cells with sphingomyelinase (rHlb) abrogated rHla-mediated increases in $[Ca^{2+}]_i$ in airway epithelial cells. Calcium-dependent fluorescence intensities (normalized to the mean fluorescence level during the initial 6 min of each recording) were monitored in Indo-1 loaded 16HBE14o- (**A**) or S9 cells (**B**), respectively. Cells were pre-treated in the absence or presence of rHlb as controls (PBS + PBS; rHlb + PBS) or additionally treated with 2000 ng/mL rHla (PBS + rHla; rHlb + rHla) during the experiment. Pre-incubation of cells with rHlb prevented rHla-mediated increases in $[Ca^{2+}]_i$ (rHlb + rHla, circles) compared with those cells that were not pre-exposed to rHlb (PBS + rHla, triangles). Mean traces ($\pm$ S.D. at 2 min intervals) of $n = 9$ (16HBE14o-, **A**) or $n = 5$ (S9, **B**) independent cells preparations are shown.

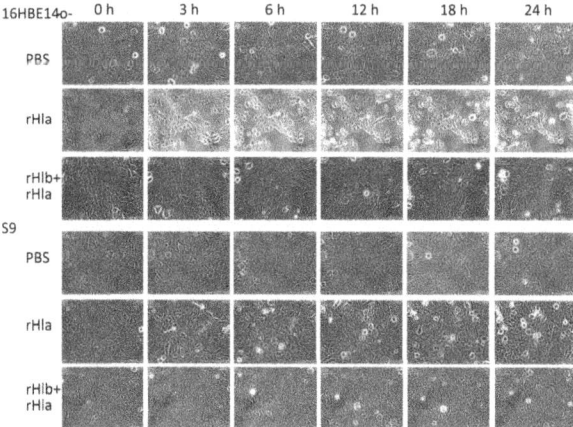

Figure 3. Pre-incubation of cells with sphingomyelinase (rHlb) prevented rHla-mediated formation of microscopically visible gaps in airway epithelial cell layers. Cellular changes, like loss of contacts to neighboring cells, changes in cell shape, and formation of paracellular gaps (white arrows) induced by *S. aureus* alpha-toxin (rHla, added at 0 h) in initially confluent cell layers of human airway epithelial model cells (16HBE14o- or S9 cells) were inhibited by pre-incubation of cells with *S. aureus* beta-toxin (rHlb), a sphingomyelinase. Treatments of cells with rHlb alone showed no cellular changes (data not shown). Phase contrast images were taken at different time points after adding PBS (vehicle control, PBS), 2000 ng/mL rHla, or rHla in the continued presence of 5000 ng/mL rHlb, respectively. Images were taken at the indicated times from time-lapse movies (Biostation IM, Nikon) monitoring the cells over 24 h.

2.4. Effects of Pre-Treatment of Airway Epithelial Cells with Sphingomyelinase (rHlb) on rHla-Mediated Hypo-Phosphorylation of Cofilin

Earlier studies have shown that rHla-treatment of airway epithelial model cells resulted in hypo-phosphorylation of the actin depolymerizing factor cofilin [19], which is likely to be the most important part in the chain of events leading to rHla-mediated changes in cell shape and paracellular gap formation. In this study, we tested the impact of rHla on pSer3-phosphorylation of cofilin with or without pre-treatment of cells with rHlb. As reported previously [19], and again shown in Figure 4, treatment of human airway epithelial cells (16HBE14o- and S9 cells) with 2000 ng/mL rHla significantly decreased the levels of pSer3-cofilin in 16 HBE14o- cells, as well as in S9 cells (Figure 4B,D, dots). After adding rHla to the 16HBE14o- cell culture, the level of pSer3-cofilin declined, reaching a minimum after 2 h that was maintained for the remaining experimental time (Figure 4B). In S9 cells, the decrease of cofilin phosphorylation was transient and started to recover between 2 and 4 h of rHla-exposure (Figure 4D). These results are in accordance with those reported previously [19]. When cells had been pre-incubated with 5000 ng/mL *S. aureus* rHlb, no decline in cofilin phosphorylation was observed in either cell type (Figure 4B,D, diamond). Total cofilin abundance (normalized to β-actin), although somewhat variable in cells, was not significantly affected by any of these treatments.

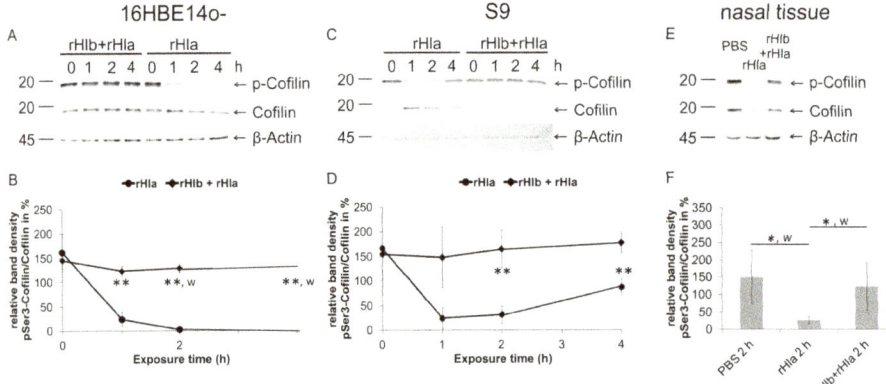

Figure 4. Pre-incubation of cells with sphingomyelinase (rHlb) prevented rHla-mediated hypo-phosphorylation of cofilin in airway epithelial cells and nasal tissue. Ser3-cofilin phosphorylation was monitored in whole cell extracts of 16HBE14o- cells (**A**,**B**) and S9 cells (**C**,**D**) that had been pre-treated (1 h) with 5000 ng/mL rHlb (rHlb) or not (PBS, rHla) and subsequently treated with rHla (rHla, 2000 ng/mL) either in the absence or continued presence of 5000 ng/mL rHlb by semi-quantitative Western blotting. In addition, primary human nasal tissue samples (**E**,**F**) were pre-treated (1 h) with 5000 ng/mL rHlb (rHlb) or not (PBS, rHla) and subsequently treated with PBS (vehicle control) or 2000 ng/mL rHla (rHla), either in the absence or continued presence of 5000 ng/mL rHlb for 2 h. Representative pSer3-cofilin-, cofilin, and β-actin Western blot signals are shown (**A**,**C**,**E**: positions of molecular mass standards in kDa are indicated). Means ± S.D. of $n = 3$ (**B**,**D**) or $n = 5$ (**F**) independent preparations. Individual means were tested for significant differences using Student's *t*-test or Welch's *t*-test (*w*): * $p < 0.05$, ** $p < 0.01$.

When freshly prepared human nasal tissue was exposed to 2000 ng/mL rHla for 2 h, a significant decline in pSer3-phosphoration of cofilin was observed compared with untreated control tissue (Figure 4F). However, when nasal tissue was pre-incubated with 5000 ng/mL rHlb, rHla-mediated hypo-phosphorylation of cofilin was absent (Figure 4F). This indicates that the results obtained using 16HBE14o- or S9 airway model cells are of physiological relevance.

3. Discussion

Besides sphingomyelin (SM), the most common and important phospholipids in the plasma membrane of eukaryotic cells are phosphatidylcholine (PC), phosphatidylserine (PS), and phosphatidylethanolamine (PEA) [44]. The choline containing lipids, SM and PC, have been implicated in the formation of lipid rafts [38,45,46]. SM has been discussed as an important factor mediating the deleterious effects of *S. aureus* alpha-toxin (Hla) [21,47] on host cells [35,39]. However, it was not clear which of the sequential step(s) of forming functional transmembrane pores in the plasma membranes of host cells (monomer binding, monomer heptamerization, and pre-pore assembly or unfolding of the stem loops of each of the monomers to from the functional transmembrane pore) were facilitated by sphingomyelin. We used *S. aureus* β-hemolysin (Hlb), another secreted virulence factor of *S. aureus* that functions as a neutral sphingomyelinase [42,43] (Figure S1) as a tool to deplete the pool of sphingomyelin in the outer leaflet of the plasma membranes of airway epithelial model cells.

To investigate whether SM is necessary for binding Hla monomers to the plasma membrane or formation of heptameric pores in the plasma membrane, immortalized human airway epithelial cells (16HBE14o-, S9), as well as freshly isolated human nasal tissue, were pre-treated with Hlb and subsequently exposed to rHla. As shown in Figure 1, the removal of SM (rHlb + rHla) had no effect on the abundance of Hla monomers in the protein extracts from cells when compared with the samples obtained from cells that had not been pre-treated with Hlb (Figure 1B,E,H). This indicates that Hla monomer binding was not affected by the presence or absence of SM. Western blot signals associated with Hla heptamers, however, showed a completely different picture (Figure 1C,F,I). Because heptamer formation by membrane-bound alpha-toxin monomers is a very rapid process [48], the abundance of Hla heptamers in rHla-treated airway epithelial model cells steadily increased over the time of Hla exposure in those cells that were not pre-treated with Hlb (Figure 1C,F; rHla). In cells that had been pre-treated with sphingomyelinase (rHlb), however, the abundance of Hla heptamers was significantly lower or entirely absent and did not change over time during the incubation period (Figure 1C,F; rHlb + rHla). The latter finding was confirmed when freshly isolated human airway tissue was used (Figure 1I), indicating that the lack of Hla heptamer formation is a general feature of SM-depleted airway epithelial cells. These results indicate that the presence of SM in plasma membranes of airway epithelial cells is dispensable for the attachment of toxin monomers to the cell surface, but essential for the heptamerization process and the formation of the pre-pore. Without the formation of a pre-pore there should be no unfolding of the stem loops of the seven subunits and no formation of a functional transmembrane pore. Thus, our expectation was that none of the usual Hla-mediated cell physiological changes would occur upon Hla exposure of our eukaryotic model cells if these had been pre-treated with sphingomyelinase.

In the absence of SM in airway epithelial cell membranes, rHla heptamerization was almost completely suppressed. The residual multimer formation may be due to incomplete SM degradation during the pre-incubation of cells with rHlb. Alternatively, choline-containing lipids like PC may be able to replace SM to a certain extent in mediating Hla heptamer formation [35]. Another potential explanation would be that Hla monomers assemble spontaneously and at a low rate without any assistance from lipids. This conclusion is supported by the observation that rHla monomers maintained for 10 min in aqueous solution at room temperature (Figure 1, pos con) sometimes show spontaneous heptamerization without the need for any additional reagents.

While these options remain to be tested, we focused on the question whether the removal of SM from host cell membranes and almost complete suppression of heptamer formation may be able to suppress the cell physiological effects that are normally seen in Hla-treated cells upon pore-formation [15,18,19]. We chose to measure the time course of changes in $[Ca^{2+}]_i$, the formation of paracellular gaps, as well as the hypo-phosphorylation of cofilin upon addition of rHla to airway epithelial model cells that had or had not been pre-treated with rHlb.

Exposure of Indo-1-loaded 16HBE14o- cells or S9 cells to rHla resulted in a slow increase in Ca^{2+}-mediated dye fluorescence, whose onset was delayed for 3 to 8 min, probably due to the time required for generating Hla pore-mediated calcium influx that exceeded the capacity of endogenous calcium extrusion mechanisms in these cells (Figure 2). These results match those that have been reported previously [15,18]. Pre-treatment of cells with sphingomyelinase (rHlb), however, completely abolished this cellular response to rHla-exposure (Figure 2, traces rHlb + rHla), indicating that sphingomyelin depletion prevents Hla from forming functional transmembrane pores. The few residual heptamers that seemed to form in rHlb-treated cells (Figure 1C,F) may allow some influx of calcium ions into the cytosol of these cells that is, however, not large enough to out-perform the endogenous calcium extrusion mechanisms.

Monitoring of cell shape changes and paracellular gap formation in confluent cultures of 16HBE14o-, as well as S9 cells during exposure to rHla using time lapse microscopy, we could confirm that treatment of cells with rHla induced loosening of the cells from each other and from the culture dish (Figure 3, traces rHla). These effects were much more pronounced in 16HBE14o- cells than in S9 cells, a finding that confirmed the results of previous studies [18,19]. However, when cells had been pre-treated with sphingomyelinase (rHlb), these effects were completely suppressed, and the cultures maintained their confluent appearance over the entire experimental period despite the continued presence of rHla (Figure 3, traces rHlb + rHla).

The rHla-mediated changes in shell shape and paracellular gap formation are associated with and most likely caused by the disruption of the original architecture of the actin cytoskeleton in rHla-treated airway epithelial model cells [19]. We monitored the level of pSer3-phosphorylation of the actin depolymerizing factor cofilin that has been shown to be downregulated upon rHla exposure of cells [19]. As shown in Figure 4 (traces with dots), we could confirm the previous findings that cofilin hypo-phosphorylation occurred under rHla-treatment in airway epithelial model cells with a sustained effect in 16HBE14o- and a transient effect in S9 cells. However, when cells had been pre-treated with rHlb before the start of the rHla-experiment there was no indication of any change in pSer3-phosphorylation in cofilin over the experimental period of 4 h (Figure 4, traces with diamonds). Parallel experiments using freshly isolated human nasal tissue gave similar results (Figure 4F). This indicates that pre-treatment of airway epithelial cells with sphingomyelinase (rHlb) prevents the activation of signaling pathways and any of the detrimental cell physiological effects usually induced by rHla exposure.

An interesting question was whether the attenuation of Hla-mediated cell damage by sphingomyelinase pre-treatment is a phenomenon limited to airway epithelial cells, or a more general effect in eukaryotic cells. To test this we used sheep erythrocyte agar plates and pre-treated the red blood cells in the agar matrix with sphingomyelinase (rHlb) before exposing them to rHla. Exposing cells to sphingomyelinase (rHlb) results in some changes in cell integrity (Figure S3B,C), known as "incomplete hemolysis" [9]. As shown in Figure S3C, these cells did not show the typical complete hemolysis that occurs in erythrocytes exposed to rHla (Figure S3D). We concluded that pre-treatment of sheep erythrocytes with sphingomyelinase (*S. aureus* rHlb) suppresses rHla-mediated hemolysis. This indicates that the removal of sphingomyelin from the outer leaflet of the plasma membranes of eukaryotic cells may generally save eukaryotic cells from the deleterious actions of *S. aureus* alpha-toxin. Such a conclusion is consistent with the recent observation that cells lacking sphingomyelin synthase 1 (SGMS1) are resistant against Hla-virulence [25].

4. Materials and Methods

4.1. Chemicals and Reagents

Indo-1/AM was obtained from Invitrogen (Karlsruhe, Germany). Trypsin (with EDTA) was purchased from PAN-Biotech (Aidenbach, Germany). Lipid standards were ordered from Sigma (Steinheim, Germany). Antibodies (Ab) were obtained from these sources: Hla-Ab (S7531) and β-Actin-Ab (AC-15, A5441) from Sigma (Steinheim, Germany), p-Cofilin1-Ab (hSer3-R, sc-12912-R), cofilin-Ab (E-8, sc-376476), goat anti-rabbit IgG-HRP (sc-2004) and goat anti-mouse IgG-HRP (sc-2005) from Santa Cruz (Heidelberg, Germany). All other chemicals were reagent grade and obtained from Roth (Karlsruhe, Germany).

4.2. Expression and Purification of Recombinant Staphylococcus aureus Hla and Hlb

Recombinant alpha-toxin (rHla) and recombinant beta-toxin (rHlb) were prepared and purified as described previously [49]. The purity of rHla and rHlb was assessed by SDS-PAGE. The concentration of rHla routinely used was 2000 ng/mL (60 nmol/L), for reasons discussed previously [18]. Exposure of airway epithelial cells to such a concentration of rHla for 2 h results in cell death of less than 15% of all cells [17]. Lower concentrations of rHla may induce pore formation as well, but the physiological responses are much less pronounced and can hardly be detected. As we have previously shown, treatment of cells with 200 ng/mL rHla does not induce MAP kinase activation [49] or intracellular calcium accumulation [18].

4.3. Cell Culture

Immortalized human airway epithelial cells (16HBE14o- or S9) were obtained from Karl Kunzelmann (Regensburg, Germany, 16HBE14o-) or from ATCCLGC Standards (Wesel, Germany, S9) and cultured on 10 cm Cell+ plates (Sarstedt, Numbrecht, Germany) at 37 °C and gassed with 5% CO_2. Eagle's MEM (PAN-Biotech, Aidenbach, Germany) containing 10% FBS superior 168 (Biochrom, Berlin, Germany) and 1% (w/v) penicillin/streptomycin (PAA 169 Laboratories, Cölbe, Germany) was routinely used as cell culture medium and changed every 3–4 days. Shortly before cells formed confluent monolayers, they were passaged on new cell culture plates or directly used in the experiments. All cell cultures were checked for *Mycoplasma* contamination on a regular basis.

4.4. Freshly Prepared Human Airway Tissue

Primary human airway tissue was isolated from the ethmoid sinus uncinate process in chronic rhinosinusitis patients with macroscopically normal mucosa undergoing surgery for the removal of nasal polyps. The sheet of epithelial and underlying connective tissue was lifted from the bone, rinsed several times in cell culture medium (see above), and transferred to the lab. Tissue was sliced in a vertical direction (0.5 mm thickness) to maintain the original tissue structure. The experiments were approved by the ethics committee of the university hospitals in Greifswald (BB95/10; September 1, 2010) and Münster (2017-120-f-S). Informed written consent was obtained from all donors.

4.5. Sample Preparation for Western Blotting

Immortalized human airway epithelial cells (16HBE14o- or S9) were incubated in the presence or absence of 5000 ng/mL recombinant *S. aureus* Hlb (sphingomyelinase) for 1 h. Afterwards, cells were treated with 2000 ng/mL recombinant Hla for 0, 1, 2, or 4 h or with phosphate buffered saline (PBS) (control). Primary human airway tissue was treated with 5000 ng/mL rHlb for 1 h and subsequently with 2000 ng/mL rHla for 2 h or with PBS (control). After incubation of cells or tissue in the presence or absence of rHla, the culture medium was carefully aspirated and the material was washed using 5 mL PBS. To each 10 cm-plate of cultured cells, 400 μL lysis buffer (100 mmol/L KCl, 20 mmol/L NaCl, 2 mmol/L $MgCl_2$, 0.96 mmol/L NaH_2PO_4, 0.84 mmol/L $CaCl_2$, 1 mmol/L EGTA, 0.5% (v/v) Tween 20, 25 mmol/L HEPES (free acid), pH 7.2 containing 10 mmol/L each of aprotinin, leupeptin, and pepstatin, as well as 100 mmol/L PMSF and 0.33 mmol/L ortho-vanadate) was added. Cells were scraped off the cell culture plate using a cell scraper. The suspension of cytosolic extract and particulate matter was transferred to a 1.5 mL Eppendorf-reaction tube and immediately transferred on ice. The samples were homogenized on ice using a T8-Ultraturrax (IKA Labortechnik, Staufen, Germany) for 30 s, each, and combined with an equal volume of SDS sample-buffer, mixed, and frozen at −80 °C [19].

4.6. Semi-Quantitative Western Blotting

Proteins were separated by SDS/PAGE (10% or 13% gels) in a minigel apparatus (BioRad, Munich, Germany) and transferred to nitrocellulose membrane (HP40, Roth, Karlsruhe, Germany) [19]. Western blotting for the quantification of total or phosphorylated proteins was performed using (phospho-) specific antibodies, HRP-linked secondary antibodies (1:6000), and enhanced luminescence reagents (Biozym, Oldendorf, Germany). Signals were recorded using a Fusion FX7-SL gel imager (Vilbert Lourmat, Eberhardzell, Germany). Band signal intensities were assessed by densitometry using Phoretix 1 D (Nonlinear Dynamics, Newcastle upon Tyne, UK). To correct for potential minor differences in exposure time, the mean density of all bands on each gel image was used to normalize the densities of individual bands of the same gel. Signal intensities of the phosphorylated forms of proteins were normalized to the signals obtained using antibodies against the respective core proteins or against the β-actin band densities. Relative band densities were used to calculate means and standard deviations of different experiments.

4.7. Intracellular Calcium Concentrations

Changes in intracellular calcium concentration ($[Ca^{2+}]_i$) were monitored in airway epithelial cells using the calcium sensitive indicator dye indo-1, as described previously [15,18]. The cell suspension was split in equal aliquots after the dye loading procedure. Cells in one portion were treated with 5000 ng/mL rHlb, and cells in the other portion with phosphate buffered saline (PBS) during the 30 min recovery period. Subsequently, cells were washed 3 times (2 min at $600\times$ g, each) and finally resuspended in 1.5 mL HEPES-buffered saline. Portions of 300 μl cell suspension were transferred to each well of a 96-well-plate (black flat-bottomed microplate, Biozym, Oldendorf, Germany) and treated with 2000 ng/mL rHla or PBS as control after a pre-run of 12 min. Calcium concentrations in the samples were determined using the Infinite M200Pro microplate reader (Tecan, Crailsheim, Germany) equipped with the software package I-control V1.11 (Tecan, Grödig, Austria, 2014) at a constant temperature (37 °C). Excitation wavelength was set to 338 nm with a slit width of 9 nm, emission wavelength was set to 400 nm with a slit width of 20 nm. Fluorescence data were recorded in intervals of 12 s. All fluorescence intensity values during the individual measurements were normalized to the average fluorescence intensities during the initial 6 min measuring period and expressed in % of these pre-run intensities.

4.8. Time Lapse Microscopy

Airway epithelial cells (16HBE14o-, S9) were cultured in 35 mm μ-cell culture plates (Ibidi, Planegg, Germany) in medium as described above until they reached confluence. The medium was renewed one day before the plate was transferred to the time lapse-microscope (Biostation II, Nikon Instruments, Düsseldorf, Germany). The microscope chamber was thermostatically controlled at 37 °C and gassed with 5% CO_2 in air during the experiment. Images of cells were taken every 3 min over 24 h and combined to time lapse-movies of 30 s duration.

4.9. Data Presentation and Statistics

Data are presented as means and S.D. of n experiments on different cell/tissue preparations. Significant differences in the series of means were detected by ANOVA. Individual means were tested for significant differences to the appropriate controls using Student's *t*-test (used if variances were equal) or Welch's *t*-test (w, used if variances were not equal). Significant differences of means were assumed at $p < 0.05$.

Supplementary Materials: The following are available online at http://www.mdpi.com/2072-6651/11/2/126/s1, Figure S1: Incubation of cells with sphingomyelinase (rHlb) reduced sphingomyelin abundance in PMs of airway epithelial cells. Phospholipids isolated from 16HBE14o- or S9 cells were separated using high performance thin-layer chromatography; Figure S2: (A) Incubation of 16HBE14o- cells with sphingomyelinase (rHlb) does not readily affect $[Ca^{2+}]_i$. Means $\pm$ S.D. (at each full min) of six (PBS) or four (rHlb) experiments, respectively, are shown. (B) Pre-incubation of 16HBE14o- cells with rHlb does not render the cells unable to respond to calcium influx-inducing agents other than rHla; Figure S3: Pre-incubation of sheep erythrocytes with sphingomyelinase (rHlb) prevented Hla-mediated hemolysis.

Author Contributions: Conceptualization, S.Z. and J.-P.H. Investigation, S.Z., N.M., A.N., and C.M. Resources, A.G.B. and J.-P.H. Writing—Original Draft Preparation, S.Z. and J.-P.H. Writing—Review and Editing, S.Z. and J.-P.H.

Funding: This research received no external funding.

Acknowledgments: The authors thank Karoline Gäbler for conducting the calcium measurements reported in Figure S2A and Elvira Lutjanov and Katrin Harder for excellent technical assistance. We acknowledge support for the article processing charge from the DFG (German Research Foundation, 393148499) and the open access publication fund of the University of Greifswald. We thank three anonymous reviewers for their valuable comments on the earlier version of the manuscript.

Conflicts of Interest: The authors declare no conflict of interest.

References

1. Evans, S.E.; Xu, Y.; Tuvim, M.J.; Dickey, B.F. Inducible innate resistance of lung epithelium to infection. *Annu. Rev. Physiol.* **2010**, *72*, 413–435. [CrossRef] [PubMed]
2. Knowles, M.R.; Boucher, R.C. Mucus clearance as a primary innate defense mechanism for mammalian airways. *J. Clin. Invest.* **2002**, *109*, 571–577. [CrossRef] [PubMed]
3. Kong, C.; Neoh, H.M.; Nathan, S. Targeting *Staphylococcus aureus* toxins: A potential form of anti-virulence therapy. *Toxins* **2016**, *8*, 72. [CrossRef] [PubMed]
4. Hildebrandt, J.-P. Pore-forming virulence factors of *Staphylococcus aureus* destabilize epithelial barriers-effects of alpha-toxin in the early phases of airway infection. *AIMS Microbiol.* **2015**, *1*, 11–36. [CrossRef]
5. Lowy, F.D. *Staphylococcus aureus* infections. *N. Engl. J. Med.* **1998**, *339*, 520–532. [CrossRef] [PubMed]
6. Holtfreter, S.; Nguyen, T.T.; Wertheim, H.; Steil, L.; Kusch, H.; Truong, Q.P.; Engelmann, S.; Hecker, M.; Völker, U.; van Belkum, A.; et al. Human immune proteome in experimental colonization with *Staphylococcus aureus*. *Clin. Vaccine Immunol.* **2009**, *16*, 1607–1614. [CrossRef]
7. Bubeck Wardenburg, J.; Schneewind, O. Vaccine protection against *Staphylococcus aureus* pneumonia. *J. Exp. Med.* **2008**, *205*, 287–294. [CrossRef] [PubMed]
8. Bhakdi, S.; Tranum-Jensen, J. Alpha-toxin of *Staphylococcus aureus*. *Microbiol. Rev.* **1991**, *55*, 733–751.
9. Glenny, A.T.; Stevens, M.F. *Staphylococcus* toxins and antitoxins. *J. Pathol. Bacteriol.* **1935**, *40*, 201–210. [CrossRef]
10. Dinges, M.M.; Orwin, P.M.; Schlievert, P.M. Exotoxins of *Staphylococcus aureus*. *Clin. Microbiol. Rev.* **2000**, *13*, 16–34. [CrossRef]
11. von Hoven, G.; Rivas, A.J.; Neukirch, C.; Klein, S.; Hamm, C.; Qin, Q.; Meyenburg, M.; Fuser, S.; Saftig, P.; Hellmann, N.; et al. Dissecting the role of ADAM10 as a mediator of *Staphylococcus aureus* alpha-toxin action. *Biochem. J.* **2016**, *473*, 1929–1940. [CrossRef] [PubMed]
12. Baaske, R.; Richter, M.; Möller, N.; Ziesemer, S.; Eiffler, I.; Müller, C.; Hildebrandt, J.-P. ATP release from human airway epithelial cells exposed to *Staphylococcus aureus* alpha-toxin. *Toxins* **2016**, *8*, 365. [CrossRef] [PubMed]
13. Suttorp, N.; Seeger, W.; Dewein, E.; Bhakdi, S.; Roka, L. Staphylococcal alpha-toxin-induced PGI2 production in endothelial cells: Role of calcium. *Am. J. Physiol.* **1985**, *248*, C127–C134. [CrossRef] [PubMed]
14. Walev, I.; Martin, E.; Jonas, D.; Mohamadzadeh, M.; Müller-Klieser, W.; Kunz, L.; Bhakdi, S. Staphylococcal alpha-toxin kills human keratinocytes by permeabilizing the plasma membrane for monovalent ions. *Infect. Immun.* **1993**, *61*, 4972–4979. [PubMed]
15. Eichstaedt, S.; Gäbler, K.; Below, S.; Müller, C.; Kohler, C.; Engelmann, S.; Hildebrandt, P.; Völker, U.; Hecker, M.; Hildebrandt, J.-P. Effects of *Staphylococcus aureus*-hemolysin A on calcium signalling in immortalized human airway epithelial cells. *Cell Calcium* **2009**, *45*, 165–176. [CrossRef] [PubMed]

16. Kloft, N.; Busch, T.; Neukirch, C.; Weis, S.; Boukhallouk, F.; Bobkiewicz, W.; Cibis, I.; Bhakdi, S.; Husmann, M. Pore-forming toxins activate MAPK p38 by causing loss of cellular potassium. *Biochem. Biophys. Res. Commun.* **2009**, *385*, 503–506. [CrossRef] [PubMed]

17. Eiffler, I.; Behnke, J.; Ziesemer, S.; Müller, C.; Hildebrandt, J.-P. *Staphylococcus aureus* alpha-toxin-mediated cation entry depolarizes membrane potential and activates p38 MAP kinase in airway epithelial cells. *Am. J. Physiol. Lung Cell. Mol. Physiol.* **2016**, *311*, L676–L685. [CrossRef] [PubMed]

18. Hermann, I.; Räth, S.; Ziesemer, S.; Volksdorf, T.; Dress, R.J.; Gutjahr, M.; Müller, C.; Beule, A.G.; Hildebrandt, J.-P. *Staphylococcus aureus* hemolysin A disrupts cell-matrix adhesions in human airway epithelial cells. *Am. J. Respir. Cell Mol. Biol.* **2015**, *52*, 14–24. [CrossRef] [PubMed]

19. Ziesemer, S.; Eiffler, I.; Schönberg, A.; Müller, C.; Hochgräfe, F.; Beule, A.G.; Hildebrandt, J.-P. *Staphylococcus aureus* alpha-toxin induces actin filament remodeling in human airway epithelial model cells. *Am. J. Respir. Cell Mol. Biol.* **2018**, *58*, 482–491. [CrossRef] [PubMed]

20. Bernstein, B.W.; Bamburg, J.R. ADF/cofilin: A functional node in cell biology. *Trends Cell Biol.* **2010**, *20*, 187–195. [CrossRef]

21. Jayasinghe, L.; Miles, G.; Bayley, H. Role of the amino latch of staphylococcal alpha-hemolysin in pore formation: A co-operative interaction between the N terminus and position 217. *J. Biol. Chem.* **2006**, *281*, 2195–2204. [CrossRef] [PubMed]

22. Song, L.; Hobaugh, M.R.; Shustak, C.; Cheley, S.; Bayley, H.; Gouaux, J.E. Structure of staphylococcal alpha-hemolysin, a heptameric transmembrane pore. *Science* **1996**, *274*, 1859–1866. [CrossRef] [PubMed]

23. Montoya, M.; Gouaux, E. Beta-barrel membrane protein folding and structure viewed through the lens of alpha-hemolysin. *Biochim. Biophys. Acta* **2003**, *1609*, 19–27. [CrossRef]

24. Wilke, G.A.; Bubeck Wardenburg, J. Role of a disintegrin and metalloprotease 10 in *Staphylococcus aureus* alpha-hemolysin-mediated cellular injury. *Proc. Natl. Acad. Sci. USA* **2010**, *107*, 13473–13478. [CrossRef] [PubMed]

25. Virreira Winter, S.; Zychlinsky, A.; Bardoel, B.W. Genome-wide CRISPR screen reveals novel host factors required for *Staphylococcus aureus* alpha-hemolysin-mediated toxicity. *Sci. Rep.* **2016**, *6*, 24242. [CrossRef] [PubMed]

26. Richter, E.; Harms, M.; Ventz, K.; Gierok, P.; Chilukoti, R.K.; Hildebrandt, J.-P.; Mostertz, J.; Hochgräfe, F. A multi-omics approach identifies key hubs associated with cell type-specific responses of airway epithelial cells to staphylococcal alpha-toxin. *PLoS ONE* **2015**, *10*, e0122089. [CrossRef] [PubMed]

27. Liang, X.; Ji, Y. Alpha-toxin interferes with integrin-mediated adhesion and internalization of *Staphylococcus aureus* by epithelial cells. *Cell. Microbiol.* **2006**, *8*, 1656–1668. [CrossRef] [PubMed]

28. Liang, X.; Ji, Y. Involvement of alpha5beta1-integrin and TNF-alpha in *Staphylococcus aureus* alpha-toxin-induced death of epithelial cells. *Cell. Microbiol.* **2007**, *9*, 1809–1821. [CrossRef] [PubMed]

29. Maharaj, I.; Fackrell, H.B. Rabbit erythrocyte band 3: A receptor for staphylococcal alpha toxin. *Can. J. Microbiol.* **1980**, *26*, 524–531. [CrossRef]

30. Pany, S.; Vijayvargia, R.; Krishnasastry, M.V. Caveolin-1 binding motif of alpha-hemolysin: Its role in stability and pore formation. *Biochem. Biophys. Res. Commun.* **2004**, *322*, 29–36. [CrossRef] [PubMed]

31. Hildebrand, A.; Pohl, M.; Bhakdi, S. *Staphylococcus aureus* alpha-toxin. Dual mechanism of binding to target cells. *J. Biol. Chem.* **1991**, *266*, 17195–17200. [PubMed]

32. Watanabe, M.; Tomita, T.; Yasuda, T. Membrane-damaging action of staphylococcal alpha-toxin on phospholipid-cholesterol liposomes. *Biochim. Biophys. Acta* **1987**, *898*, 257–265. [CrossRef]

33. Tomita, T.; Watanabe, M.; Yasuda, T. Influence of membrane fluidity on the assembly of *Staphylococcus aureus* alpha-toxin, a channel-forming protein, in liposome membrane. *J. Biol. Chem.* **1992**, *267*, 13391–13397. [PubMed]

34. Galdiero, S.; Gouaux, E. High resolution crystallographic studies of alpha-hemolysin-phospholipid complexes define heptamer-lipid head group interactions: Implication for understanding protein-lipid interactions. *Protein Sci.* **2004**, *13*, 1503–1511. [CrossRef] [PubMed]

35. Valeva, A.; Hellmann, N.; Walev, I.; Strand, D.; Plate, M.; Boukhallouk, F.; Brack, A.; Hanada, K.; Decker, H.; Bhakdi, S. Evidence that clustered phosphocholine head groups serve as sites for binding and assembly of an oligomeric protein pore. *J. Biol. Chem.* **2006**, *281*, 26014–26021. [CrossRef] [PubMed]

36. Simons, K.; Ehehalt, R. Cholesterol, lipid rafts, and disease. *J. Clin. Invest.* **2002**, *110*, 597–603. [CrossRef] [PubMed]

37. Abrami, L.; van der Goot, F.G. Plasma membrane microdomains act as concentration platforms to facilitate intoxication by aerolysin. *J. Cell Biol.* **1999**, *147*, 175–184. [CrossRef] [PubMed]

38. Kulma, M.; Herec, M.; Grudzinski, W.; Anderluh, G.; Gruszecki, W.I.; Kwiatkowska, K.; Sobota, A. Sphingomyelin-rich domains are sites of lysenin oligomerization: Implications for raft studies. *Biochim. Biophys. Acta* **2010**, *1798*, 471–481. [CrossRef]

39. Schwiering, M.; Brack, A.; Stork, R.; Hellmann, N. Lipid and phase specificity of alpha-toxin from *S. aureus*. *Biochim. Biophys. Acta* **2013**, *1828*, 1962–1972. [CrossRef]

40. Bernheimer, A.W.; Avigad, L.S.; Kim, K.S. Staphylococcal sphingomyelinase (beta-hemolysin). *Ann. N. Y. Acad. Sci.* **1974**, *236*, 292–306. [CrossRef]

41. Huseby, M.; Shi, K.; Brown, C.K.; Digre, J.; Mengistu, F.; Seo, K.S.; Bohach, G.A.; Schlievert, P.M.; Ohlendorf, D.H.; Earhart, C.A. Structure and biological activities of beta toxin from *Staphylococcus aureus*. *J. Bacteriol.* **2007**, *189*, 8719–8726. [CrossRef] [PubMed]

42. Walev, I.; Weller, U.; Strauch, S.; Foster, T.; Bhakdi, S. Selective killing of human monocytes and cytokine release provoked by sphingomyelinase (beta-toxin) of *Staphylococcus aureus*. *Infect. Immun.* **1996**, *64*, 2974–2979. [PubMed]

43. Wiseman, G.M.; Caird, J.D. The nature of staphylococcal beta hemolysin. I. Mode of action. *Can. J. Microbiol.* **1967**, *13*, 369–376. [CrossRef] [PubMed]

44. Skowron, M.; Zakrzewski, R.; Ciesielski, W. Application of thin-layer chromatography image analysis technique in quantitative determination of sphingomyelin. *J. Anal. Chem.* **2016**, *71*, 808–813. [CrossRef]

45. Huitema, K.; van den Dikkenberg, J.; Brouwers, J.F.; Holthuis, J.C. Identification of a family of animal sphingomyelin synthases. *EMBO J.* **2004**, *23*, 33–44. [CrossRef] [PubMed]

46. Slotte, J.P. Molecular properties of various structurally defined sphingomyelins - correlation of structure with function. *Prog. Lipid Res.* **2013**, *52*, 206–219. [CrossRef] [PubMed]

47. Gouaux, J.E.; Braha, O.; Hobaugh, M.R.; Song, L.; Cheley, S.; Shustak, C.; Bayley, H. Subunit stoichiometry of staphylococcal alpha-hemolysin in crystals and on membranes: A heptameric transmembrane pore. *Proc. Natl. Acad. Sci. USA* **1994**, *91*, 12828–12831. [CrossRef] [PubMed]

48. Thompson, J.R.; Cronin, B.; Bayley, H.; Wallace, M.I. Rapid assembly of a multimeric membrane protein pore. *Biophys. J.* **2011**, *101*, 2679–2683. [CrossRef] [PubMed]

49. Below, S.; Konkel, A.; Zeeck, C.; Müller, C.; Kohler, C.; Engelmann, S.; Hildebrandt, J.-P. Virulence factors of *Staphylococcus aureus* induce Erk-MAP kinase activation and c-Fos expression in S9 and 16HBE14o- human airway epithelial cells. *Am. J. Physiol. Lung Cell Mol. Physiol.* **2009**, *296*, L470–L479. [CrossRef] [PubMed]

Communication

Bioinformatics and Functional Assessment of Toxin-Antitoxin Systems in *Staphylococcus aureus*

Gul Habib, Qing Zhu and Baolin Sun *

Division of Molecular Medicine, Hefei National Laboratory for Physical Sciences at Microscale, the CAS Key Laboratory of Innate Immunity and Chronic Disease, School of Life Sciences, University of Science and Technology of China, Hefei 230027, China; gulhabib@mail.ustc.edu.cn (G.H.); zhuqq@mail.ustc.edu.cn (Q.Z.)
* Correspondence: sunb@ustc.edu.cn; Tel.: +86-551-6360-6748; Fax: +86-551-6360-7438

Received: 14 October 2018; Accepted: 11 November 2018; Published: 14 November 2018

Abstract: *Staphylococcus aureus* is a nosocomial pathogen that can cause chronic to persistent infections. Among different mediators of pathogenesis, toxin-antitoxin (TA) systems are emerging as the most prominent. These systems are frequently studied in *Escherichia coli* and *Mycobacterial* species but rarely explored in *S. aureus*. In the present study, we thoroughly analyzed the *S. aureus* genome and screened all possible TA systems using the Rasta bacteria and toxin-antitoxin database. We further searched *E. coli* and *Mycobacterial* TA homologs and selected 67 TA loci as putative TA systems in *S. aureus*. The host inhibition of growth (HigBA) TA family was predominantly detected in *S. aureus*. In addition, we detected seven pathogenicity islands in the *S. aureus* genome that are enriched with virulence genes and contain 26 out of 67 TA systems. We ectopically expressed multiple TA genes in *E. coli* and *S. aureus* that exhibited bacteriostatic and bactericidal effects on cell growth. The type I Fst toxin created holes in the cell wall while the TxpA toxin reduced cell size and induced cell wall septation. Besides, we identified a new TA system whose antitoxin functions as a transcriptional autoregulator while the toxin functions as an inhibitor of autoregulation. Altogether, this study provides a plethora of new as well as previously known TA systems that will revitalize the research on *S. aureus* TA systems.

Keywords: *Staphylococcus aureus*; toxin-antitoxin systems; HigBA; pathogenicity islands

Key Contribution: This study reveals 67 toxin-antitoxin systems, 13 TA families classified in five TA types, and seven pathogenicity islands in *Staphylococcus aureus*.

---

1. Introduction

Staphylococcus aureus is a commensal pathogen that can cause a diverse array of infections and syndromes such as toxic shock syndrome, bacteremia, endocarditis, osteomyelitis, pneumonia, soft-tissue infection and many more [1]. *S. aureus* can detect and respond to diverse environmental stimuli such as nutrient starvation and stress to increase its fitness by altering the expression of numerous genes such as the toxin-antitoxin (TA) system and two-component system [2,3]. A typical TA system consists of two components: a toxin that can disrupt a cellular process and an antitoxin that functions as an antidote for the toxin. TA systems have been involved in virulence and persistent infections in *Escherichia coli* [4], *Mycobacterium* [5], and *Salmonella* [6]. They are classified into six types based on the mechanism of interaction between toxin and antitoxin [7]. The type I TA system has been characterized by an RNA-RNA interaction of the antitoxin antisense RNA with the toxin mRNA, while the type II TA system exhibits protein-protein interaction between toxin and antitoxin. In the type III system, the activity of the toxic protein is inhibited by the interaction with the antitoxin RNA, while toxin and antitoxin proteins in the type IV TA system compete for the same target [4,7,8]. The antitoxin protein in the type V TA system cleaves the toxin mRNA [9]. The newly identified type VI TA system differs from others in a way that the antitoxin functions as an adaptor to facilitate toxin proteolysis [10,11]. The abundance of TA loci has been considered to have evolved through horizontal gene transfer, by which TA systems successfully integrate into the bacterial genome and are involved in genetic stability as well as pathogenic potency [12]. Under unfavorable conditions such as starvation, bacteriophage attack, heat shock, and antibiotic stresses, the unstable antitoxin is degraded, and as a result, the toxin is freed to target vital cellular processes such as translation, transcription, and replication [7,10]. TA systems have been extensively studied in *E. coli* [4] and *Mycobacterium* [5], while the study on TA systems in *S. aureus* is less advanced. *S. aureus* possesses a number of TA genes in its genome, however just three type II and two type I TA systems have been studied [3]. The rest of TA systems still await discovery and analysis. Here, we have performed an extensive screening and assessment of TA loci in the genome of methicillin-resistant *S. aureus* MW2 utilizing available software and databases. We screened TA systems with Rasta bacteria [13], the toxin-antitoxin database (TADB) [14], and position-specific iterative basic local alignment search tool (PSI-BLAST) searches supported by the "guilt by association" principle and classified them into different families and types. HigBA was the most abundant TA family with 15 systems, followed by six GCN5-related acetyltransferases (GNATs) and five RelBE TA systems. Other TA families include MazEF, VapBC, HicBA, CcdBA, ParDE, HipBA, transporter TA systems, abortive infection (Abi) TA systems, and some with a domain of unknown function. Further, we ectopically expressed some toxin and antitoxin genes to assess their predicted functions. Most of the type II toxins did inhibit cell growth while their antitoxins did not show any inhibitory effects. We overexpressed RelE, MazF, and HigB toxins that exhibited growth inhibition. We detected seven type I TA systems including previously characterized type I toxins Fst and TxpA [15]. Transmission electron microscopy (TEM) analysis of the bacterial cell wall revealed that the Fst toxin broke the cell wall while TxpA and holin toxins increased cell wall septation. We also detected a new TA system and found that its toxin and antitoxin did not show any growth inhibition upon overexpression. In this new TA system, toxin and antitoxin can be co-transcribed, and the antitoxin can bind to the promoter and autoregulate its operon. The toxin protein interfered with the antitoxin function and inhibited binding of the antitoxin to the promoter. In summary, our present study has greatly expanded the plethora of TA systems in *S. aureus* and revealed a variety of genes involved in bacterial cellular processes.

2. Results and Discussion

2.1. Comprehensive Screening of Toxin-Antitoxin Systems in S. aureus

The typical toxin-antitoxin (TA) system consists of a stable toxin and an unstable antitoxin and is widely distributed in bacterial genomes. The toxin and antitoxin form a conjugative pair of genes that are co-transcribed, and are involved in various cellular processes, and importantly counterbalance each other. The fraction of the TA system encoded by a bacterial genome varies from one pathogen to another. *Mycobacterium* contains 88 [5], *E. coli* contains 40 [4], *Salmonella typhimurium* carries 27 [6,16], and *Pseudomonas aeruginosa* has 26 TA systems [17]. To date, three type II TA systems have been well characterized in *S. aureus*, namely Axe1-Txe1, Axe2-Txe2, and MazEF. These type II TA systems show similarity to *E. coli* Phd-Doc, YefM-YoeB, and MazEF systems, respectively [3].

In the present work, we combined multiple approaches and explored the *S. aureus* genome for putative TA loci. We screened for TA genes and predicted a total of 67 putative TA systems with Rasta bacteria, TADB, and PSI-BLAST. To make the classification more accurate, we used the TA systems of *E. coli* [4], *Klebsiella pneumoniae* [18], and *Mycobacterium* [5,19] as modal and manually curated each TA system and classified them into different families. We found type I, II, III, IV, and type V TA systems in *S. aureus* on the basis of software prediction, conserved domain analysis, and orthologues searches from Kyoto Encyclopedia of Genes and Genomes (KEGG). Details of the screening method and TA systems is given (Figure 1A,B), and the selected TA families are provided in Table 1.

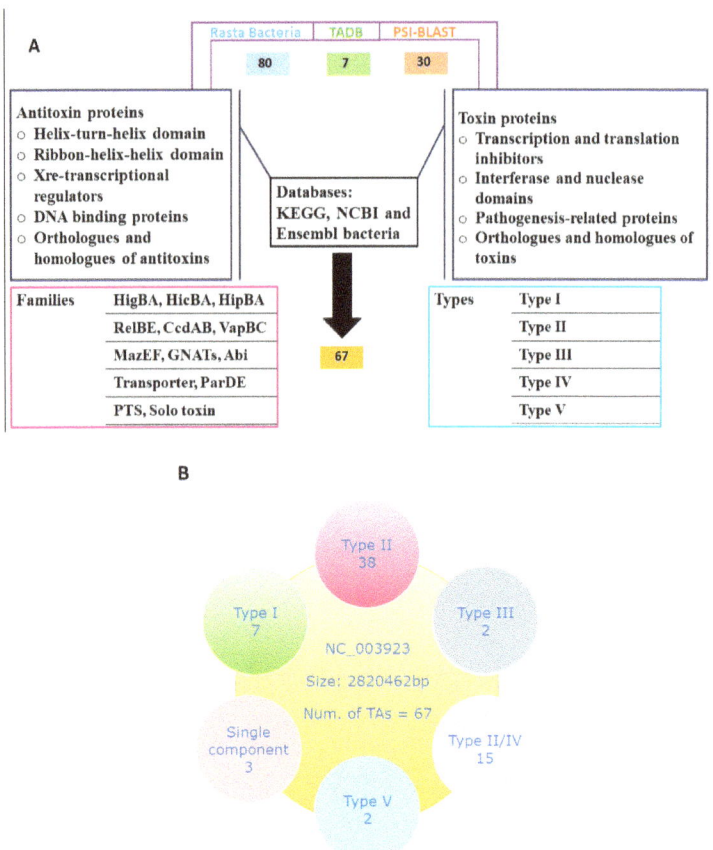

Figure 1. Screening of TA systems. (**A**) Preliminary screening of TA loci in *S. aureus* genome was performed with Rasta bacteria and toxin-antitoxin database (TADB) and followed by position-specific iterative basic local alignment search tool (PSI-BLAST) searches. In total, 80 putative TA systems are predicted by Rasta bacteria, seven by TADB, and 30 by PSI-BLAST. TA systems were manually curated by conserved domains such as helix-turn-helix (HTH), ribbon-helix-helix (RHH) and transcriptional regulators for antitoxin, and conserved domains such as interferases, nucleases, transcription and translation inhibitors and pathogenesis-related proteins for toxins. All selected TA system orthologues and homologs were analyzed, and finally, 67 putative TA systems were selected. (**B**) The selected 67 TA systems were classified into different families and types. Type II TA systems were abundantly detected followed by putative type II/IV TA systems.

Table 1. List of all selected TA systems and their predicted families. The TA systems were arranged in order of their proximity to the TA family. The length of toxin and antitoxin proteins is given in amino acids (aa).

Family	Toxin Genes	Length (aa)
TxpA/Fst/Holin (type I toxins)	MW1888-TxpA	44
	MW2354-Fst	84
	MW0911-TxpA	35
	MW1440-TxpA	34
	MW0405	64
	MW1434	81
	MW1381-Holin	100

Table 1. *Cont.*

Family	Toxin/Antitoxin Genes	Length (aa) Toxin/Anti-toxin
HigBA (type II)	MW1419/MW1418	84/57
	MW1725/MW1724	59/117
	MW2494/MW2493	76/106
	MW1228/MW1227	79/77
	MW1413/MW1412	62/66
	MW0749/MW0748	74/48
	MW1338/MW1337	59/116
	MW1928/MW1927	106/53
	MW1397/MW1396	92/110
	MW1822/MW1821	91/154
	MW0789/MW0788	98/128
	MW1919/MW1918	72/134
	MW2464/MW2463	72/275
	MW0907/MW0906	72/189
	MW1056/MW1055	44/62
RelBE (type II)	MW2329/MW2330	88/83
	MW2380/MW2381	88/85
	MW1230/MW1231	80/155
	MW0324/MW0325	129/67
	MW1394/MW1395	131/133
GNATs (type II)	MW2441/MW2440	94/144
	MW1089/MW1088	70/203
	MW0641/MW0640	180/99
	MW0924/MW0923	160/71
	MW2449/MW2450	163/56
	MW2497/MW2498	99/185
Nucleases (type II) (CcdAB/ParDE)	MW1311/MW1312	73/166
	MW0461/MW0462	87/130
	MW1403/MW1402	104/101
	MW1733/MW1732	85/159
VapBC (type II)	MW1070/MW1071	263/224
	MW0302/MW0303	266/315
	MW0572/MW0573	211/295
MazEF (type II)	MW1992/MW1993	120/56
	MW1492/MW1493	114/144
HicBA (type II)	MW0075/MW0076	109/83
	MW0771/MW0772	72/94
Transporters (type II/IV)	MW1042/MW1043	61/77
	MW1045/MW1046	48/65
	MW2414/MW2413	77/102
	MW1722/MW1723	100/147
	MW2025/MW2026	77/134
	MW1779/MW1780	121/68
	MW0350/MW0349	69/83
Transporters (type V/II)	MW2507/MW2506	91/122
	MW0268/MW0269	617/164
Abi (type III/IV)	MW2234/MW2233	245/104
	MW0420/MW0421	89/131
HipBA (type II)	MW0980/MW0981	91/179
PTS (type II/IV)	MW1437/MW1436	133/107
	MW1004/MW1005	84/129
	MW0777/MW0776	78/67
	MW1150/MW1149	94/391
	MW1196/MW1195	83/34
	MW1292/MW1291	89/102
	MW1432/MW1431	71/87
	MW1164/MW1165	275/130
	MW0101	50
	MW0036	103
	MW1827	128

2.2. Distribution of TA Genes in Different Families

The toxin gene codes for a poison protein and antitoxin is the cognate antidote. The toxin and antitoxin sequence domains are the determinants for families and TA types. The antitoxin mostly contains a DNA-binding domain such as helix-turn-helix (HTH) or ribbon-helix-helix (RHH) of the transcriptional regulator. For example, the RelBE family in *E. coli* is the most diverse family and most of the RelB antitoxins bind to the promoter region and autoregulate the TA operon [20]. In *Mycobacterium*, VapBC is the most abundant TA family, and the VapBC protein complex can autoregulate the TA operon [5]. When we screened *S. aureus* genome, we observed the enriched profusion of the host inhibition of growth (HigBA) TA family, which is a sub-family of RelBE super-family [14]. The HigBA TA system has the unique genetic annotation in type II TA systems, i.e., the toxin gene is upstream of the antitoxin, which makes it different from the rest of type II TA families [21,22]. HigB is a ribonuclease in function followed by HigA antitoxin, and they can form a heterotetrameric complex. The antitoxin contains an HTH-Xre domain and can bind to DNA, while the toxin targets 50S ribosomal subunit [22]. To date, no HigBA TA system has been reported in *S. aureus*. In this study, we identified 15 HigBA TA systems in the *S. aureus* genome and classified them into three groups by pairwise alignment to its close members (Figure 2A–C).

In all three groups, the HigB toxin is followed by the HigA antitoxin, which satisfies the first requirement in predicating the HigBA TA system (Table 1). Such enriched abundance of HigBA TA systems has not been reported in other bacteria, but possible homologs can be detected. For example, homologues of HigB toxins (MW1228, MW1928, and MW1413) are prevalent in *Lactobacillus ceti*, *Enterococcus faecium*, *Streptococcus pneumoniae*, *E. coli*, *P. aeruginosa*, *Bacillus cereus*, and *Staphylococcus* species (Figure 3A–C). We aligned HigB toxin MW1228 with homologues from *L. ceti*, *E. faecium*, *S. pneumoniae*, *E. coli*, *P. aeruginosa*, and *B. cereus*, and detected a highly conserved domain of 35 residues (13–48 aa) at the N-terminus (Figure 3A). We did not detect a similar domain in other HigB toxins such as MW1928 and MW1413, but they also shared a unique domain with their homologues. When we aligned the homologues of MW1928 and MW1413, we detected a highly conserved domain of 17 residues (10–27 aa) in MW1928 and 11 residues (20–31 aa) in MW1413 homologues (Figure 3B,C). Although the homologues of the HigBA TA family are widely detected in many bacteria, mostly remain uncharacterized except in *Proteus vulgaris*, *Mycobacterium* [19], *Vibrio cholera* [23], and *P. aeruginosa*. In *P. aeruginosa*, the HigB toxin functions as RNase and reduces biofilm formation, pyocyanin and pyochelin production, and swarming motility [22]. The endonuclease toxin HigB cleaved AAA rich sequences as well as single monomer (A) in the coding region of *P. vulgaris* [24]. Interestingly, the HigB target coincided with amino acid lysine (AAA), one of the most frequent amino acid in the GC-less and AT-rich *Staphylococcus* species. Therefore, the enriched profusion of the HigBA system highlights the target specificity and functional sites in *S. aureus*.

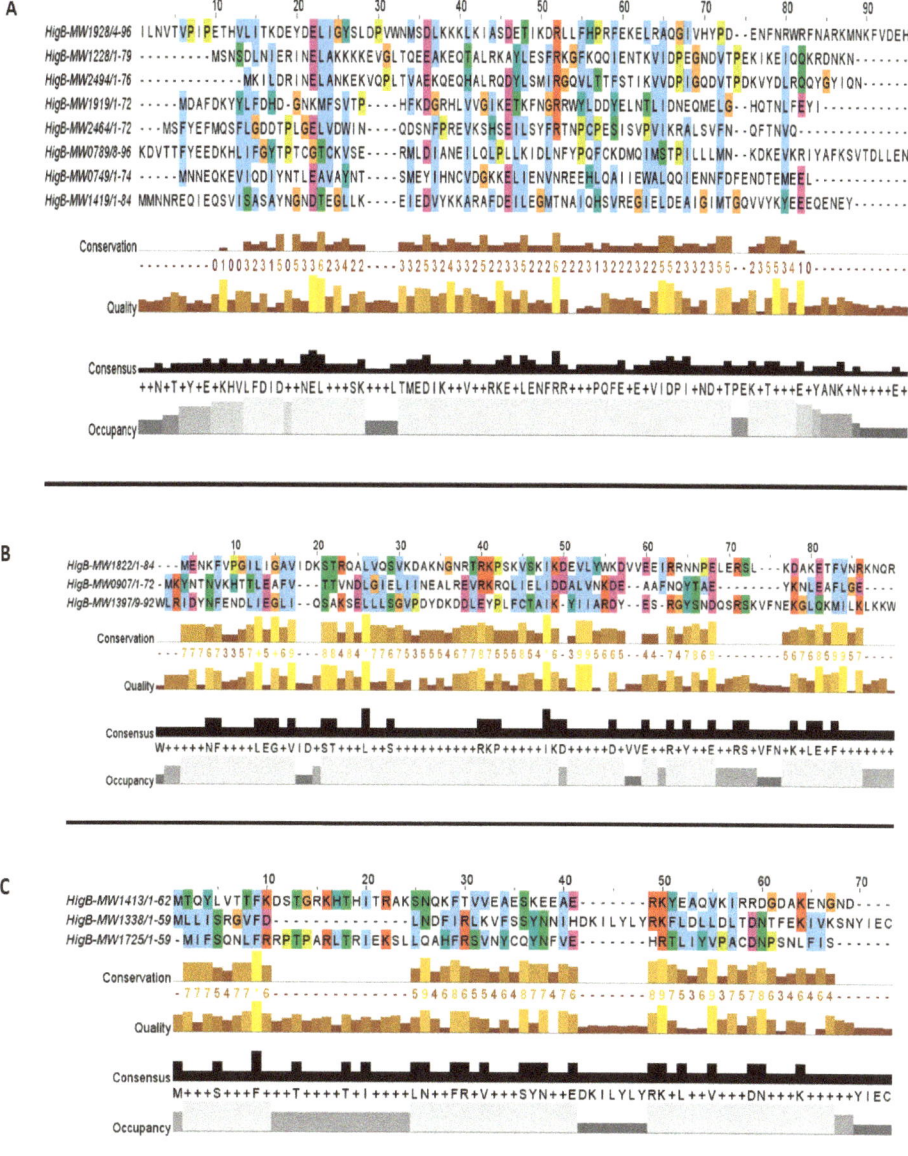

Figure 2. Multiple sequence alignment of HigB toxins. (**A–C**) The HigB sequences were aligned by MUSCLE software across the *S. aureus* genome. HigB toxins that were short in length and showed divergence in their family were aligned together in a sub-family as shown in (**B,C**).

The RelBE super-family has sub-families, namely YefM-YoeB, MqsAR, PrlF-YhaV, and RelBE in *E. coli* [14]. The RelBE sub-family antitoxin RelB is degraded upon a stringent response that leads to the activation of the RelE toxin and inhibition of growth via targeting translation. Upon stringent response, RelB is degraded and subsequently leads to the activation of RelE, which targets translation and results in growth inhibition. The YefM-YoeB sub-family has been well characterized in *S. aureus* [3],

S. quorum [25], *S. suis* [26], *Streptomyces* [27], *E. coli,* and *S. pneumoniae* [28]. The YoeB toxin can form a complex with the YefM antitoxin and target translation. In this study, we detected five RelBE TA systems (Figure 4A) that can be further classified into sub-families depending on their functions. For example, we characterized a new TA system of the RelBE super-family that showed resemblance with YefM-YoeB sub-family, and was involved in virulence control in *S. aureus* [20].

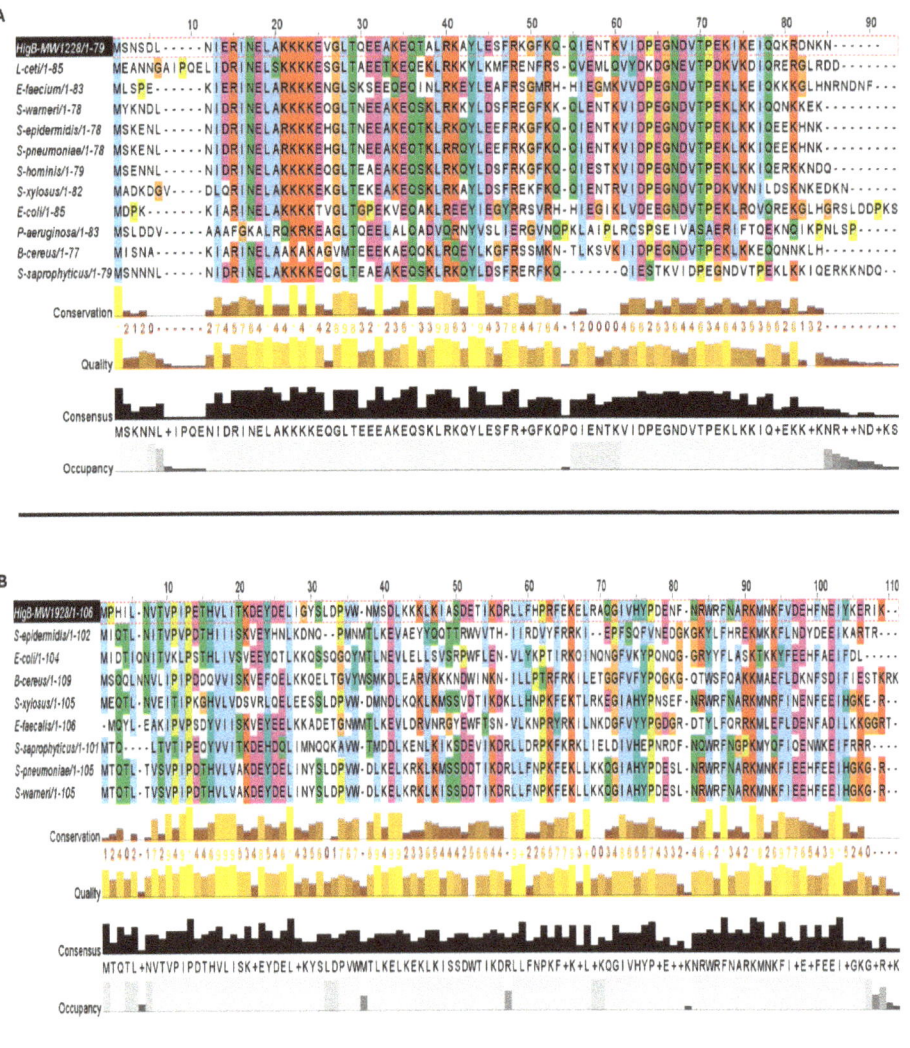

Figure 3. *Cont.*

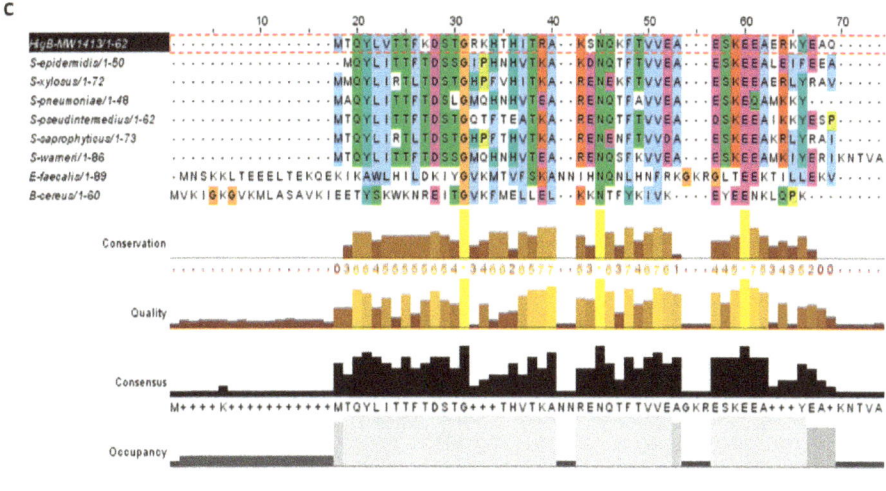

Figure 3. Homologs of the HigB toxins. (**A**) The HigB toxin MW1228 showed high conservation in *L. ceti*, *E. faecium*, *S. pneumoniae*, *E. coli*, *P. aeruginosa*, and *B. cereus*. The N-terminal 35 residues (13–48 aa) was the highly conserved region in all of the selected homologues. (**B**) The toxin MW1928 was the largest protein (106 aa) in the HigBA family and displayed a highly conserved region at N-terminal (10–27 aa) and a less conserved region at C-terminal (95–109 aa). (**C**) The toxin MW1413 is a small protein (62 aa) in the HigBA family and highly conserved in *Staphylococcus* species such as *S. epidermidius*, *S. xylosus*, *S. pseudointermedius*, *S. saprophyticus*, *S. warneri*, and *S. pneumoniae*. MW1413 was less conserved in *E. faecalis* and *B. cereus*.

The GNATs-like TA system is the new addition of the TA family that can prevent peptide bond formation by acetylating tRNA and inhibiting translation in *Salmonella* and *Klebsilla* [18,29]. The TacT toxin inhibits translation via acetylating the initiator tRNA, and induces persister cell formation [29]. Six loci were detected in *S. aureus* that showed resemblance with GNATs TA systems (Figure 4B), and their toxins showed acetyltransferase-like domains.

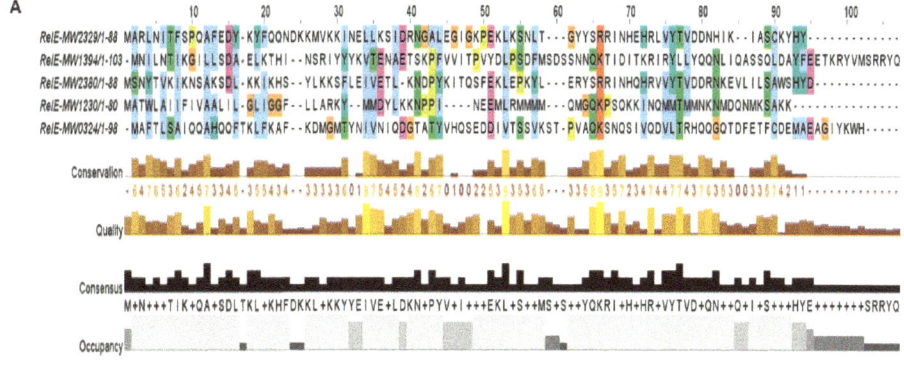

Figure 4. *Cont.*

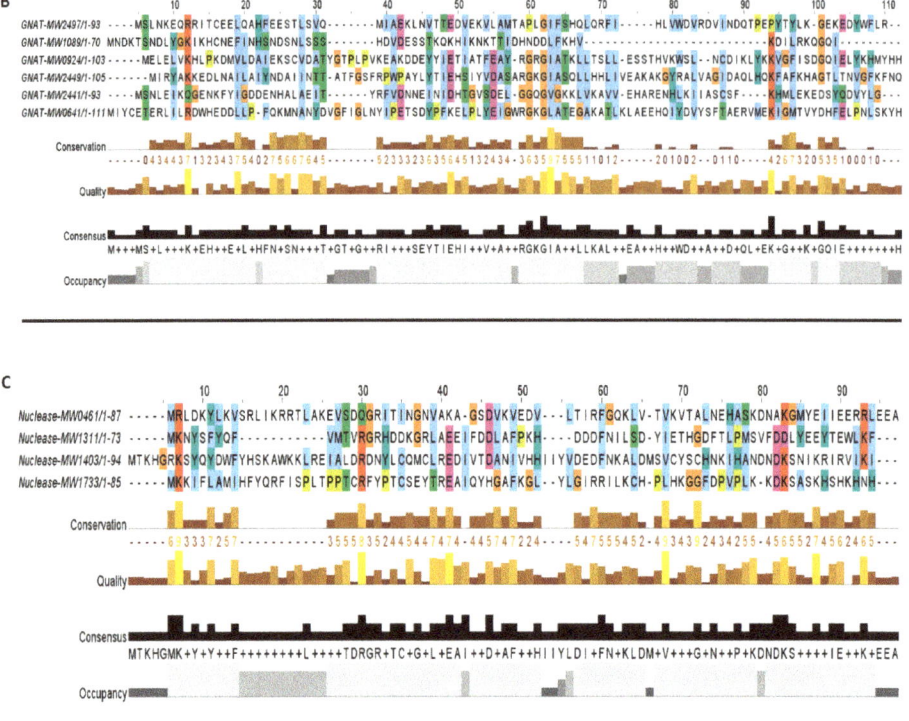

Figure 4. The predicted toxins were aligned by MUSCLE and sorted by pairwise alignment to close matching protein. (**A**) The RelE toxins MW2329, MW2380, and MW1230 were shorter in length and showed homology towards YefM-YoeB sub-family, while the MW0324 and MW1394 were longer in length and displayed less conservation, possibly making a new sub-family. (**B**) The GNATs toxins showed a functional site of 28 residues (41–68 aa) that contained the most conserved region in this family. (**C**) A moderate consensus site of 20 residues (30–49 aa) was detected in the nuclease family. The nuclease toxins MW1311, MW1403, and MW1733 were closely related in sequence alignment, while MW0461 was less conserved.

Four TA loci showed resemblance with DNA gyrase inhibitors (nucleases) and were predicted to be either the ParDE or the CcdBA TA system (Figure 4C). The par toxin targets DNA gyrase and inhibits DNA replication in *E. coli* and *M. tuberculosis* [4,5]. The ParE-ParD components have been reported to form a heterotetrameric complex via the ParD N-terminal RHH domain that can increase the dimerization of ParD, while the C-terminal domain has a specific ParE toxin-binding site [30]. The CcdB toxin can bind to DNA gyrase, form a gyrase-DNA complex, and create a double-strand break into DNA. The CcdA antitoxin can resolve the complex by binding to the toxin protein and dislodge CcdB from gyrase [31]. So far, these loci have not been studied in detail in *S. aureus* and characterization of these systems will reveal the mechanism of toxin action.

The first persistence-associated gene *hipA* was first detected in 1983 [32] and later studied in *E. coli* as the HipBA TA system [33]. We found one HipBA-like system in *S. aureus* that is different in sequence and length, but its orthologue search revealed its relatedness with the HipBA family. Here we named it as the first HipBA TA system in *S. aureus*. In the *hip* operon, HipA is a toxin that phosphorylates glutamyl-tRNA synthetase, while HipB is the antitoxin that counteracts the HipA toxicity [33].

The VapBC TA system is the most abundant TA system in prokaryotes and is found copiously in *E. coli*, *Mycobacterium*, *Shigella*, *Sulfolobus*, and *Leptospira* [2,4,5,34], while there is no report of *vapBC* loci in *S. aureus*. The toxin VapC targets translation by cleaving mRNA or tRNA. The VapB antitoxin is usually a transcriptional regulator that has a DNA binding domain and forms a complex with VapC. VapBC can respond to different stress conditions such as hypoxia, antibiotics stress, and latent infections [34,35]. Collectively, we detected 3 VapBC TA systems in *S. aureus* (Table 1), which have not been characterized so far.

The second most abundant family that consists of nine TA systems belongs to the transporter TA family in *S. aureus*. These TA systems are either located in secretion systems or contain conserved domains associated with transport pathways. For example, toxin MW0268 and antitoxin MW0269 are located in the type VII secretion system while toxin MW2507 is an effector protein that showed proximity towards the type VII secretion system secretory proteins EsxA and EsxB. Similarly, toxins MW1042 and MW1722 are associated with ATP-binding cassette transporter permeases, while toxin MW2414 has orthologue homology with TerC protein that has been implicated in the efflux of tellurium ions in *E. coli* [36]. The toxins MW1779 and MW0350 are related to sodium and hydrogen antiporter systems. These putative TA systems were grouped into the transporter TA family, and their toxin proteins were aligned together that were less conserved and more divergent (Figure 5A). We believed that these systems might be involved in inhibition and regulation of secretion systems and transport pathways.

The MazEF family has been characterized by the toxin MazF that acts as an endoribonuclease targeting mRNA, 16S and 23S ribosomal RNAs in *Mycobacterium* and *E. coli* [5,10]. Components of the MazEF TA system are commonly called global translation inhibitors [37], and are activated upon nutrient or oxidative stress, DNA damaging response, and heat shock [5]. The MazEF TA systems have been studied in *E. coli*, *M. tuberculosis*, *Clostridium difficile*, and *S. aureus* [2–5]. In the present study, we detected one new and one previously studied MazEF systems. The new MazF toxin showed a high degree of conservation in other bacteria including *E. faecium*, *M. tuberculosis*, and *lactobacillus buchneri* (Figure 5B).

When we mined the *S. aureus* genome, we discovered two HicBA TA systems (Figure 5C,D). In the HicBA TA system, *hicA* codes for the toxin that cleaves mRNA and reduces the rate of translation, while *hicB* rescues the bacteriostatic effects induced by *hicA* [38]. The transcription of *hicBA* is induced upon carbon and amino acid starvation and depends on lon protease in *E. coli* [39].

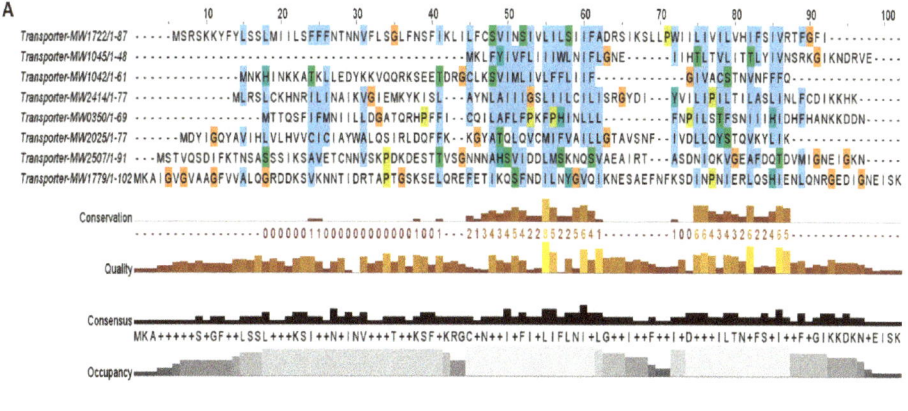

Figure 5. *Cont.*

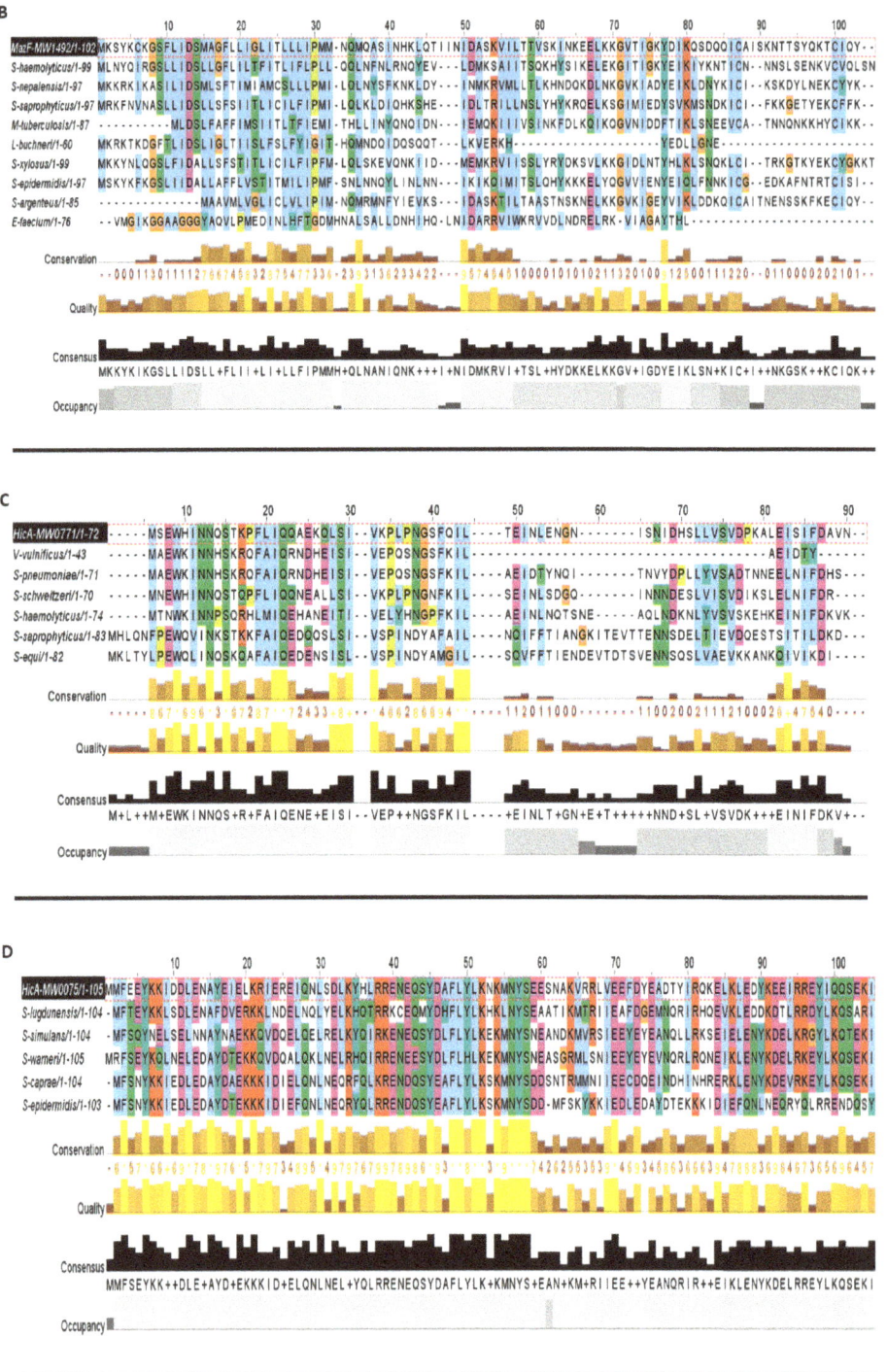

Figure 5. *Cont.*

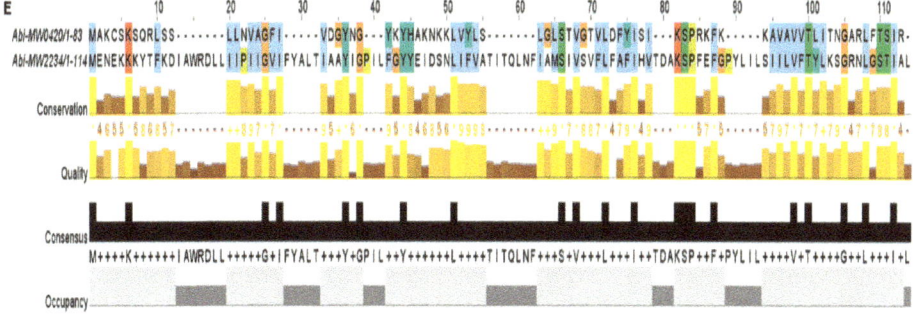

Figure 5. The toxins were aligned by MUSCLE and sorted by pairwise alignment to close matching protein. (**A**) The transporter family toxins displayed less conservation among all TA families because each of the protein is associated with different secretion systems and transport pathways. The 15 residues (48–62 aa) are the highly consensus sites in this family. (**B**) Homologs of MazF toxin MW1492 were aligned that showed a high degree of conservation in *M. tuberculosis*, *L. buchneri*, *E. faecium*, and *Staphylococcus* species. The highly conserved region contains 25 residues (8–32 aa) at N-terminus. (**C**) The HicA toxin MW0771 homologs were detected in *Vibrio vulnificus*, *S. pneumoniae*, and *Staphylococcus* species. The 20 residues (11–30 aa) are the highly consensus sites in the homologs of MW00771 protein. (**D**) The homologs of MW0075 could not be detected in *M. tuberculosis*, *L. buchneri*, *E. faecium*, *V. vulnificus*, *S. pneumoniae* and *B. cereus*. The MW0075 toxin is a highly conserved protein in *Staphylococcus* species. (**E**) The two Abi toxins were aligned by pairwise alignment that showed 16% sequence similarity. Both Abi toxin homologs could be detected in different bacterial species.

Apart from the type II TA system, two loci were predicted as the abortive infection system (Abi) TA system that can probably constitute the type III or IV TA system (Figure 5E). The TA system ToxIN in *Pectobacterium atrosepticum* was the first TA system described that also acted as the abortive infection system. This system can lead to altruistic cell death and viral replication inhibition [40,41]. The Abi TA system has not been discovered in *S. aureus* so far and awaits further investigation

The *S. aureus* genome contains 11 TA systems that neither show any similarity with the reported TA families nor contain conserved domain. Therefore, we kept them as putative TA systems (PTS). Among these PTS, 3 TA systems can be three components (MW1430-1431-1432, MW1149-1150-1151, and MW0776-0777-0778) or two components including toxin and antitoxin as shown in Table 1. The third component can be either a second antitoxin or transcriptional regulator that probably alters the expression of TA genes. Three of the PTS were classified as a solo toxin (MW0101, MW0036, and MW1827) that can probably constitute different types of TA systems or might function differently (Table 1). In general, three-component TA systems and some solo toxins have been studied in detail in cyanobacteria [42]. Further research on these PTS systems may reveal their classification and function in *S. aureus*. In conclusion, we compared *S. aureus* TA families with the reported TA families of *E. coli*, *Mycobacterium*, *P. aeruginosa*, and *S. typhimurium*, and our results could advance our understanding of the distribution of TA families in bacteria (Figure 6).

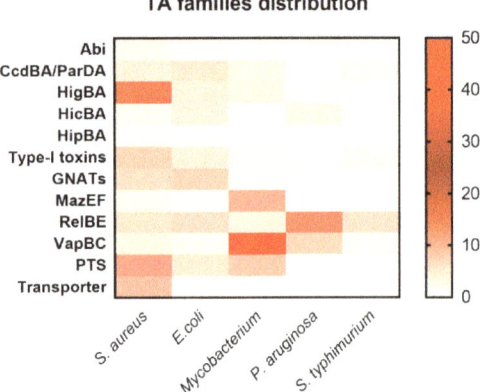

Figure 6. Distribution of TA families in *S. aureus, E. coli, Mycobacterium, P. aruginosa,* and *S. typhimurium.* The selected TA families were compared with the reported TA families of other bacteria. The highest number of TA systems in *Mycobacterium* is 50 VapBC, 10 MazEF, 8 PTS, and 3 HigBA TA systems. *P. aruginosa* has 13 RelBE, 7 VapBC, and 3 HicBA systems. *E. coli* has 7 GNATs, 6 RelBE, 5 nuclease toxins, and 4 each HigBA and HicBA systems. *S. typhimurium* has 6 RelBE, 2 VapBC, and 2 nuclease toxins. Seven type I toxins were found in *S. aureus,* 4 in *E. coli,* 2 in *S. typhimurium,* and 1 in *Mycobacterium.* The HipBA TA system was the rarest family that has not been discovered in any bacteria except *E. coli, P. aeruginosa,* and *S. aureus.*

2.3. Pathogenicity Islands TA Systems

Pathogenicity islands (PIs) are unique regions of DNA and consist of virulence genes and mobile genetic elements. We detected 7 PIs in *S. aureus* genomes by using PIs database [12]. These islands are enriched with TA-like genes. The roles of TA systems in PIs include but are not limited to successful integration, maintenance and stability of islands, and defense against phages [12,43]. The PI phi-Sa-2 and VSa-gamma have 11 and 4 TA systems (Table 2). We predicted that PIs-TA systems are the most functional TA systems in *S. aureus.* This prediction is supported by previous reports that mechanistically related TA systems can display different functions depending on their localization in regions such as PIs [43]. In total, we detected 26 TA systems out of 67 that are located within pathogenicity island (PI) regions with an abundance of virulence genes in the neighborhood.

Table 2. The predicted seven PIs in the genome of *S. aureus.* Each PI has different number of TA systems and unique virulence genes.

PIs	Size (bp)	Number of TA Systems	Representative Toxins from TA Systems	Virulence Genes in PIs
Phi-Sa-2	45,550	11	MW1381, MW1394, MW1397, MW1403, MW1413, MW1419, MW1432, MW1434, MW1437, MW1440, MW1492	Panton–Valentine leukocidin
VSa-gamma	20,881	4	MW1042, MW1045, MW1056, MW1070	Alpha hemolysin, phenol soluble modulin
Phi-Sa-3	43,598	3	MW1888, MW1919, MW1928	Staphylokinase
VSa-beta	31,977	3	MW1725, MW1733, MW1779	Proteases
VSa-alpha	34,481	3	MW0302, MW0324, MW0405	–
VSa-3	14,380	2	MW0749, MW0771	Enterotoxins, exotoxin
VSa-4	2710	0		–

2.4. Functional Assessment of the Type I TA System

The type I TA system toxins are called bacterial time bombs because these toxins target the bacterial cell wall and induce cell death. Importantly, type I toxins are considered to be the most critical peptides to eradicate *S. aureus* cells [3,44]. Here we have detected seven type I TA systems that have toxin domains such as TxpA, Fst, and holin (Table 1). We have also revealed two type I TA systems Fst and TxpA that were previously reported [3,15]. To assess the activity of three type I toxins that target cell wall, we constructed different recombinant vectors as shown in Materials and Methods. The toxin was ectopically overexpressed in *S. aureus* and the OD_{600} value was determined. The expression of toxins inhibited cell growth, confirming the deadly nature of type I toxins (Figure 7A). Further, the cellular toxicity assay was performed on anhydrotetracycline (ATC) inducible plates. The results indicated that cell growth was inhibited and cannot be restored (Figure 7B), demonstrating that type I toxins are bactericidal and can induce cell death in *S. aureus*.

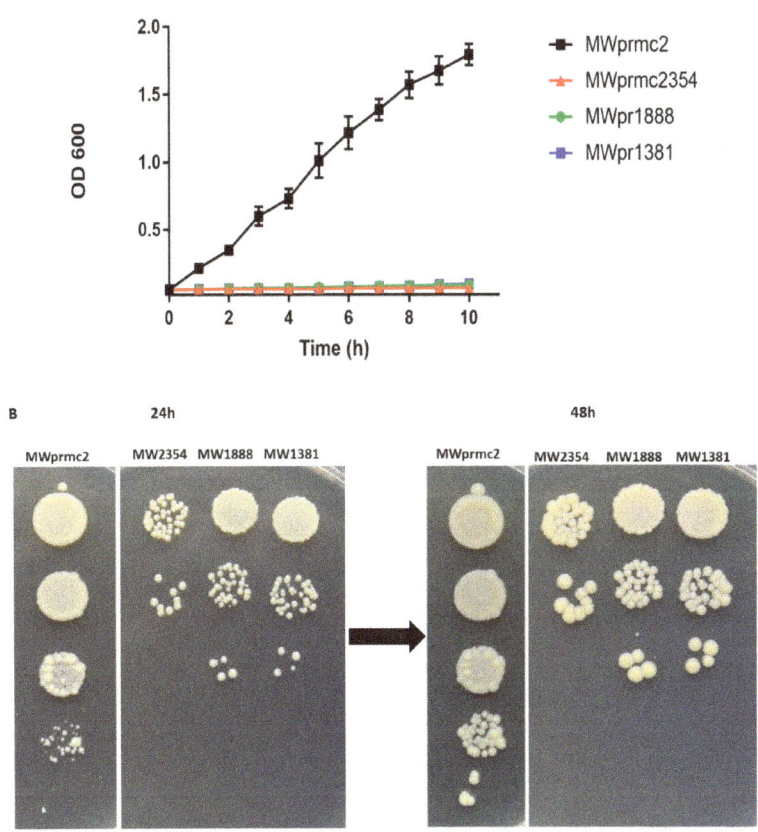

Figure 7. Overexpression of type I toxins. (**A**) Normal growth was observed in MWprmc2 that contains an empty pRMC2 vector while complete growth inhibition was observed in MWprmc2354, MWprmc1888 and MWprmc1381 that contain Fst, TxpA, and holin toxins, respectively. (**B**) The bactericidal nature of three type I toxins was determined by ATC inducible plates, and the results showed that cells expressing Fst, TxpA, and holin toxins were unable to restore growth as compared to control MWprmc2.

2.5. Transmission Eelectron Microscopy Analysis of the Bacterial Cell Wall

Type I toxins are short peptides and apoptotic in nature and target cell wall [45]. We ectopically expressed three type I toxins for 1 h, harvested the cells, and analyzed the cell wall by transmission electron microscope (TEM). As a control, the *S. aureus* with empty pRMC2 vector was used (Figure 8A). The type I toxins targeted the cell wall differently. The MWprmc2354 (Fst toxin) created holes and broke the cell wall, and its lysis activity was intense. Apart from this, empty cells with no cellular entities were observed, and cells unable to withstand the toxin effects and ultimately cell wall remaining remnants were detected with ectopic expression of Fst (Figure 8B). The ectopic expression of MWprmc1888 (TxpA toxin) and MWprmc1381 (holin toxin) led to neither holes nor breakage, and instead increased wall septation. Two cells intersecting their cell walls and septum formation were ostensible with the TxpA (Figure 8C) and holin (Figure 8D) toxins as shown by TEM micrographs (Figure 8A–D).

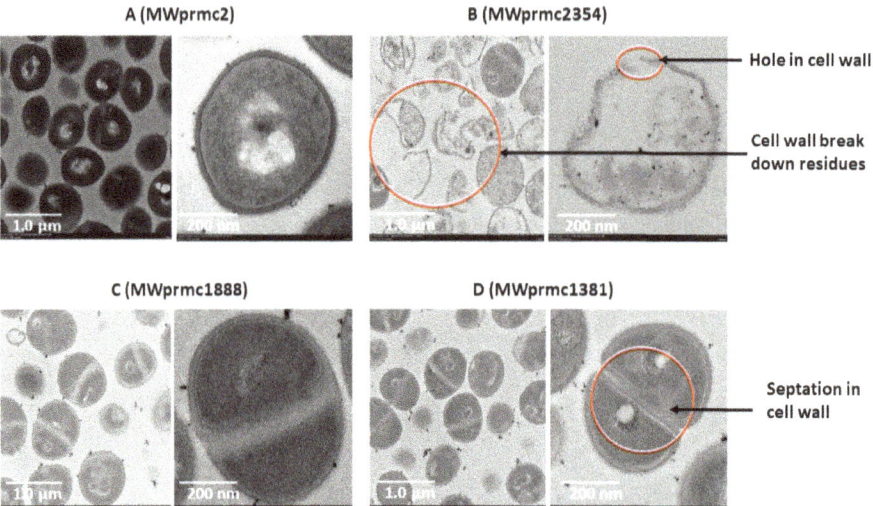

Figure 8. The bacterial cell wall was analyzed by TEM. Single cell micrographs were captured at magnification ×25.0 K and scale bar 200 nm while the group of cells micrographs were captured at magnification ×8.0 K and scale bar 1.0 µm. (**A**) Mwprmc2 was an empty vector and used as control. (**B**) The Fst toxin (MWprmc2354) broke down the cell wall as well as created holes. (**C,D**) The TxpA (MWprmc1888) and holin (MWprmc1381) toxins increased cell wall septation and reduced cell size compared to the control MWprmc2 with the empty vector.

2.6. Response and Function of Cellular Proteases in S. aureus

Most of TA systems contain a liable antitoxin that is either degraded by cellular proteases or suppressed to enhance toxin expression under stress conditions, suggesting that the antitoxin can be a positive or a negative regulator of the toxin. We detected six different types of Clp proteases; ClpA, ClpB, ClpC, ClpP, ClpQ, and ClpX in *S. aureus*. Among them, ClpC and ClpX form complexes with ClpP and are involved in the protein homeostasis by either conditional degradation or disposing of the unwanted proteins [46]. In the present study, we observed the upregulation of *clpP* towards type I toxins, while *clpX* remained steady (Figure 9A,B). Our data indicated that *clpP* was activated when the toxin was ectopically expressed. This result suggests that the proteases play an essential role in keeping the toxin level in the balanced state in order not to severely damage the bacterial cells. The response and function of proteases towards TA systems are an exciting topic and need further study.

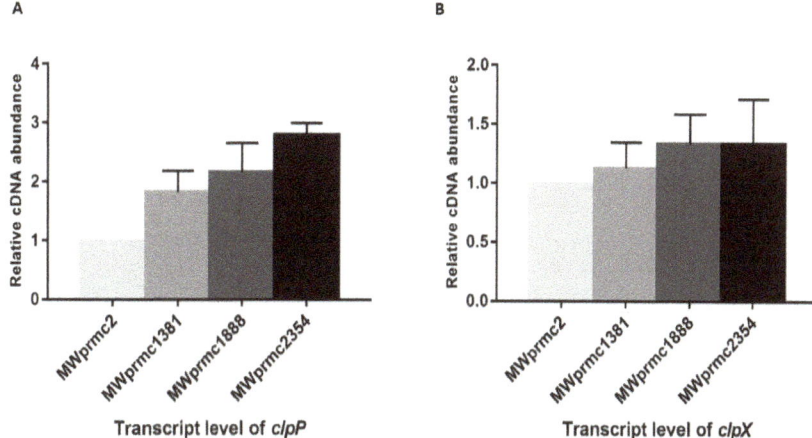

Figure 9. The transcript levels of *clpP* and *clpX* were determined by quantitative real-time PCR (qRT-PCR). The transcript levels were compared with MWprmc2 containing empty pRMC2 vector. (**A**) The ectopic expression of MWprmc2354 toxin activated *clpP* ($p \leq 0.06$) when compared to other type I toxins MWprmc1888 and MWprmc1381 ($p \leq 0.08$). (**B**) The ectopic expression of type I toxins did not alter the transcript level of *clpX*.

2.7. Functional Assessment of Type II TA Systems

The type II TA system is well studied in *E. coli*, *Mycobacterium*, *Pseudomonas*, and *Staphylococcus*. After screening the TA loci in *S. aureus*, we randomly selected five type II TA systems from different families, i.e., MW2329-MW2330 and MW2380-MW2381 TA systems from the RelBE family, MW1992-MW1993 TA system from the MazEF family, MW1419-MW1418 TA system from the HigBA family, and MW1436-MW1437 TA system from the PTS family. We aimed to confirm whether the predicted toxin or antitoxin can inhibit cell growth or not. We ectopically expressed toxins and antitoxins in *E. coli* BL21 and monitored the bacterial growth. All the toxins inhibited cell growth except MW1437, while all the antitoxins were similar to control BLpet with empty pET28a plasmid (Figure 10A,B). We conclude that most of the type II TA system toxins target metabolic pathways and induce bacteriostatic conditions, contrasting to type I toxins with bactericidal activities.

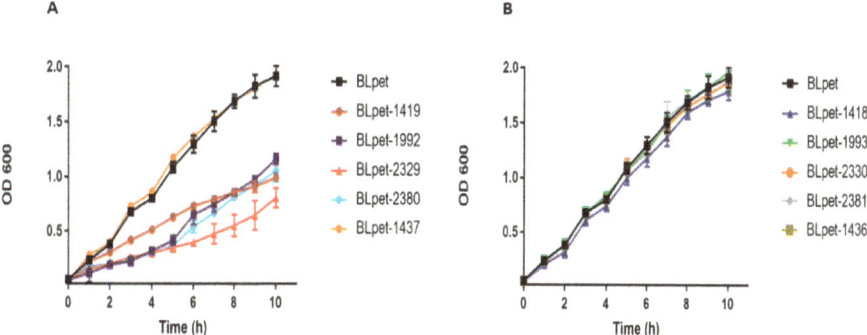

Figure 10. Expression of type II toxins and antitoxins in *E. coli* BL21. (**A**) The ectopic expression of RelE toxins (BLpet-2329 and BLpet-2380), HigB toxin (BLpet-1419), and MazF toxin (BLpet-1992) inhibited cell growth while BLpet-1437 has no inhibitory effects on cell growth. (**B**) RelB antitoxins (BLpet-2330 and BLpet-2381), HigA antitoxin (BLpet-1418), MazE antitoxin (BLpet-1993), and BLpet-1436 did not inhibit cell growth.

The MW1436-MW1437 TA locus was predicted to be either a type II or a type IV TA system and classified as PTS (Table 1). Locus MW1437 codes for a toxin while MW1436 codes for an antitoxin. The two genes are separated by a 12 bp intergenic region and are located in the phi-Sa-2 PI region within the neighborhood of one of the type I TA system (MW1434) (Figure 11). The toxin protein has a Zn-dependent protease-like domain, while the antitoxin is an HTH domain-containing transcriptional regulator. The ectopic expression of both toxin BLpet-1437 and antitoxin BLpet-1436 were growth-supportive, and no cellular toxicity was observed (Figure 10A,B).

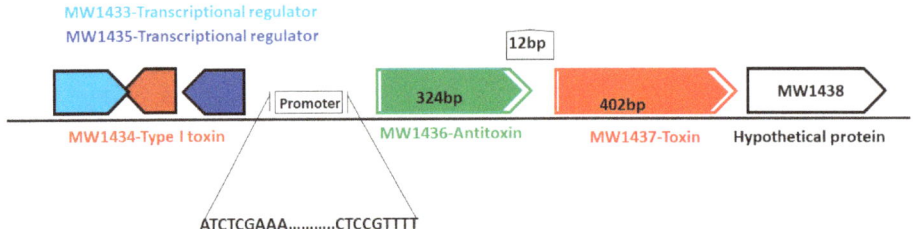

Figure 11. Schematic representation of a new TA system from phi-Sa-2 pathogenicity island. The toxin-coding gene is 402 bp, the antitoxin-coding gene is 324 bp in length, and they are located close to the type I TA system and transcriptional regulator.

We did not detect the in vivo protein–protein interaction between toxin (BLpet-1437) and antitoxin (BLpet-1436) proteins (data not shown). Further, we evaluated the DNA-binding ability of antitoxin and toxin proteins. Firstly, we determined the 98 bp promoter region of this operon using SoftBerry software. We incubated the toxin and antitoxin separately as well as together with the 98 bp biotin-labeled promoter probe. Our gel-shift assay results indicated that the antitoxin can bind to the probe and probably autoregulate this operon (Figure 12A, lanes 5–7), while the toxin could not bind to the probe (Figure 12A, lanes 2–4). Interestingly, the addition of the toxin protein to antitoxin and the probe mixture inhibited antitoxin–probe complex formation, suggesting that the toxin protein can inhibit the binding of antitoxin to the promoter region and prevent autoregulation (Figure 12A, lanes 8 and 9). To further determine whether the two genes can be co-transcribed, we performed RT-PCR using RNA, cDNA, and DNA as templates and detected the 700 bp co-transcript of MW1436-MW1437 (Figure 12B, lanes 3 and 4). Our data also showed that *relBE* genes MW2329-MW2330 were co-transcribed (Figure 12B, lanes 6 and 7). In summary, although the toxin of this TA system did not inhibit cell growth, it may be involved in some other functions such as autoregulation of the neighbor genes or stability of the phi-Sa-2 island.

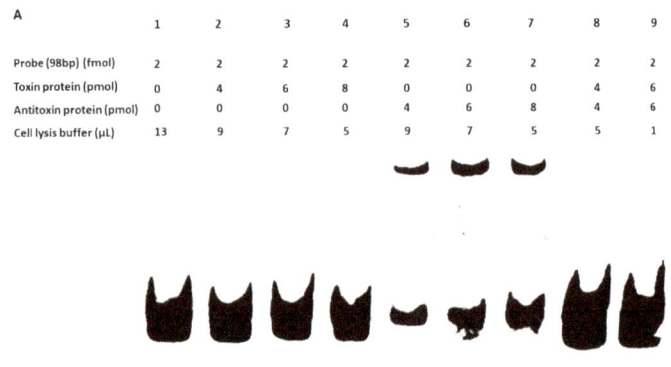

Figure 12. *Cont.*

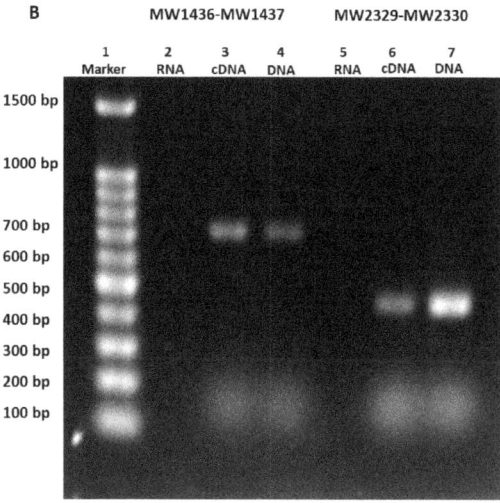

Figure 12. (**A**) Gel-shift assay was performed to detect the binding of toxin and antitoxin to the promoter. The up-shift represents the binding of the antitoxin protein to the 98 bp probe (lanes 5–7), while the toxin protein was unable to bind (lanes 2–4). No shift was observed when antitoxin and toxin proteins were mixed together with the 98 bp probe (lanes 8 and 9). (**B**) The TA genes MW1436-MW1437, and MW2329-M2330 were co-transcribed. 1 = marker, 2, 5 = RNA, 3, 6 = cDNA, and 4, 7 = DNA. The co-transcript product (lanes 3 and 4) represents the 700 bp size of MW1436-MW1437, and lanes 6 and 7 represent the 400 bp size of MW2329-MW2330.

3. Conclusions

We comprehensively screened all the possible TA loci in *S. aureus* genome and predicted previously unidentified 67 TA systems, and classified them into different families and types. Our data substantially expanded the collection of TA systems in *S. aureus* and revealed the nature of type I and type II toxins. We detected a new TA system whose toxin did not inhibit cell growth but prevented cognate antitoxin binding to the promoter. The presence of these chromosomally encoded TA systems suggests that these loci may be involved in *S. aureus* physiology and genome stability. More importantly, our prediction may open a new window for further study of these uncharacterized TA systems, especially, their association with *S. aureus* pathogenicity and virulence.

4. Materials and Methods

4.1. Bacterial Strains and Growth Conditions

The bacterial strains used in this study are listed in Table S1 of the Supplementary Material. S. aureus was grown in tryptic soy broth (TSB) medium and E. coli was grown in lysogeny broth (LB). pRMC-2, an ATC inducible shuttle vector, and pET28a, an isopropyl β-D-1-thiogalactopyranoside (IPTG) inducible plasmid, were used in S. aureus and E. coli BL21, respectively. For plasmid maintenance, each medium was supplemented with appropriate antibiotic (ampicillin 150 μg/mL, chloramphenicol 15 μg/mL, and kanamycin 50 μg/mL). Cells were washed with normal saline. Normal growth conditions for S. aureus and E. coli were 37 °C and 220 rpm. Growth curves were determined by measuring optical density at 600 nm using a BioTek microplate reader.

4.2. Screening of TA Systems in S. aureus

Initially, we identified all putative TA genes in *S. aureus* by Rasta bacteria and TADB. Rasta bacteria cutoff score >53 was set and 80 TA systems were selected, and seven TA systems were predicted by TADB. Further, we performed PSI-BLAST searches using *Mycobacterium* and *E. coli* known TA systems as modal and selected 30 more TA systems in *S. aureus*. To improve the screening strategy, we analyze the predicted TA domains in databases such as KEGG, NCBI, and Ensembl bacteria, and analyzed the orthologues and homologs. Then the selected TA systems were classified into different families and types. The approach is summarized in Figure 1A.

4.3. Construction of Vectors

TA genes were amplified by PCR and inserted into pRMC2 vectors via KpnI-SacI-EcoRI and pET28a via NdeI-BamHI-EcoRI-XhoI restriction enzymes, resulting in the recombinant vectors MWprmc2354, MWprmc1888, MWprmc1381, BLpet2329, BLpet2330, BLpet2380, BLpet2381, BLpet1992, BLpet1993, BLpet1418, and BLpet1419. MWprmc2 and BLpet were empty vectors and used as controls in *S. aureus* and *E. coli*, respectively. All the constructed vectors were confirmed by DNA sequencing. The primers used for the construction of vectors are provided in Table S1 of the Supplementary Material.

4.4. Cellular Toxicity Assay

The constructed vectors were transformed into *E. coli* Trans T1 and BL21 by heat shock at 42 °C for 1 min. The pRMC2-vectors were initially transformed into *E. coli* Trans T1 by heat shock and then to *S. aureus* RN4220 and *S. aureus* MW2 by electroporation. Further, the 24 h culture was diluted 200 times and grown at 37 °C for 2 h. When the culture reached OD_{600} = 0.45–0.48, ATC (200 ng/mL) or IPTG (0.1 mM) was added, and the growth pattern was monitored for 10 h at 1 h intervals. The bactericidal nature of type I toxins was assessed on ATC plates. *S. aureus* was grown as described above to OD_{600} = 0.85–0.90. Cells (normally 5×10^8 CFU/mL) were serially diluted in water, and 10 μL from each dilution was spotted on ATC inducible plates. The plates were incubated at 37 °C and observed for the restoration of growth after 24 h and 48 h intervals.

4.5. RNA Extraction and qRT-PCR

S. aureus was grown in TSB medium under specified conditions such as ATC induction for 1 h. Then, 1 mL of culture was centrifuged at 12,000 rpm for 2 min and the cells were washed and suspended in RNAiso Plus. Cells were treated with 0.1 mm silica beads in RNA isolation tube by the FastPrep-24 Automated system for 40 s twice. Trichloromethane was added in the RNA isolation tube and kept at 4 °C for 5 min, and then centrifuged at 12,000 rpm at 4 °C for 20 min. The up-layer was taken and mixed with an equal volume of isopropanol and kept at −20 °C for 2 h. Then the tubes were centrifuged at 12,000 rpm for 30 min, washed with 75% ethanol, air dried and dissolved in water (50 μL). RNase-free DNase I was used for residual DNA removal, and then RNAiso Plus and trichloromethane were added and centrifuged at 12,000 rpm at 4 °C for 20 min. The up-layer supernatant was transferred to a new tube and mixed with an equal volume of isopropanol and kept at −20 °C for 2 h. The tubes were centrifuged at 12,000 rpm at 4 °C for 30 min, and RNA was collected and washed with 75% ethanol and air dried. RNA concentration and quality were determined by Beckman Coulter and NanoDrop. The PrimeScript First Strand cDNA synthesis kit and SYBR Premix Ex Taq kit (TaKaRa) were used for RT and real-time PCR, respectively. The gene signal was normalized with *gyrA* or *hu* cDNA abundance [47]. All the qRT-PCR primers were designed by Beacon designer 7 and synthesized by General Biosystems China and listed in Table S1 of the Supplementary Material.

4.6. Preparation of Samples for Transmission Electron Microscopy

The type I toxins were ectopically expressed for 1 h in *S. aureus*, and cells were harvested and washed with phosphate buffered saline (PBS). The samples were then fixed with 2% paraformaldehyde and 2.5% glutaraldehyde in PBS buffer at 4 °C overnight, and washed three times in PBS buffer, then fixed with 2% Osmium tetroxide in PBS buffer at 4 °C overnight, and washed three times in PBS buffer. After fixation, the samples were dehydrated by a graded series of ethanol (30%, 50%, 70%, 80%, 90%, 95%, and 100%) for 20 min at each step, and transferred to absolute acetone for 20 min. The samples were placed in 1:1 mixture of Spurr resin and absolute acetone at room temperature for 1 h, and then transferred to 3:1 mixture of Spurr resin and absolute acetone at room temperature for 3 h, and to Spurr resin mixture overnight. Then the samples were placed in capsules contained embedding medium and heated at 70 °C overnight. The specimen sections were stained by uranyl acetate and alkaline lead citrate for 5–10 min, respectively, and images were captured with a Hitachi HT-7700 transmission electron microscope.

4.7. Gel-Shift Assay

Toxin and antitoxin expression vectors BLpet1437 and BLpet1436 were transferred into *E. coli* BL21 cells, and the transformants were grown in LB medium at 37 °C to OD_{600} = 0.45–0.48, and induced with IPTG (0.1 mM) at 30 °C for 5 h. Kanamycin 50 μg/mL was added into LB medium to maintain the plasmid. Cells were harvested, washed with normal saline, and lysed by sonication for 15 min in lysis buffer (50 mM Tris-HCl, pH 8.0, 300 mM NaCl). The proteins were purified by Ni^{2+}-NTA histidine affinity column and concentrated by Millipore protein concentration column. Then, the biotin label primer was synthesized and the 98 bp promoter region was amplified by PCR and incubated with toxin and antitoxin proteins at 37 °C for 1 h. The 15μL reaction mixture contains probe (2 fmol), each toxin and antitoxin protein (4–8 pmol), and lysis buffer. The reaction was stopped by adding gel-shift loading dye and heated at 95 °C for 10 min, and then loaded on 4% native polyacrylamide gel and run in 0.5× Tris-borate-EDTA buffer at 400 mA for 30 min as previously described [20]. DNA bands were transferred onto nylon membrane and detection was performed with chemiluminescent detection kit. Image Quant LAS 4000 mini imaging system was used to detect the membrane.

4.8. Statistical Analysis

The nucleotides and protein sequences were obtained from the NCBI database and analyzed by Vector NTI software. For protein sequence alignment, MUSCLE [48] software was used and visualized by Jalview. Graph pad prism and Origin pro 9 was used for data analysis. Primers were designed by Beacon designer 7. Welch *t*-test was used for statistical significance.

Supplementary Materials: The following is available online at http://www.mdpi.com/2072-6651/10/11/473/s1, Table S1: List of all bacterial strains, vectors, and primers used in this study. The lower case letters represent restriction enzyme sites.

Author Contributions: B.S. and G.H. designed the project; G.H. and Q.Z. performed the experiments; G.H. analyzed the data and wrote the manuscript; All authors proofread the manuscript.

Funding: Gul Habib was supported by the World Academy of Sciences Ph.D. Presidential Fellowship Program. This study was funded by the Strategic Priority Research Program of the Chinese Academy of Sciences (XDB29020000).

Acknowledgments: We thank Zeyu Jin and Jiade Zhu for technical assistance. We are grateful to the Network on Antimicrobial Resistance in *Staphylococcus aureus* (NARSA) for providing the staphylococcal strains.

Conflicts of Interest: The authors declare no conflict of interest.

References

1. Baba, T.; Takeuchi, F.; Kuroda, M.; Yuzawa, H.; Aoki, K.; Oguchi, A.; Nagai, Y.; Iwama, N.; Asano, K.; Naimi, T.; et al. Genome and virulence determinants of high virulence community-acquired mrsa. *Lancet* **2002**, *359*, 1819–1827. [CrossRef]

2. Fernandez-Garcia, L.; Blasco, L.; Lopez, M.; Bou, G.; Garcia-Contreras, R.; Wood, T.; Tomas, M. Toxin-antitoxin systems in clinical pathogens. *Toxins* **2016**, *8*, 227. [CrossRef] [PubMed]

3. Schuster, C.F.; Bertram, R. Toxin-antitoxin systems of staphylococcus aureus. *Toxins* **2016**, *8*, 140. [CrossRef] [PubMed]

4. Yamaguchi, Y.; Park, J.H.; Inouye, M. Toxin-antitoxin systems in bacteria and archaea. *Annu. Rev. Genet.* **2011**, *45*, 61–79. [CrossRef] [PubMed]

5. Sala, A.; Bordes, P.; Genevaux, P. Multiple toxin-antitoxin systems in mycobacterium tuberculosis. *Toxins* **2014**, *6*, 1002–1020. [CrossRef] [PubMed]

6. De la Cruz, M.A.; Zhao, W.; Farenc, C.; Gimenez, G.; Raoult, D.; Cambillau, C.; Gorvel, J.P.; Meresse, S. A toxin-antitoxin module of salmonella promotes virulence in mice. *PLoS Pathog.* **2013**, *9*, e1003827. [CrossRef] [PubMed]

7. Page, R.; Peti, W. Toxin-antitoxin systems in bacterial growth arrest and persistence. *Nat. Chem. Biol.* **2016**, *12*, 208–214. [CrossRef] [PubMed]

8. Wen, Z.; Wang, P.; Sun, C.; Guo, Y.; Wang, X. Interaction of type iv toxin/antitoxin systems in cryptic prophages of escherichia coli k-12. *Toxins* **2017**, *9*, 77. [CrossRef] [PubMed]

9. Wang, X.; Lord, D.M.; Cheng, H.Y.; Osbourne, D.O.; Hong, S.H.; Sanchez-Torres, V.; Quiroga, C.; Zheng, K.; Herrmann, T.; Peti, W.; et al. A new type v toxin-antitoxin system where mrna for toxin ghot is cleaved by antitoxin ghos. *Nat. Chem. Biol.* **2012**, *8*, 855–861. [CrossRef] [PubMed]

10. Kedzierska, B.; Hayes, F. Emerging roles of toxin-antitoxin modules in bacterial pathogenesis. *Molecules* **2016**, *21*, 790. [CrossRef] [PubMed]

11. Aakre, C.D.; Phung, T.N.; Huang, D.; Laub, M.T. A bacterial toxin inhibits DNA replication elongation through a direct interaction with the beta sliding clamp. *Mol. Cell* **2013**, *52*, 617–628. [CrossRef] [PubMed]

12. Yoon, S.H.; Park, Y.K.; Kim, J.F. Paidb v2.0: Exploration and analysis of pathogenicity and resistance islands. *Nucleic Acids Res.* **2015**, *43*, D624–D630. [CrossRef] [PubMed]

13. Sevin, E.W.; Barloy-Hubler, F. Rasta-bacteria: A web-based tool for identifying toxin-antitoxin loci in prokaryotes. *Genome Biol.* **2007**, *8*, R155. [CrossRef] [PubMed]

14. Xie, Y.; Wei, Y.; Shen, Y.; Li, X.; Zhou, H.; Tai, C.; Deng, Z.; Ou, H.Y. Tadb 2.0: An updated database of bacterial type ii toxin-antitoxin loci. *Nucleic Acids Res.* **2018**, *46*, D749–D753. [CrossRef] [PubMed]

15. Fozo, E.M.; Makarova, K.S.; Shabalina, S.A.; Yutin, N.; Koonin, E.V.; Storz, G. Abundance of type i toxin-antitoxin systems in bacteria: Searches for new candidates and discovery of novel families. *Nucleic Acids Res.* **2010**, *38*, 3743–3759. [CrossRef] [PubMed]

16. Lobato-Marquez, D.; Moreno-Cordoba, I.; Figueroa, V.; Diaz-Orejas, R.; Garcia-del Portillo, F. Distinct type i and type ii toxin-antitoxin modules control salmonella lifestyle inside eukaryotic cells. *Sci. Rep.* **2015**, *5*, 9374. [CrossRef] [PubMed]

17. Andersen, S.B.; Ghoul, M.; Griffin, A.S.; Petersen, B.; Johansen, H.K.; Molin, S. Diversity, prevalence, and longitudinal occurrence of type ii toxin-antitoxin systems of pseudomonas aeruginosa infecting cystic fibrosis lungs. *Front. Microbiol.* **2017**, *8*, 1180. [CrossRef] [PubMed]

18. Yeo, C.C. Gnat toxins of bacterial toxin-antitoxin systems: Acetylation of charged trnas to inhibit translation. *Mol. Microbiol.* **2018**, *108*, 331–335. [CrossRef] [PubMed]

19. Schuessler, D.L.; Cortes, T.; Fivian-Hughes, A.S.; Lougheed, K.E.; Harvey, E.; Buxton, R.S.; Davis, E.O.; Young, D.B. Induced ectopic expression of higb toxin in mycobacterium tuberculosis results in growth inhibition, reduced abundance of a subset of mrnas and cleavage of tmrna. *Mol. Microbiol.* **2013**, *90*, 195–207. [PubMed]

20. Wen, W.; Liu, B.; Xue, L.; Zhu, Z.; Niu, L.; Sun, B. Autoregulation and virulence control by the toxin-antitoxin system savrs in staphylococcus aureus. *Infect. Immun.* **2018**, *86*, IAI-00032. [CrossRef] [PubMed]

21. Armalyte, J.; Jurenas, D.; Krasauskas, R.; Cepauskas, A.; Suziedeliene, E. The higba toxin-antitoxin module from the opportunistic pathogen acinetobacter baumannii—Regulation, activity, and evolution. *Front. Microbiol.* **2018**, *9*, 732. [CrossRef] [PubMed]

22. Wood, T.L.; Wood, T.K. The higb/higa toxin/antitoxin system of pseudomonas aeruginosa influences the virulence factors pyochelin, pyocyanin, and biofilm formation. *Microbiologyopen* **2016**, *5*, 499–511. [CrossRef] [PubMed]

23. Budde, P.P.; Davis, B.M.; Yuan, J.; Waldor, M.K. Characterization of a higba toxin-antitoxin locus in vibrio cholerae. *J. Bacteriol.* **2007**, *189*, 491–500. [CrossRef] [PubMed]

24. Hurley, J.M.; Woychik, N.A. Bacterial toxin higb associates with ribosomes and mediates translation-dependent mrna cleavage at a-rich sites. *J. Biol. Chem.* **2009**, *284*, 18605–18613. [CrossRef] [PubMed]

25. Nolle, N.; Schuster, C.F.; Bertram, R. Two paralogous yefm-yoeb loci from staphylococcus equorum encode functional toxin-antitoxin systems. *Microbiology* **2013**, *159*, 1575–1585. [CrossRef] [PubMed]

26. Zheng, C.; Xu, J.; Ren, S.; Li, J.; Xia, M.; Chen, H.; Bei, W. Identification and characterization of the chromosomal yefm-yoeb toxin-antitoxin system of streptococcus suis. *Sci. Rep.* **2015**, *5*, 13125. [CrossRef] [PubMed]

27. Sevillano, L.; Diaz, M.; Yamaguchi, Y.; Inouye, M.; Santamaria, R.I. Identification of the first functional toxin-antitoxin system in streptomyces. *PLoS ONE* **2012**, *7*, e32977. [CrossRef] [PubMed]

28. Nieto, C.; Cherny, I.; Khoo, S.K.; de Lacoba, M.G.; Chan, W.T.; Yeo, C.C.; Gazit, E.; Espinosa, M. The yefm-yoeb toxin-antitoxin systems of escherichia coli and streptococcus pneumoniae: Functional and structural correlation. *J. Bacteriol.* **2007**, *189*, 1266–1278. [CrossRef] [PubMed]

29. Cheverton, A.M.; Gollan, B.; Przydacz, M.; Wong, C.T.; Mylona, A.; Hare, S.A.; Helaine, S. A salmonella toxin promotes persister formation through acetylation of trna. *Mol. Cell* **2016**, *63*, 86–96. [CrossRef] [PubMed]

30. Dalton, K.M.; Crosson, S. A conserved mode of protein recognition and binding in a pard-pare toxin-antitoxin complex. *Biochemistry* **2010**, *49*, 2205–2215. [CrossRef] [PubMed]

31. De Jonge, N.; Garcia-Pino, A.; Buts, L.; Haesaerts, S.; Charlier, D.; Zangger, K.; Wyns, L.; De Greve, H.; Loris, R. Rejuvenation of ccdb-poisoned gyrase by an intrinsically disordered protein domain. *Mol. Cell* **2009**, *35*, 154–163. [CrossRef] [PubMed]

32. Moyed, H.S.; Bertrand, K.P. Hipa, a newly recognized gene of escherichia coli k-12 that affects frequency of persistence after inhibition of murein synthesis. *J. Bacteriol.* **1983**, *155*, 768–775. [PubMed]

33. Germain, E.; Castro-Roa, D.; Zenkin, N.; Gerdes, K. Molecular mechanism of bacterial persistence by hipa. *Mol. Cell* **2013**, *52*, 248–254. [CrossRef] [PubMed]

34. Cooper, C.R.; Daugherty, A.J.; Tachdjian, S.; Blum, P.H.; Kelly, R.M. Role of vapbc toxin-antitoxin loci in the thermal stress response of sulfolobus solfataricus. *Biochem. Soc. Trans.* **2009**, *37*, 123–126. [CrossRef] [PubMed]

35. Ramage, H.R.; Connolly, L.E.; Cox, J.S. Comprehensive functional analysis of mycobacterium tuberculosis toxin-antitoxin systems: Implications for pathogenesis, stress responses, and evolution. *PLoS Genet.* **2009**, *5*, e1000767. [CrossRef] [PubMed]

36. Orth, D.; Grif, K.; Dierich, M.P.; Wurzner, R. Variability in tellurite resistance and the ter gene cluster among shiga toxin-producing escherichia coli isolated from humans, animals and food. *Res. Microbiol.* **2007**, *158*, 105–111. [CrossRef] [PubMed]

37. Pandey, D.P.; Gerdes, K. Toxin-antitoxin loci are highly abundant in free-living but lost from host-associated prokaryotes. *Nucleic Acids Res.* **2005**, *33*, 966–976. [CrossRef] [PubMed]

38. Li, G.; Shen, M.; Lu, S.; Le, S.; Tan, Y.; Wang, J.; Zhao, X.; Shen, W.; Guo, K.; Yang, Y.; et al. Identification and characterization of the hicab toxin-antitoxin system in the opportunistic pathogen pseudomonas aeruginosa. *Toxins* **2016**, *8*, 113. [CrossRef] [PubMed]

39. Jorgensen, M.G.; Pandey, D.P.; Jaskolska, M.; Gerdes, K. Hica of escherichia coli defines a novel family of translation-independent mrna interferases in bacteria and archaea. *J. Bacteriol.* **2009**, *191*, 1191–1199. [CrossRef] [PubMed]

40. Dy, R.L.; Przybilski, R.; Semeijn, K.; Salmond, G.P.; Fineran, P.C. A widespread bacteriophage abortive infection system functions through a type iv toxin-antitoxin mechanism. *Nucleic Acids Res.* **2014**, *42*, 4590–4605. [CrossRef] [PubMed]

41. Short, F.L.; Akusobi, C.; Broadhurst, W.R.; Salmond, G.P.C. The bacterial type iii toxin-antitoxin system, toxin, is a dynamic protein-rna complex with stability-dependent antiviral abortive infection activity. *Sci. Rep.* **2018**, *8*, 1013. [CrossRef] [PubMed]

42. Kopfmann, S.; Roesch, S.K.; Hess, W.R. Type ii toxin-antitoxin systems in the unicellular cyanobacterium synechocystis sp. Pcc 6803. *Toxins* **2016**, *8*, 228. [CrossRef] [PubMed]

43. Feng, Y.; Chen, C.J.; Su, L.H.; Hu, S.; Yu, J.; Chiu, C.H. Evolution and pathogenesis of staphylococcus aureus: Lessons learned from genotyping and comparative genomics. *FEMS Microbiol. Rev.* **2008**, *32*, 23–37. [CrossRef] [PubMed]

44. Pinel-Marie, M.L.; Brielle, R.; Felden, B. Dual toxic-peptide-coding staphylococcus aureus rna under antisense regulation targets host cells and bacterial rivals unequally. *Cell Rep.* **2014**, *7*, 424–435. [CrossRef] [PubMed]

45. Sayed, N.; Nonin-Lecomte, S.; Rety, S.; Felden, B. Functional and structural insights of a staphylococcus aureus apoptotic-like membrane peptide from a toxin-antitoxin module. *J. Biol. Chem.* **2012**, *287*, 43454–43463. [CrossRef] [PubMed]

46. Donegan, N.P.; Thompson, E.T.; Fu, Z.; Cheung, A.L. Proteolytic regulation of toxin-antitoxin systems by clppc in staphylococcus aureus. *J. Bacteriol.* **2010**, *192*, 1416–1422. [CrossRef] [PubMed]

47. Valihrach, L.; Demnerova, K. Impact of normalization method on experimental outcome using rt-qpcr in staphylococcus aureus. *J. Microbiol. Methods* **2012**, *90*, 214–216. [CrossRef] [PubMed]

48. Edgar, R.C. Muscle: Multiple sequence alignment with high accuracy and high throughput. *Nucleic Acids Res.* **2004**, *32*, 1792–1797. [CrossRef] [PubMed]

Article

Erianin against *Staphylococcus aureus* Infection via Inhibiting Sortase A

Ping Ouyang [†], Xuewen He [†], Zhong-Wei Yuan [†], Zhong-Qiong Yin, Hualin Fu, Juchun Lin, Changliang He, Xiaoxia Liang, Cheng Lv, Gang Shu, Zhi-Xiang Yuan, Xu Song, Lixia Li and Lizi Yin *

College of Veterinary Medicine, Sichuan Agriculture University, Chengdu 610000, China;
ouyang.ping@live.cn (P.O.); xuewen-he@hotmail.com (X.H.); yuanzhongwei_sicau@163.com (Z.-W.Y.);
yinzhongq@163.com (Z.-Q.Y.); fuhl2005@sohu.com (H.F.); juchunlin@126.com (J.L.); hecl@sicau.edu.cn (C.H.);
liangxiaoxia@sicau.edu.cn (X.L.); lvcheng1980@163.com (C.L.); dyysg2005@sicau.edu.cn (G.S.);
zhixiang-yuan@hotmail.com (Z.-X.Y.); songx@sicau.edu.cn (X.S.); lilixia905@163.com (L.L.)
* Correspondence: yinlizi@sicau.edu.cn; Tel.: +86-170-9284-8186; Fax: +86-288-629-2116
† These authors contributed equally.

Received: 7 August 2018; Accepted: 20 September 2018; Published: 23 September 2018

Abstract: With continuous emergence and widespread of multidrug-resistant *Staphylococcus aureus* infections, common antibiotics have become ineffective in treating these infections in the clinical setting. Anti-virulence strategies could be novel, effective therapeutic strategies against drug-resistant bacterial infections. Sortase A (srtA), a transpeptidase in gram-positive bacteria, can anchor surface proteins that play a vital role in pathogenesis of these bacteria. SrtA is known as a potential antivirulent drug target to treat bacterial infections. In this study, we found that erianin, a natural bibenzyl compound, could inhibit the activity of srtA in vitro (half maximal inhibitory concentration—IC_{50} = 20.91 ± 2.31 µg/mL, 65.7 ± 7.2 µM) at subminimum inhibitory concentrations (minimum inhibitory concentrations—MIC = 512 µg/mL against *S. aureus*). The molecular mechanism underlying the inhibition of srtA by erianin was identified using molecular dynamics simulation: erianin binds to srtA residues Ile182, Val193, Trp194, Arg197, and Ile199, forming a stable bond via hydrophobic interactions. In addition, the activities of *S. aureus* binding to fibronectin and biofilm formation were inhibited by erianin, when co-culture with *S. aureus*. In vivo, erianin could improve the survival in mice that infected with *S. aureus* by tail vein injection. Experimental results showed that erianin is a potential novel therapeutic compound against *S. aureus* infections via affecting srtA.

Keywords: sortase A; *Staphylococcus aureus*; erianin; inhibitor; molecular mechanism

Key Contribution: We confirmed that erianin could inhibit srtA in vitro, and explored mechanism of action using molecular dynamics simulation for the first time. We also found that erianin can reduce the activities of *S. aureus* binding to fibronectin and biofilm formation, and improve the survival in *S. aureus* infected mice.

---

1. Introduction

Staphylococcus aureus (*S. aureus*), a gram-positive bacterium, is an important opportunistic pathogen in human and animals [1]. *S. aureus* can cause a range of diseases when the host has weakened immunity. Methicillin-resistant *S. aureus* (MRSA) is a resistant strain in *S. aureus*. Antibiotics have limited or no effects on MRSA infection and contribute to increased antimicrobial resistance. MRSA is seriously threatening public health worldwide, with higher morbidity and mortality rates than non-resistant *S. aureus* strains and high therapy costs [2]. New antibiotics and new therapeutic strategies are urgently needed. In recent decade, antivirulent treatment strategy has become an immediate research focus on the treatment bacteria-medicated diseases [3,4].

Modern research found that virulence factors play vital roles in bacterial pathogenesis. In *S. aureus*, the virulence factors assist the bacteria to adhere the surface of host mucosal surface, destroy the red blood cells and leukocytes, and evade host's immune defenses [1]. During the infection, *S. aureus* firstly adheres to the surface of host organ tissues via its surface proteins. *S. aureus* without surface proteins cannot adhere to the host cells [1], it soon recognized by the host's immune system [5,6]. These extracellular associated proteins secreted by the bacteria are covalently anchored to the cell wall peptidoglycan by transpeptidase to become true surface proteins [7]. sortase A (srtA) is one of the primary surface-anchored transpeptidases in *S. aureus* [8]. In 2000, Mazmanian et al. reported that the *S. aureus* with deleted *srtA* gene had no influence on the bacterial growth, but the mutant decreased the number of surface leucine, proline, any amino acid, threonine and glycine (LEXTG)-containing proteins and reduced the ability of the bacteria to cause renal abscesses and acute infection in mouse models [9]. In a rat endocarditis model, *S. aureus srtA* mutant showed low pathogenicity [10]. Previous studies found that *S. aureus* with *srtA* gene deletion lost its ability to bind IgG, fibronectin, and fibrinogen and had reduced survival in macrophages [9,11]. The catalytic center of srtA comprises a set of amino acids residues (His120, Cys184 and Arg197) [12]. His120 and Cys184 maintain the activity of srtA [13]. Arg197 can effectively cleavage the T-G peptide bond of LPXTG-containing proteins, which are substrates of srtA [14]. SrtA is a potentially promising target for treating *S. aureus* infections.

In our group, we mainly focus on finding new compounds from traditional Chinese medicines (TCM) against *S. aureus* virulence factors. In previous researches, we found natural molecules can inhibit the α-hemolysin in vitro [15–18]. For more virulence factors, we have screened natural compounds targeting the srtA by detecting the inhibition rate of enzyme activity. We have identified several natural molecules from TCM herbs, and found that erianin had relatively high inhibitory activities. Erianin is a bibenzyl compound (Figure 1). Erianinis is also an important bioactive constituent of *Dendrobium chrysotoxum* [19]. Erianin has exhibited pharmacological antitumor by inhibiting angiogenesis [20–22], endothelial metabolism [23], and inflammation and by inducing the cells arrest, apoptosis, and autophagy [24,25]. Erianin has also exhibited antioxidant activity [19] and anti-benign prostatic hyperplasia [26]. On our knowledge, there are no studies focused on erianin inhibited the srtA in *S. aureus*. In this study, we evaluated the influence erianin on *S. aureus* srtA and the molecular mechanism of erianin binding to srtA.

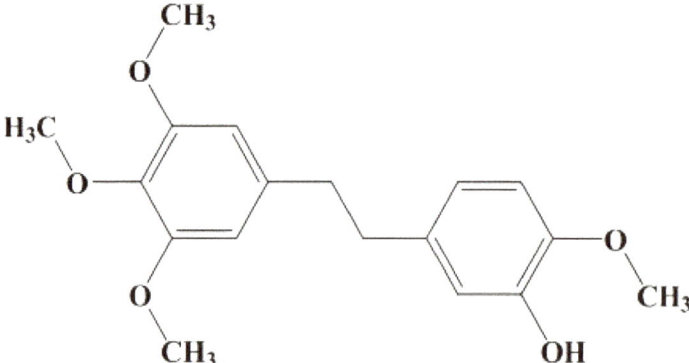

Figure 1. Chemical structure of erianin (CAS No. 95041-90-0).

2. Results

2.1. The Activity of Erianin against S. aureus SrtA

The fluorescent peptide Dabcyl-QALPETGEE-Edans is a substrate for srtA. Purified srtA can cleave the peptide, thus increasing the fluorescence intensity. The inhibitory activity of erianin was showed with IC_{50}, which is the concentration that decreases the 50% fluorescence intensity relative to the negative control group (without erianin). The results of different concentrations of erianin against *S. aureus* srtA are presented as the percentage of inhibitory activity in Figure 2. The IC_{50} of erianin against *S. aureus* SrtA was calculated to be 20.91 ± 2.31 μg/mL (65.7 ± 7.2 μM).

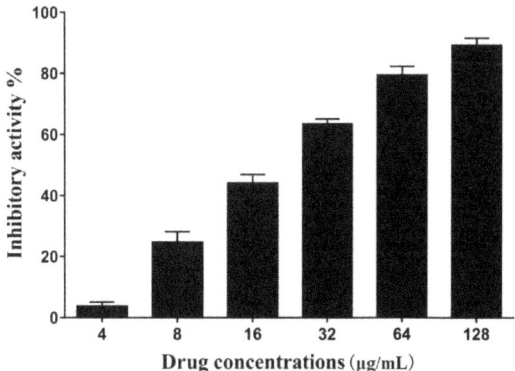

Figure 2. Inhibitory effects of erianin (different levels) against srtA from *Staphylococcus aureus* Newman D2C in vitro.

2.2. The Influence of Erianin on S. aureus Growth

The minimum inhibitory concentrations (MICs) of erianin against *S. aureus* strain ATCC25904 (Newman D2C) and ΔsrtA strain were same (512 μg/mL). A growth curve assay was performed using strain Newman D2C with (8–64 μg/mL) or without erianin, and srtA mutant strain. The growth curves showed that erianin had no inhibitory effects on the growth of *S. aureus* at concentrations of 8–64 μg/mL (Figure 3).

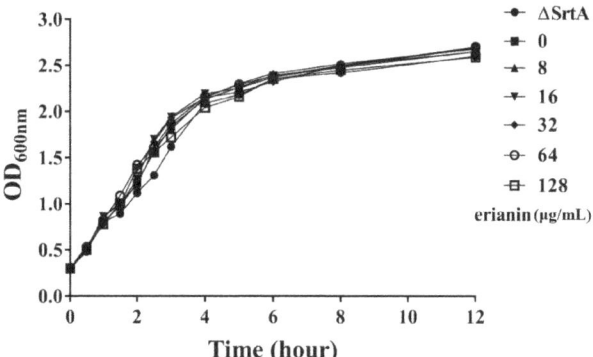

Figure 3. Growth curves of *Staphylococcus aureus* strain Newman D2C with or without erianin and gene knockout mutant ΔsrtA in brain-heart infusion (BHI). Symbol ●, ■, ▲, ▼, ◆, ○ and □ represent gene knockout mutant ΔsrtA, *Staphylococcus aureus* strain Newman D2C culture with 0, 4, 16, 32, 64 and 128 μg/mL erianin, respectively.

2.3. Erianin Reduced the Activity of S. aureus Adhesion to Fibrinogen

During the early phase of *S. aureus* infection, surface proteins can adhere to host cells and invade tissues to escape the host immune defense. SrtA, a primary anchoring enzyme in gram-positive bacteria, anchors many surface proteins, such as protein A, fibronectin binding proteins, and collagen-binding proteins. Fibronectin binding proteins can invade cells via binding fibronectin/fibrinogen in host cells [27–29]. Erianin decrease the catalytic activity of srtA, and then surface proteins (fibronectin binding proteins) will reduce. Fg-binding assays were used to test the inhibition of binding ability of *S. aureus* srtA by erianin. The capacity of *S. aurues* Newman D2C treated with different concentrations erianin and ΔsrtA to adhere to an Fg-coat surface were examined. The results were shown in Figure 4. The adhering capacity significantly decreased in the ΔsrtA, with an Fg-adhesion rate of 3.3 ± 1.5%. The adhesion rate of *S. aurues* Newman D2C treated with 8 µg/mL erianin was lower (91.8 ± 4.1%) than that of wild type (WT) group. The Newman D2C strain treated with 64 µg/mL erianin showed a binding rate of 16.1 ± 2.9%. These results showed that erianin reduced the capacity of *S. aureus* to adhere to Fg in a dose-response manner.

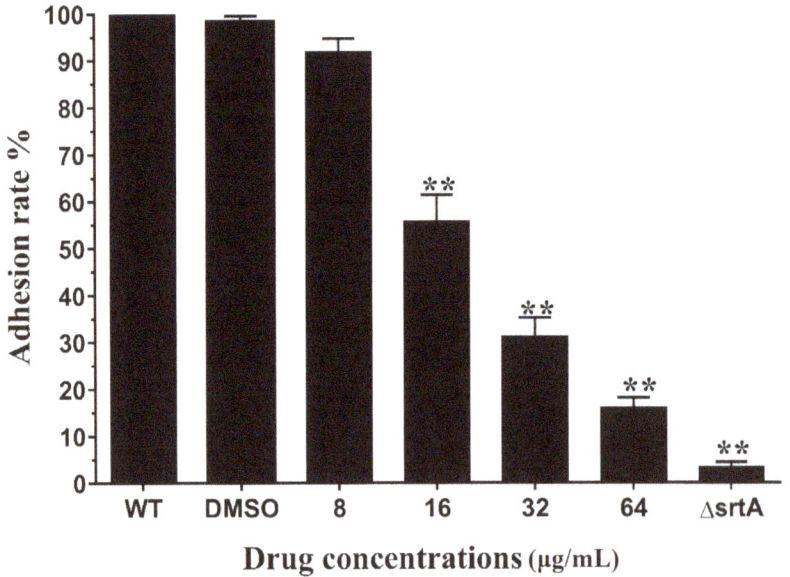

Figure 4. Adhesion rate of bacterial cells to Fg. erianin reduced the adhere of wild type (WT) to Fg in a dose dependent manner. Each result was derived from three independent experiments and presented as the mean ± SEM. ** $p < 0.01$ vs. the WT group.

2.4. Erianin Decreased Biofilm Formation

The formation of bacterial biofilm is related to persistent infection in human and animals, an important factor in the failure of antibiotics [30,31]. Some surface proteins (fibronectin binding proteins, *S. aureus* surface protein C and *S. aureus* surface protein G) make a direct role in *S. aureus* biofilm formation [32]. We detected the *S. aureus* biofilm formation in the absence or presence different concentrations of erianin. *S. aureus* biofilm formations were decreased in the presence of erianin (Figure 5). When *S. aureus* treated with 64 µg/mL erianin declined significantly in the biofilm formation. Erianin affected the biofilm formation could be via inhibiting the activity of *S. aureus* srtA.

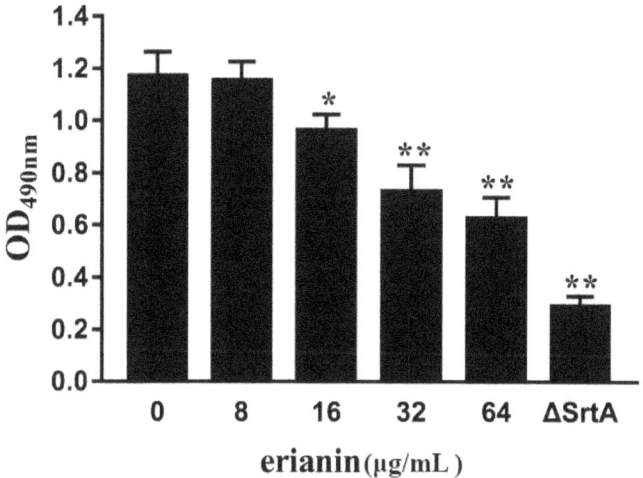

Figure 5. Erianin reduces the biofilm formation of *S. aureus*. Three independent experiments were performed to obtain stable results. * $p < 0.05$ vs. the WT group, ** $p < 0.01$ vs. the WT group.

2.5. Mechanism Underlying the Binding of Erianin to SrtA

Molecular docking and molecular dynamics simulations were used to explore the potential binding sites and mechanism between erianin and srtA via the AutoDock vina 1.1.2 and Amber12 software package. Based on the docking results, the mechanism of erianin binding to srtA was determined using 20-ns molecular dynamics simulations. The root mean square deviation (RMSD) values of the protein backbone were used to explore the dynamic stability of the models and ensure the rationality of the sampling strategy. RMSD values of the protein backbone were calculated based on the starting structure along the simulation time and plotted as shown in Figure 6A. We found that the protein structure of srtA and its binding with erianin were stabilized during the simulation.

To reveal the flexibility of the residues of the whole protein in the srtA-erianin complex and free srtA, the root mean square fluctuations (RMSF) of the residues were calculated. Different flexibilities of all residues in the srtA protein with or without of erianin were shown in Figure 6B. The fluctuation patterns of the srtA-erianin complex and free srtA were different during the final 20 ns of the simulation. All of the residues in the binding site of srtA-erianin complex showed a certain degree of flexibility compared with free srtA, with RMSF of <3 Å. These results indicated that the residues in the srtA-erianin complex were more rigid.

2.6. Identification of the Site of Erianin Binding to SrtA

The Molecular Mechanics Generalized Born Surface Area (MMGBSA) method was used to calculate the binding free energy of the residues surrounding the binding site to explain their contribution to the entire binding system. The binding free energies of each residue included Van der Waals (ΔEvdw), solvation (ΔEsol), electrostatic (ΔEele) and total contributions (ΔEtotal) (Figure 6C). The Arg197 residue of the srtA-erianin complex has a strong electrostatic contribution, with the ΔEele of <−3.0 kcal/mol. There was cation-π interaction between SrtA and erianin, because Arg197 is close to the phenyl group of erianin, and electrostatic interactions exist in this complex (Figure 6D). In addition, residue Trp194, with the ΔEvdw of <−1.5 kcal/mol, has an appreciable Van der Waals interactions with erianin, leading to formation of a hydrogen bond between srtA and erianin. The Van der Waals interactions were a primary source of the decomposed energy, except Arg197 and Trp194, possibly via hydrophobic interactions (Ala104, Ile182, Val193, and Ile199). These results suggested that these six residues were key residues for erianin, especially Val193 and Arg197 with ΔEtotal of <−2.0 kcal/mol.

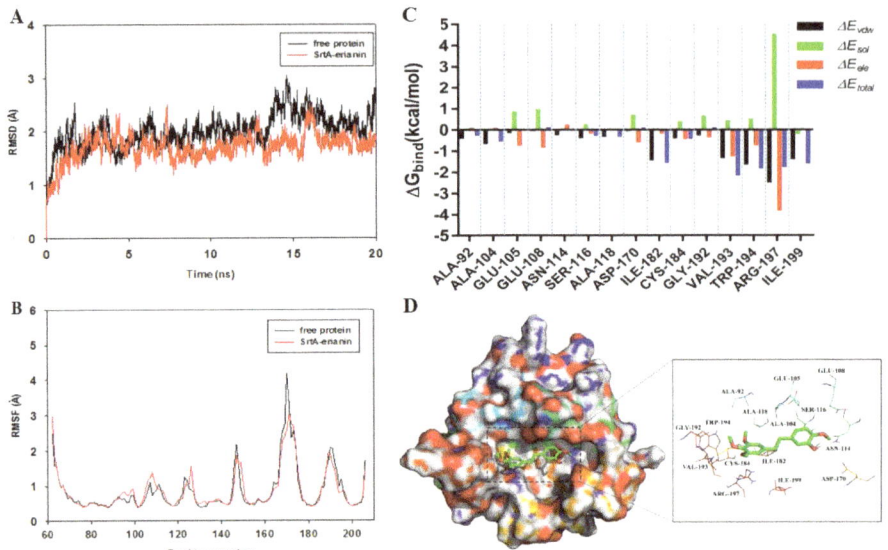

Figure 6. (**A**) The root-mean-square deviations (RMSDs) of all the atoms of srtA-erianin complex with respect to its initial structure as function of time; (**B**) RMSF of residues of the whole protein in srtA-erianin complex and free SrtA during the 20 ns simulation; (**C**) Decomposition of the binding energy (ΔG_{bind}) on a per-residue basis in the srtA-erianin complex; (**D**) The predicted binding mode of erianin in srtA binding pocket obtained from MD simulation.

2.7. Confirmation of the Molecular Basis of Erianin against SrtA

To validate these theoretical results of the binding of erianin-srtA complex, we obtained two srtA-mutants (V193A-srtA and R197A-srtA) by point mutation. The total binding free energies (ΔG_{bind}) of the interaction between WT-srtA, srtA mutants and erianin were calculated using an MM-GBSA approach. According to the calculation results (Table 1), the binding energy of WT-srtA was bigger than the mutants. That means that WT-srtA had the strongest ability to bind to erianin. The ΔG_{bind} and the number of binding sites between erianin and the two mutants were measured using fluorescence spectroscopy quenching, and the results were obtained by computational methods (Table 1). We found that these results showed consistent between the fluorescence spectroscopy quenching and MM-GBSA approach. To further confirm the simulation results, V193A-srtA and R197A-srtA were constructed for fluorescence resonance energy transfer (FRET) assay. The results of FRET assay (Figure 7) showed that the activity of erianin against V193A-srtA and R197A-srtA was significantly reduced compared with WT-srtA. These results showed that the information from the MD simulation on the srtA-erianin complex is reliable. Erianin inhibit the biological activity of srtA by binding to the active site region (residues of Ala92/Ala104/Ser116/Ala118/Ile182/Val193/Trp194/Arg197/Ile199).

Table 1. The binding free energy (kcal/mol) of WT-Erianin, V193A-Erianin and R197A-Erianin systems based on computational method and the values of the binding constants (K_A) based on the fluorescence spectroscopy quenching.

Proteins	WT-SrtA	V193A-SrtA	R197A-SrtA
The binding energy	-24.1 ± 2.3	-23.8 ± 2.2	-18.9 ± 2.0
K_A (1×10^4) L mol^{-1}	44.5 ± 7.2	43.9 ± 6.7	39.6 ± 4.8

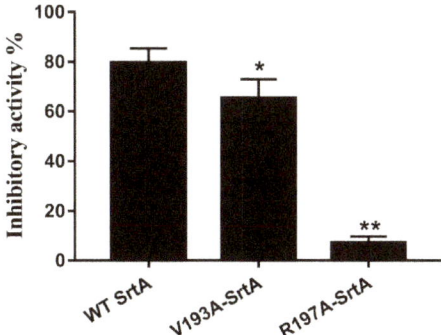

Figure 7. Inhibitory effects of erianin against WT-srtA and srtA mutants. WT-srtA and srtA mutants (V193A-srtA and R197A-srtA) were incubated with 64 μg/mL erianin, and the catalytic activity of recombinant srtA was determined as described in Figure 2. The error bars show the standard deviations (SD). * $p < 0.05$, ** $p < 0.01$ compared with WT-srtA.

2.8. Erianin Protected Mice from Fatal S. aureus Infection

The protective effects of erianin in vivo were determined by survival rate of mice infected with *S. aureus*. All mice inoculated with Newman D2C strain (WT group) died within 8 days. However, the mortality rate of mice infected with ΔsrtA strain was 10% (Figure 8). The *S. aureus*-inoculated mice were treated with erianin (50 mg/kg, three times per day) via subcutaneous injection. Nine days after infected, three mice still survive in the WT + erianin group (Figure 8). These results showed that erianin can increase the survival rate of mice infected with *S. aureus*. In addition, erianin cannot lysis rabbit red cells at the concentrations of 0–512 μg/mL (data not shown).

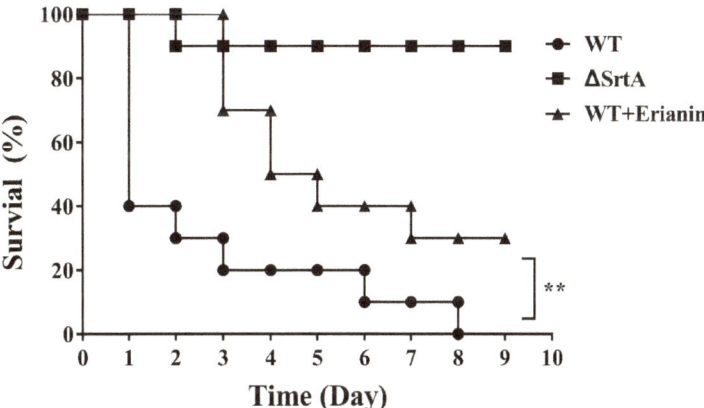

Figure 8. Effects of erianin on survival rates after 9-days monitoring of mice. Survival percentage of BALB/c mice ($N = 10$) after challenged with intravenous injection of 2×10^9 CFU of *S. aureus* (WT) and ΔSrtA. Treatment with erianin (50 mg/kg, three times a day) was initiated 2 h after inoculation and again at 8-hour intervals. Survival statistics were calculated by Log-rank (Mantel–Cox) test. The statistical significance determined as follows: WT vs. ΔSrtA, ** $p < 0.001$.

3. Discussion

Virulence inhibitors are deemed new treatment strategies for bacterial infections, especially for multidrug-resistant bacterial [3,4,33]. Traditional antibiotics produce great pressures on bacterial growth during the process of administration, whereas virulence inhibitors produce low survival pressure on bacterial reproduction. In *S. aureus*, surface proteins and extracellular toxins are key virulence factors and play important roles in the adhesion, colonization, and destruction to host cells [1]. SrtA is a common and conserved transpeptidase found in many gram-positive bacteria [28]. SrtA anchors LEXTG-containing proteins to the gram-positive bacterial cell surface. In *S. aureus*, there are approximately 20 LEXTG-containing proteins, including fibronrctin-binding proteins (FnbpA and FnbpB), protein A (Spa), serine-aspartate repeat proteins (SdrC, ScdD and ScdE), collagen-binding protein (Cna) and clumping factors (ClfA and ClfB) [34]. These proteins play important roles in adhesion, biofilm formation, colonization, and evasion of the host innate immune defense in the pathogenicity of *S. aureus* [26]. SrtA inhibitors could affect the anchoring of many proteins to cell surface, a strategy which is better than targeting single surface protein for altering bacterial virulence [30]. Many natural and synthesized compounds capable of inhibiting the bioactivity of srtA in vivo and in vitro have been identified [35].

Our results showed that erianin (Figure 1) was a potential srtA inhibitor. In this study, erianin inhibited the activity of srtA and adhesion of *S. aureus* to fibrinogen at concentrations below the MIC without influencing the hemolysis ability of bacterial culture medium (data not shown). The IC_{50} of erianin against *S. aureus* srtA was 20.91 ± 2.31 μg/mL (65.7 ± 7.2 μM). We also evaluated the effect of erianin using a mouse infection model and found that it reduced their mortality rate. Erianin would be an assistance drug to against *S. aureus* infection. However, the exact mechanisms of erianin against *S. aureus* infection in vitro are not clear. TCM herbs exhibited the pharmacodynamics targeted different pathways with multi-components. Erianin is possible to act on the host by inhibiting the phlogistic pathways and improving the immunity, or interfere with several virulence factors of bacteria.

The molecular mechanism of erianin binding to srtA was also revealed using the molecular simulation methods. The results showed that erianin tightly binds to the residues in the active center of srtA. One previous study reported that the residues His120, Cys184 and Arg197, the β6/β7 loop and β7/β8 loop are essential for the catalysis by srtA [12]. The β6/β7 loop recognizes the substrate (LPXTG-containing proteins) of srtA [36,37]. Erianin binds to the β6/β7 loop via Van der Waals interaction (Ala104), hydrophobic interactions, and electrostatic interactions (Ala118). A comparison of free srtA and erianin-srtA complexes with respect to RMSF values of the residues indicated that erianin affected the flexibility of β6/β7 loop. Consequently, a combination of erianin and srtA would influence the recognition of the substrate, which is the first step of the process of protein anchoring by srtA. The mobility of β7/β8 loop in srtA provides a binding site for Lipid II [38,39]. The methoxy group of phenyl ring forms a hydrogen bond with Val193 (binding site of Lipid II in srtA) and places the β7/β8 loop into a closed state in the erianin-srtA complex. The thiol group of Cys184 bond to the carbonyl group of Thr (LPXTG-containing proteins) and form a thioester acyl enzyme intermediate [40]. Arg197 facilitates the cleavage of Thr and Gly through the ionization of Cys184 by forming a stabilized tetrahedral oxyanion transition state and providing the required activation energy [41]. In the docking model of erianin-SrtA complex, the phenyl group of erianin was placed near Arg197, thus facilitating electrostatic interactions between them. Furthermore, cation–π interactions were observed between srtA and erianin. The phenyl rings of erianin formed a hydrophobic binding packet in the active site of SrtA with the alky side chains of Ala92, Ala104, Asn114, Ser116, Ile182, Cys184, Trp194 and Ile199, which further stabilized the active center.

Oh et al. (2006) reported that (Z)-3-(2,5-dimethoxyphenyl)-2-(4-methoxyphenyl) acrylonitrile can inhibit the activity of srtA [42]. This compound has the same basic structural backbone of erianin. Erianin (IC_{50} = 20.91 ± 2.31 μg/mL, 65.7 ± 7.2 μM) has weaker activity in inhibiting srtA than (Z)-3-(2,5-dimethoxyphenyl)-2-(4-methoxyphenyl) acrylonitrile (IC_{50} = 2.7 μg/mL). Because of the structure of erianin, the stereohindrance due to trimethoxy groups make the ligand unavailable for srtA residues, and the hydroxyl enhances the hydrophilicity of erianin. The double bond and the cyano can increase the inhibitory activity [43].

These results suggested that erianin inhibited the anchoring of srtA by preventing the access and binding of the T-G peptide chain and Lipid II of the surface proteins to the bioactivity center.

These results showed that erianin has the potential to treat *S. aureus* infections. Su et al. have reported that erianin basically had no cytotoxicity on human normal liver cell line L02 [21], although the toxicology of erianin will be research before it used as a therapeutic option against S. aureus infections in vivo.

4. Materials and Methods

4.1. Expression and Purification of WT-SrtA, V193A-SrtA and R197A-SrtA

S. aureus strain Newman D2C (ATCC25904) was commercially obtained from the American Type Culture Collection (ATCC) and used in this study. The Newman D2C strain can produce srtA without hemolysins and coagulase. The ΔsrtA strain, which was constructed from Newman D2C, was graciously provided by Professor Deng, College of Veterinary Medicine, Jilin University [5]. The *strA* fragment was cloned from the *S. aureus* NewmanD2C genomic DNA by PCR. The *strA* fragment (the sequence express only residues 60–206) was cloned into the pGEX-6P-1 vector (GE Amersham), and made the recombinant plasmid pG-srtAΔN59 [44]. The mutations V193A and R197A were obtained from site-directed mutagenesis on the recombinant plasmid pG-srtAΔN59 using the QuickChange site-directed mutagenesis kit (Stratagene, La Jolla, CA, USA) according the manufacturer's protocol. Electroporation was used to transfect the recombinant plasmid pG-srtAΔN59 and the mutant constructs into *Escherichia coli* strain BL21 (Invitrogen, Carlsbad, CA, USA). The transformed *Escherichia coli* were grown at 37 °C in LB broth with ampicillin (100 μg/mL). When bacteria showed an initial logarithmic growth phase initially ($OD_{600\,nm}$ = 0.6–0.8), isopropyl β-D-1-thiogalactopyr-anoside (IPTG, 1 mM) was added into the culture medium to induce the target protein. The bacteria were grown at 16 °C for 12–16 h. The recombinant proteins (WT-srtA, V193A-srtA and R197A-srtA) were released from the cells by sonication and dissolved in the reaction buffer [41]. These proteins solution were added into a GST-affinity column (2 mL glutathione Sepharose 4B; GE Amersham). These recombinant proteins were bounded to the GST column. The unbound proteins were washed using the reaction buffer. Proteins were concentrated by molecular size elution column and detected by SDS-PAGE. BCA protein assay kit (Pierce, Thermo Fisher Scientific, Shanghai, China) was used for determining the concentration of proteins. All expression vectors were confirmed via DNA sequencing. The mutagenic primer pairs employed to produce the three mutants are listed in Table 2.

Table 2. Oligonucleotide primers used in this study.

Primer Name	Oligonucleotide (5-3) [a]
PsrtA59F	GCGGGATCCCCGGAATTCCAAGCTAAACCTCAAATTCC
PsrtA59R	CCGCTCGAGTTATTTGACTTCTGTAGCTACAA
V193A-srtA-F	TGAAAAGACAGGCGCTTGGGAAAAAC
V193A-srtA-R	TTCCCAGCGCCTGTCTTTTCATTGTAATCAT
R197A-srtA-F	GACAGGCGTTTGGGAAAAAGCGAAAATCTTTGTAGCTACAG
R197A-srtA-R	CTGTAGCTACAAAGATTTTCGCTTTTTCCCAAACGCCTGTC

[a] Restriction endonuclease recognition sites or mutated codons are underlined.

4.2. Determination of Mutant and WT SrtA Activity

Erianin, which was purchased from Chengdu Herbpurify Co., Ltd., (Chengdu, China), was dissolved in Dimethyl sulfoxide (DMSO, Sigma, St. Louis, MO, USA). The solution was stored at 4 °C before use. The activities of erianin against WT-srtA, V193A-srtA and R197A-srtA were detected using the FRET assays. FRET assays were performed by disrupting a synthetic substrate peptide Dabcyl-QALPETGEE-Edans (GL Biochem, Shanghai, China) according to the protocols which have been published [7,14]. All reactions were performed in the black 96-well plate. Briefly, 300 μL of the reaction volume contained with reaction buffer, synthetic substrate peptide, recombinant proteins (WT-srtA, V193A-srtA and R197A-srtA), and different concentrations of erianin. The negative control contained all of the above components, except erianin. First, the mixture without synthetic substrate peptide was incubated at 37 °C for 30 min; then, the synthetic substrate peptides were added into the reaction system and incubated for 60 min at 37 °C. The sample fluorescence was analyzed at an emission wavelength of 495 nm and an excitation wavelength of 350 nm. We also checked the fluorescences of erianin co-culture with recombinant proteins, and synthetic substrate peptide. Each experiment was tested in triplicate to ensure reproducibility.

4.3. Susceptibility Testing and Growth Curve Assay

The minimal inhibitory concentrations (MICs) of erianin against *S. aureus* were measured using the broth microdilution method recommended by the Clinical and Laboratory Standards Institute. MIC was defined as the lowest concentrations of erianin that inhibited *S. aureus* growth. The negative control contained DMSO without erianin. For the growth curves, overnight cells cultures were grown in fresh brain-heart infusion (BHI) broth (Sigma) by diluted 1:100. When the $OD_{600\,nm}$ of the culture were reached 0.3, they were resuspended in a solution containing different concentrations of erianin (8, 16, 32, 64 and 128 μg/mL). DMSO was used as negative control group. The solutions were incubated with constant shaking (200 rpm) at 37 °C for different durations. The OD value was measured by UV-spectrophotometer (Agilent Technologies, Santa Clara, CA, USA) at 600 nm.

4.4. Fibrinogen-Binding Assay

S. aureus cultures in the logarithmic phase were diluted to an initial $OD_{600\,nm}$ of 0.05 in BHI broth. Different concentrations of erianin were added to the *S. aureus* Newman D2C cultures. The Newman ΔsrtA strain was used as the positive control. The Newman D2C strain without erianin was as the wild type (WT) group. DMSO was used as negative control group. The mixtures were incubated at 37 °C with constant shaking (200 rpm) for 2 h. The bacteria were collected by centrifugation (5000× *g* for 5 min), and washed two times with sterile PBS, and the pellets were resuspended with PBS until use.

Bovine Fibrinogen (Sigma, 20 μg/mL) was seeded onto 96-well plates (Polystyrene Costar) and incubated at 4 °C overnight for coating. After washing, the plates were blocked with 5% bovine serum albumin (BSA, Sigma) at 37 °C. After 2 h, the plates were washed three times with sterile PBS. Then, the cell suspensions (100 μL/well) were added to the plates and incubated for 2 h at 37 °C. After removing the cells suspension, the adherent bacteria were fixed with formaldehyde (25%, *v/v*) for 30 min and stained with crystal violet (12.5 mg/mL) for 10 min. After washed with double distilled water and dried, 33% acetic acid was added to dissolve crystal violet. The absorbance of the sample was subsequently measured with a microplate reader (Thermo Scientific, Waltham, MA, USA) at 570 nm. The absorbance of the negative control group was used as the 100% adherence. The adherence rate of each sample was calculated by comparing to the negative control.

4.5. Biofilm Formation Assay

S. aureus cultured overnight, and then were grown in fresh BHI broth by diluted 1:100 with erianin or DMSO at 37 °C with constant shaking (200 rpm). The Newman ΔsrtA strain was used as the positive control. The Newman D2C strain without erianin was as the wild type (WT) group. DMSO was used as negative control group. When the $OD_{600\,nm}$ of the culture are reached 0.6, 10 μL of the bacterial solution was added into 290 μL BHI broth containing 3% (*w/v*) sucrose. The mixture was placed in the 96-well flat-bottom polystyrene microliter plates, and incubated at 37 °C in anaerobic box. After 18 h, we removed lightly the liquid mixture. And then, 100 μL of 10% formaldehyde solution was used to fix the biofilm at room temperature (RT) overnight. After removing the formaldehyde, crystal violet (12.5 mg/mL) was used to stain the biofilm at RT for 30 min. After washed with double distilled water and dried, 33% acetic acid was added to dissolve crystal violet. The absorbance of the sample was subsequently measured with a microplate reader (Thermo Scientific) at 490 nm.

4.6. Binding Affinity Determination of Erianin with Mutant and WT SrtA

The binding constants (KA) of erianin with WT-srtA and mutant srtA were measured using the fluorescence-quenching method. A 280-nm excitation wavelength with a 5-nm bandpass and a 345-nm emission wavelength with a 10-nm bandpass were used for the measurements. The details of the measurements in this study have previously been described [45,46].

4.7. Molecular Docking and Molecular Dynamics

The binding mode between erianin and srtA was investigated using molecular docking method with Autodock vina 1.1.2 [47]. The three-dimensional (3D) X-ray structure of SrtA (PDB ID: 1T2P) used in this experiment was obtained from Protein Data Bank (http://www.rcsb.org/pdb/home/home.do). ChemBioDraw Ultra 12.0 and ChemBio3D Ultra 12.0 softwares were used to prepare the 3D structure of erianin. The docking input files were generated using Auto Dock Tools 1.5.6 package [48]. Ligand structures for docking were prepared by defining rotatable bonds and merging non-polar hydrogen atoms. The search grid for srtA was identified as center_x: -34.843, center_y: -17.649, and center_z: 7.103 with dimensions size_x: 12.75, size_y: 15 and size_z: 10.5. The value of exhaustiveness was set at 20 to increase the docking accuracy. Default parameters were used if parameter details were not mentioned in Vina docking. The docking result was revised using molecular dynamics (MD).

MD simulations of the selected docked positions were performed using the Amber 12 and AmberTools 13 programs [49,50]. The topologies and parameters of erianin were prepared by AnteChamber PYthon Parser interfacE (ACPYPE) [51]. Next, the force field of the ligand was prepared and labeled "leaprc.gaff" (generalized Amber force field), whereas the receptor was labeled "leaprc.ff12SB". The details of MD simulation processes in this study have previously been described [52,53].

4.8. Animal Experiments

Mice (BALB/c) weighing 20 ± 2 g were commercially obtained from Chengdu Dossy Experimental Animals Co., Ltd., (Chengdu, China). All animal studies were carried out according to the experimental practices and standards of the animal ethics committee of Sichuan Agricultural University, and the experiment protocols were approved on 23 September 2016, and supervised by the animal care committee for project identification code 20160906.

The Newman D2C strain and Newman ΔsrtA strain were inoculated in BHI broth and incubated overnight at 37 °C. The cultures were diluted to 1:100 using fresh BHI broth and inoculated at 37 °C with constant shaking (200 rpm) for 3 h. The bacteria were collected by centrifugation ($5000\times$ g for 5 min at 4 °C) and washed two times with sterile PBS. Then, they were resuspended in fresh BHI broth to obtain staphylococcal suspension. In the survival studies, mice were infected with 100 µL of staphylococcal suspensions (2×10^9 Colony-Forming Units–CFU) via tail vein injection [54]. Thirty mice were randomly divided into three groups in the survival studies: (1) mice infected with Newman D2C strain (WT group), (2) mice infected with Newman ΔSrtA strain (ΔsrtA group) and (3) mice infected with Newman D2C strain and treated with erianin (50 mg/kg, three times a day; WT + erianin group) for three days. Survival percentages of each group were recorded 9 days after infection.

4.9. Statistical Analysis

The statistical significance of the percentage of Fg-binding was analyzed using the SPSS13.0 software (SPSS Inc., Chicago, IL, USA, 2005) with the unpaired two-tailed Student's *t*-test. The significance of the survival studies was analyzed using Log-rank (Mantel-Cox). The differences were considered statistically significant when *p*-value was <0.05.

Author Contributions: L.Y., Z.Y. (Zhongqiong Yin) and H.F. conceived and designed the experiments. P.O., X.H. and Z.Y. (Zhongwei Yuan) performed the experiments. J.L., C.H., G.S., X.L., C.L. and Z.Y. (Zhixiang Yuan) contributed reagents/materials/analysis tools. X.S. and L.L. analyzed the data. L.Y., P.O., X.H. and Z.Y. (Zhongwei Yuan) wrote the paper.

Acknowledgments: The study was funded by the Shuangzhi Project of Sichuan Agricultural University (03571444, 03572452) and the General Project of Sichuan Provincial Department of Education (Grant No. 16ZB0036).

Conflicts of Interest: All authors declare that they have no competing interest.

References

1. Lowy, F.D. Staphylococcus aureus infections. *N. Engl. J. Med.* **1998**, *339*, 520–532. [CrossRef] [PubMed]

2. Dryden, M. Complicated skin and soft tissue infections caused by methicillin-resistant *Staphylococcus aureus*: Epidemiology, risk factors, and presentation. *Surg. Infect.* **2008**, *9* Suppl. 1, s3–s10. [CrossRef] [PubMed]

3. Muhlen, S.; Dersch, P. Anti-virulence strategies to target bacterial infections. *Curr. Top. Microbiol. Immunol.* **2016**, *398*, 147–183. [PubMed]

4. Dickey, S.W.; Cheung, G.Y.C.; Otto, M. Different drugs for bad bugs: Antivirulence strategies in the age of antibiotic resistance. *Nat. Rev. Drug Discov.* **2017**, *16*, 457–471. [CrossRef] [PubMed]

5. Chen, F.; Liu, B.; Wang, D.; Wang, L.; Deng, X.; Bi, C.; Xiong, Y.; Wu, Q.; Cui, Y.; Zhang, Y.; et al. Role of sortase A in the pathogenesis of *Staphylococcus aureus*-induced mastitis in mice. *FEMS Microbiol. Lett.* **2014**, *351*, 95–103. [CrossRef] [PubMed]

6. Tan, C.; Wang, J.; Hu, Y.; Wang, P.; Zou, L. *Staphylococcus* epidermidis deltasortase A strain elicits protective immunity against *Staphylococcus aureus* infection. *Anton. Leeuwenhoek* **2017**, *110*, 133–143. [CrossRef] [PubMed]

7. Mazmanian, S.K.; Ton-That, H.; Su, K.; Schneewind, O. An iron-regulated sortase anchors a class of surface protein during *Staphylococcus aureus* pathogenesis. *Proc. Natl. Acad. Sci. USA* **2002**, *99*, 2293–2298. [CrossRef] [PubMed]

8. Mazmanian, S.K.; Liu, G.; Ton-That, H.; Schneewind, O. *Staphylococcus aureus* sortase, an enzyme that anchors surface proteins to the cell wall. *Science* **1999**, *285*, 760–763. [CrossRef] [PubMed]

9. Mazmanian, S.K.; Liu, G.; Jensen, E.R.; Lenoy, E.; Schneewind, O. *Staphylococcus aureus* sortase mutants defective in the display of surface proteins and in the pathogenesis of animal infections. *Proc. Natl. Acad. Sci. USA* **2000**, *97*, 5510–5515. [CrossRef] [PubMed]

10. Weiss, W.J.; Lenoy, E.; Murphy, T.; Tardio, L.; Burgio, P.; Projan, S.J.; Schneewind, O.; Alksne, L. Effect of srtA and srtB gene expression on the virulence of *Staphylococcus aureus* in animal models of infection. *J. Antimicrob. Chemother.* **2004**, *53*, 480–486. [CrossRef] [PubMed]

11. Kubica, M.; Guzik, K.; Koziel, J.; Zarebski, M.; Richter, W.; Gajkowska, B.; Golda, A.; Maciag-Gudowska, A.; Brix, K.; Shaw, L.; et al. A potential new pathway for *Staphylococcus aureus* dissemination: The silent survival of *S. aureus* phagocytosed by human monocyte-derived macrophages. *PLoS ONE* **2008**, *3*, e1409. [CrossRef] [PubMed]

12. Suree, N.; Yi, S.W.; Thieu, W.; Marohn, M.; Damoiseaux, R.; Chan, A.; Jung, M.E.; Clubb, R.T. Discovery and structure-activity relationship analysis of *Staphylococcus aureus* sortase A inhibitors. *Bioorg. Med. Chem.* **2009**, *17*, 7174–7185. [CrossRef] [PubMed]

13. Ton-That, H.; Mazmanian, S.K.; Alksne, L.; Schneewind, O. Anchoring of surface proteins to the cell wall of *Staphylococcus aureus*. Cysteine 184 and histidine 120 of sortase form a thiolate-imidazolium ion pair for catalysis. *J. Biol. Chem.* **2002**, *277*, 7447–7452. [CrossRef] [PubMed]

14. Ton-That, H.; Liu, G.; Mazmanian, S.K.; Faull, K.F.; Schneewind, O. Purification and characterization of sortase, the transpeptidase that cleaves surface proteins of *Staphylococcus aureus* at the lpxtg motif. *Proc. Natl. Acad. Sci. USA* **1999**, *96*, 12424–12429. [CrossRef] [PubMed]

15. Ouyang, P.; Sun, M.; He, X.; Wang, K.; Yin, Z.; Fu, H.; Li, Y.; Geng, Y.; Shu, G.; He, C.; et al. Sclareol protects *Staphylococcus aureus*-induced lung cell injury via inhibiting alpha-hemolysin expression. *J. Microbiol. Biotechnol.* **2016**, *27*, 19–25. [CrossRef] [PubMed]

16. Ping, O.; Ruixue, Y.; Jiaqiang, D.; Kaiyu, W.; Jing, F.; Yi, G.; Xiaoli, H.; Defang, C.; Weimin, L.; Li, T.; et al. Subinhibitory concentrations of prim-o-glucosylcimifugin decrease the expression of alpha-hemolysin in *Staphylococcus aureus* (USA300). *Evid.-Based Complement. Altern. Med.* **2018**, *10*, 1–8. [CrossRef] [PubMed]

17. Xuewen, H.; Ping, O.; Zhongwei, Y.; Zhongqiong, Y.; Hualin, F.; Juchun, L.; Changliang, H.; Gang, S.; Zhixiang, Y.; Xu, S.; et al. Eriodictyol protects against *Staphylococcus aureus*-induced lung cell injury by inhibiting alpha-hemolysin expression. *World J. Microbiol. Biotechnol.* **2018**, *34*, 64. [CrossRef] [PubMed]

18. Ouyang, P.; Chen, J.; Sun, M.; Yin, Z.; Lin, J.; Fu, H.; Shu, G.; He, C.; Lv, C.; Deng, X.; et al. Imperatorin inhibits the expression of alpha-hemolysin in *Staphylococcus aureus* strain baa-1717 (USA300). *Anton. Leeuwenhoek* **2016**, *109*, 915–922. [CrossRef] [PubMed]

19. Ng, T.B.; Liu, F.; Wang, Z.T. Antioxidative activity of natural products from plants. *Life Sci.* **2000**, *66*, 709–723. [CrossRef]

20. Gong, Y.Q.; Fan, Y.; Wu, D.Z.; Yang, H.; Hu, Z.B.; Wang, Z.T. In vivo and in vitro evaluation of erianin, a novel anti-angiogenic agent. *Eur. J. Cancer* **2004**, *40*, 1554–1565. [CrossRef] [PubMed]

21. Su, C.; Zhang, P.; Liu, J.; Cao, Y. Erianin inhibits indoleamine 2, 3-dioxygenase -induced tumor angiogenesis. *Biomed. Pharmacother.* **2017**, *88*, 521–528. [CrossRef] [PubMed]

22. Yu, Z.; Zhang, T.; Gong, C.; Sheng, Y.; Lu, B.; Zhou, L.; Ji, L.; Wang, Z. Erianin inhibits high glucose-induced retinal angiogenesis via blocking erk1/2-regulated hif-1alpha-vegf/vegfr2 signaling pathway. *Sci. Rep.* **2016**, *6*, 34306. [CrossRef] [PubMed]

23. Gong, Y.; Fan, Y.; Liu, L.; Wu, D.; Chang, Z.; Wang, Z. Erianin induces a jnk/sapk-dependent metabolic inhibition in human umbilical vein endothelial cells. *In Vivo* **2004**, *18*, 223–228. [PubMed]

24. Sun, J.; Fu, X.; Wang, Y.; Liu, Y.; Zhang, Y.; Hao, T.; Hu, X. Erianin inhibits the proliferation of t47d cells by inhibiting cell cycles, inducing apoptosis and suppressing migration. *Am. J. Transl. Res.* **2016**, *8*, 3077–3086. [PubMed]

25. Wang, H.; Zhang, T.; Sun, W.; Wang, Z.; Zuo, D.; Zhou, Z.; Li, S.; Xu, J.; Yin, F.; Hua, Y.; et al. Erianin induces g2/m-phase arrest, apoptosis, and autophagy via the ros/jnk signaling pathway in human osteosarcoma cells in vitro and in vivo. *Cell Death Dis.* **2016**, *7*, e2247. [CrossRef] [PubMed]

26. Dekanski, J.B. Anti-prostatic activity of bifluranol, a fluorinated bibenzyl. *Br. J. pharmacol.* **1980**, *71*, 11–16. [CrossRef] [PubMed]

27. Wann, E.R.; Gurusiddappa, S.; Hook, M. The fibronectin-binding mscramm fnbpa of *Staphylococcus aureus* is a bifunctional protein that also binds to fibrinogen. *J. Biol. Chem.* **2000**, *275*, 13863–13871. [CrossRef] [PubMed]

28. Bingham, R.J.; Rudino-Pinera, E.; Meenan, N.A.; Schwarz-Linek, U.; Turkenburg, J.P.; Hook, M.; Garman, E.F.; Potts, J.R. Crystal structures of fibronectin-binding sites from *Staphylococcus aureus* fnbpa in complex with fibronectin domains. *Proc. Natl. Acad. Sci. USA* **2008**, *105*, 12254–12258. [CrossRef] [PubMed]

29. Hansenova Manaskova, S.; Nazmi, K.; van Belkum, A.; Bikker, F.J.; van Wamel, W.J.; Veerman, E.C. Synthetic lpetg-containing peptide incorporation in the *Staphylococcus aureus* cell-wall in a sortase A- and growth phase-dependent manner. *PLoS ONE* **2014**, *9*, e89260. [CrossRef] [PubMed]

30. Van Acker, H.; Van Dijck, P.; Coenye, T. Molecular mechanisms of antimicrobial tolerance and resistance in bacterial and fungal biofilms. *Trends Microbiol.* **2014**, *22*, 326–333. [CrossRef] [PubMed]

31. Flemming, H.C.; Wingender, J.; Szewzyk, U.; Steinberg, P.; Rice, S.A.; Kjelleberg, S. Biofilms: An emergent form of bacterial life. *Nat. Rev. Microbiol.* **2016**, *14*, 563–575. [CrossRef] [PubMed]

32. Tsompanidou, E.; Denham, E.L.; Sibbald, M.J.; Yang, X.M.; Seinen, J.; Friedrich, A.W.; Buist, G.; van Dijl, J.M. The sortase A substrates fnbpa, fnbpb, clfa and clfb antagonize colony spreading of *Staphylococcus aureus*. *PLoS ONE* **2012**, *7*, e44646. [CrossRef] [PubMed]

33. Rasko, D.A.; Sperandio, V. Anti-virulence strategies to combat bacteria-mediated disease. *Nat. Rev. Drug Discov.* **2010**, *9*, 117–128. [CrossRef] [PubMed]

34. Nandakumar, R.; Nandakumar, M.P.; Marten, M.R.; Ross, J.M. Proteome analysis of membrane and cell wall associated proteins from *Staphylococcus aureus*. *J. Proteome Res.* **2005**, *4*, 250–257. [CrossRef] [PubMed]

35. Cascioferro, S.; Raffa, D.; Maggio, B.; Raimondi, M.V.; Schillaci, D.; Daidone, G. Sortase A inhibitors: Recent advances and future perspectives. *J. Med. Chem.* **2015**, *58*, 9108–9123. [CrossRef] [PubMed]

36. Bentley, M.L.; Gaweska, H.; Kielec, J.M.; McCafferty, D.G. Engineering the substrate specificity of *Staphylococcus aureus* sortase A. The beta6/beta7 loop from srtb confers npqtn recognition to srtA. *J. Biol. Chem.* **2007**, *282*, 6571–6581. [CrossRef] [PubMed]

37. Bentley, M.L.; Lamb, E.C.; McCafferty, D.G. Mutagenesis studies of substrate recognition and catalysis in the sortase A transpeptidase from *Staphylococcus aureus*. *J. Biol. Chem.* **2008**, *283*, 14762–14771. [CrossRef] [PubMed]

38. Zong, Y.; Bice, T.W.; Ton-That, H.; Schneewind, O.; Narayana, S.V. Crystal structures of *Staphylococcus aureus* sortase A and its substrate complex. *J. Biol. Chem.* **2004**, *279*, 31383–31389. [CrossRef] [PubMed]
39. Marraffini, L.A.; Ton-That, H.; Zong, Y.; Narayana, S.V.; Schneewind, O. Anchoring of surface proteins to the cell wall of *Staphylococcus aureus*. A conserved arginine residue is required for efficient catalysis of sortase A. *J. Biol. Chem.* **2004**, *279*, 37763–37770. [CrossRef] [PubMed]
40. Ilangovan, U.; Ton-That, H.; Iwahara, J.; Schneewind, O.; Clubb, R.T. Structure of sortase, the transpeptidase that anchors proteins to the cell wall of *Staphylococcus aureus*. *Proc. Natl. Acad. Sci. USA* **2001**, *98*, 6056–6061. [CrossRef] [PubMed]
41. Tian, B.X.; Eriksson, L.A. Catalytic mechanism and roles of arg197 and thr183 in the *Staphylococcus aureus* sortase A enzyme. *J. Phys. Chem. B* **2011**, *115*, 13003–13011. [CrossRef] [PubMed]
42. Oh, K.B.; Oh, M.N.; Kim, J.G.; Shin, D.S.; Shin, J. Inhibition of sortase-mediated *Staphylococcus aureus* adhesion to fibronectin via fibronectin-binding protein by sortase inhibitors. *Appl. Microbiol. Biotechnol.* **2006**, *70*, 102–106. [CrossRef] [PubMed]
43. Oh, K.B.; Kim, S.H.; Lee, J.; Cho, W.J.; Lee, T.; Kim, S. Discovery of diarylacrylonitriles as a novel series of small molecule sortase A inhibitors. *J. Med. Chem.* **2004**, *47*, 2418–2421. [CrossRef] [PubMed]
44. Wang, L.; Bi, C.; Cai, H.; Liu, B.; Zhong, X.; Deng, X.; Wang, T.; Xiang, H.; Niu, X.; Wang, D. The therapeutic effect of chlorogenic acid against *Staphylococcus aureus* infection through sortase A inhibition. *Front. Microbiol.* **2015**, *6*, 1031. [CrossRef] [PubMed]
45. Bandyopadhyay, S.; Valder, C.R.; Huynh, H.G.; Ren, H.; Allison, W.S. The beta g156c substitution in the f1-atpase from the thermophilic bacillus ps3 affects catalytic site cooperativity by destabilizing the closed conformation of the catalytic site. *Biochemistry* **2002**, *41*, 14421–14429. [CrossRef] [PubMed]
46. Jurasekova, Z.; Marconi, G.; Sanchez-Cortes, S.; Torreggiani, A. Spectroscopic and molecular modeling studies on the binding of the flavonoid luteolin and human serum albumin. *Biopolymers* **2009**, *91*, 917–927. [CrossRef] [PubMed]
47. Trott, O.; Olson, A.J. Autodock vina: Improving the speed and accuracy of docking with a new scoring function, efficient optimization, and multithreading. *J. Comput. Chem.* **2010**, *31*, 455–461. [CrossRef] [PubMed]
48. Morris, G.M.; Huey, R.; Lindstrom, W.; Sanner, M.F.; Belew, R.K.; Goodsell, D.S.; Olson, A.J. Autodock4 and autodocktools4: Automated docking with selective receptor flexibility. *J. Comput. Chem.* **2009**, *30*, 2785–2791. [CrossRef] [PubMed]
49. Gotz, A.W.; Williamson, M.J.; Xu, D.; Poole, D.; Le Grand, S.; Walker, R.C. Routine microsecond molecular dynamics simulations with amber on gpus. 1. Generalized born. *J. Chem. Theory Comput.* **2012**, *8*, 1542–1555. [CrossRef] [PubMed]
50. Salomon-Ferrer, R.; Gotz, A.W.; Poole, D.; Le Grand, S.; Walker, R.C. Routine microsecond molecular dynamics simulations with amber on gpus. 2. Explicit solvent particle mesh ewald. *J. Chem. Theory Comput.* **2013**, *9*, 3878–3888. [CrossRef] [PubMed]
51. Sousa da Silva, A.W.; Vranken, W.F. Acpype-antechamber python parser interface. *BMC Res. Notes* **2012**, *5*, 367. [CrossRef] [PubMed]
52. Wang, J.; Zhou, X.; Liu, S.; Li, G.; Shi, L.; Dong, J.; Li, W.; Deng, X.; Niu, X. Morin hydrate attenuates *Staphylococcus aureus* virulence by inhibiting the self-assembly of alpha-hemolysin. *J. Appl. Microbiol.* **2015**, *118*, 753–763. [CrossRef] [PubMed]
53. Li, Z.; Jia, L.; Wang, J.; Wu, X.; Hao, H.; Xu, H.; Wu, Y.; Shi, G.; Lu, C.; Shen, Y. Design, synthesis and biological evaluation of 17-arylmethylamine-17-demethoxygeldanamycin derivatives as potent hsp90 inhibitors. *Eur. J. Med. Chem.* **2014**, *85*, 359–370. [CrossRef] [PubMed]
54. Wang, L.; Bi, C.; Wang, T.; Xiang, H.; Chen, F.; Hu, J.; Liu, B.; Cai, H.; Zhong, X.; Deng, X.; et al. A coagulase-negative and non-haemolytic strain of *Staphylococcus aureus* for investigating the roles of srtA in a murine model of bloodstream infection. *Pathog. Dis.* **2015**, *73*, ftv042. [CrossRef] [PubMed]

Review

An Eye on *Staphylococcus aureus* Toxins: Roles in Ocular Damage and Inflammation

Roger Astley [1], Frederick C. Miller [2], Md Huzzatul Mursalin [3], Phillip S. Coburn [1] and Michelle C. Callegan [1,3,4,*]

1   Department of Ophthalmology, University of Oklahoma Health Sciences Center,
     Oklahoma City, OK 73104, USA; roger-astley@ouhsc.edu (R.A.); phillip-coburn@ouhsc.edu (P.S.C.)
2   Department of Cell Biology and Department of Family and Preventive Medicine, University of Oklahoma
     Health Sciences Center, Oklahoma City, OK 73104, USA; frederick-miller@ouhsc.edu
3   Department of Microbiology and Immunology, University of Oklahoma Health Sciences Center,
     Oklahoma City, OK 73104, USA; MDHuzzatul-Mursalin@ouhsc.edu
4   Dean McGee Eye Institute, 608 Stanton L. Young Blvd., DMEI PA-418, Oklahoma City, OK 73104, USA
*   Correspondence: michelle-callegan@ouhsc.edu; Tel.: +1-405-271-3674; Fax: +1-405-271-8781

Received: 13 May 2019; Accepted: 15 June 2019; Published: 19 June 2019

Abstract: *Staphylococcus aureus* (*S. aureus*) is a common pathogen of the eye, capable of infecting external tissues such as the tear duct, conjunctiva, and the cornea, as well the inner and more delicate anterior and posterior chambers. *S. aureus* produces numerous toxins and enzymes capable of causing profound damage to tissues and organs, as well as modulating the immune response to these infections. Unfortunately, in the context of ocular infections, this can mean blindness for the patient. The role of α-toxin in corneal infection (keratitis) and infection of the interior of the eye (endophthalmitis) has been well established by comparing virulence in animal models and α-toxin-deficient isogenic mutants with their wild-type parental strains. The importance of other toxins, such as β-toxin, γ-toxin, and Panton–Valentine leukocidin (PVL), have been analyzed to a lesser degree and their roles in eye infections are less clear. Other toxins such as the phenol-soluble modulins have yet to be examined in any animal models for their contributions to virulence in eye infections. This review discusses the state of current knowledge of the roles of *S. aureus* toxins in eye infections and the controversies existing as a result of the use of different infection models. The strengths and limitations of these ocular infection models are discussed, as well as the need for physiological relevance in the study of staphylococcal toxins in these models.

Keywords: toxins; enzymes; *Staphylococcus aureus*; eye; infection; in vivo models

Key Contribution: This review highlights the current knowledge base regarding toxins of *Staphylococcus aureus* and their roles in potentially blinding ocular infections. Although specific toxins have been identified as contributing to ocular tissue damage and inflammation, the exact mechanisms of action of these toxins on ocular tissues and their roles in inciting damaging inflammation need clarification.

---

1. Introduction

Staphylococcal involvement in wound infections was established in the 1881 report by Ogston [1], who described the formation of new abscesses in guinea pigs and mice injected with pus taken from his patients. He called the clusters of bacteria he observed in the abscesses "staphyle" ("bunch of grapes" in Greek) because the arrangement of the bacterial cells resembled a cluster of grapes [2]. In 1884, Rosenbach [3] published studies on bacteria isolated from human wounds. He isolated and categorized two staphylococci by color: golden colonies were named *Staphylococcus aureus (S. aureus)*, and white colonies named *Staphylococcus albus* (now called *epidermidis*).

In 1928, *S. aureus* became infamous after the organism was identified as the cause of 18 children becoming violently ill after receiving diphtheria vaccinations in Bundaberg, Australia. Hours after inoculation, the children began to vomit, developed high fever, experienced convulsions, and fell into unconsciousness. Twelve of the children died within 36 h of vaccination. All of the surviving children developed severe abscesses at the injection site [4]. *S. aureus* was isolated from the vaccine vials [5]. The Commonwealth of Australia commissioned Burnet to study why the contamination had been so deadly [5]. Burnet demonstrated that supernatants from the *S. aureus* culture isolated from the vaccine caused hemolysis in vitro, caused skin-necrosis after intradermal inoculation, and was rapidly lethal to rabbits after intravenous injection. He concluded that a single heat-labile factor was responsible for these diverse pathological features [6]. Burnet further reported that other *S. aureus* isolates had the same pathological properties as the one isolated from the Bundaberg vaccine [7].

Burnet's contention that the diverse disease features were due to a single toxin was disputed by various other researchers [8–10]. Glenny and Stevens [11] demonstrated that the hemolysis in vitro, skin-necrosis at the site of inoculation, and rapid lethality after intravenous injection were due to one immunologically distinct toxin, which they called "α-toxin". Whether one toxin or more were responsible for these pathological features was not resolved until researchers were able to purify α-toxin [12,13] and demonstrate that α-toxin alone caused these pathological effects. Once purified toxin was available, researchers were able to study pore formation, identify α-toxin binding receptors, and examine how this single, potent toxin is able to produce such a diverse repertoire of actions.

S. aureus is a Gram-positive, nonmotile coccoid bacteria. *S. aureus* can be distinguished from other staphylococci by its classic gold pigmentation, β-hemolysis, a positive coagulase reaction, fermentation of mannitol, and a positive deoxyribonuclease test [14] (Figure 1). Although there are over 50 species of staphylococci, *S. aureus* is by far the most significant pathogen in this group [14]. *S. aureus* has an extensive array of virulence factors in its arsenal, which are depicted in Figure 2A. These virulence factors include proteins and enzymes associated with the bacterial surface which promote adherence to tissue and assist in biofilm formation, the cell wall itself which interacts with innate immune receptors to initiate acute inflammation, and a myriad of toxins and enzymes which promote bacterial spread by damaging tissue and lysing inflammatory cells.

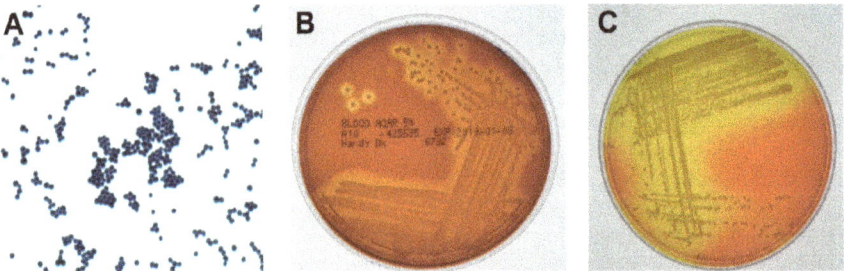

Figure 1. Phenotypes of *Staphylococcus aureus* (*S. aureus*). (**A**) Gram stain of *S. aureus*. Magnification, 100×. (**B**) *S. aureus* grown on 5% sheep blood agar overnight at 37 °C. Note the characteristic zones of β-hemolysis surrounding each colony. (**C**) *S. aureus* grown on mannitol salt agar overnight at 37 °C. *S. aureus*, unlike other staphylococci, ferments mannitol, resulting in acid production and the classic yellow halo within the deep pink agar.

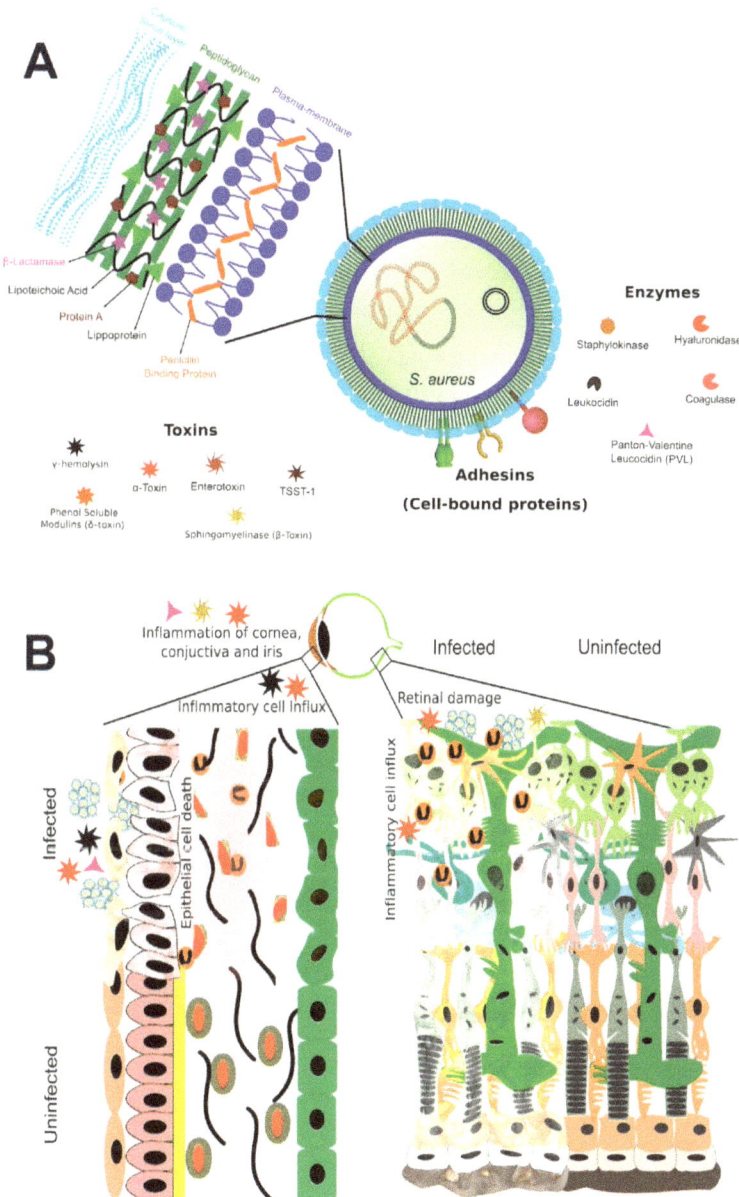

Figure 2. *S. aureus* virulence factors contribute to the pathogenesis of ocular infections. (**A**) *S. aureus* contains numerous virulence factors in its armamentarium which assists in antibiotic resistance, adherence to tissue, interactions with innate immunity, and spread of infection. (**B**) *S. aureus* toxins which have been analyzed in experimental infection models include α-toxin, β-toxin, γ-toxin, and Panton–Valentine leukocidin (PVL). The contribution of these toxins to ocular infections is strain-and model-dependent, as discussed below.

Although this collection of virulence factors renders *S. aureus* a formidable pathogen, this organism is part of the normal human flora. *S. aureus* nasal carriage among the general adult population varies depending on the geographical location [15]. In the Netherlands, the *S. aureus* carriage rate was reported to be 24–25.2% [16,17]. In the United States from 2001–2004, 30.4% of the population was reported to carry *S. aureus* [18]. Among healthy Japanese, 35.7% of individuals were carriers [19]. Saxena et al. [20] reported the carriage rate in India to be 29.4%. In Kenya, 18.3% of health care workers were *S. aureus* carriers [21]. In Mexico, 37.1% of healthy adults carried *S. aureus* [22]. Taken collectively, a quarter to one-third of individuals are carriers for *S. aureus* [23]. The most common site of colonization is the nasal mucosa, but *S. aureus* can also colonize the throat and the perineum [15,16,24,25].

Antibiotic resistance in *S. aureus* has emerged in waves, as *S. aureus* has counteracted the introduction of antibiotics as the primary treatment for bacterial diseases [26]. The first wave began after two patients were treated with penicillin in 1941 [27] and 1942 [27,28]. The first report of antibiotic resistance to penicillin was in 1944 [29], and by 1948, 59% of *S. aureus* cultures were reported to be resistant to penicillin [30], which is now know to be due to a plasmid-encoded penicillinase. *S. aureus* in this group were most commonly of phage type 80/81 [31]. By the 1950s, this phage group had become pandemic [26]. Phage type 80/81 seemingly disappeared after the introduction of methicillin as the preferred antibiotic for the treatment of staphylococcal infections [26,32].

The second wave of antibiotic resistance emerged with the introduction of methicillin in 1959 [33]. The first isolations of methicillin resistant *S. aureus* (MRSA) were reported in 1961 [34]. These *S. aureus* clones carried the *mecA* gene, which encodes the low-affinity penicillin binding protein PBP2a that imparts broad anti-beta-lactam resistance [35]. These isolates were mainly associated with hospitals and other health care establishments and are referred to as Healthcare-Associated MRSA (HA-MRSA). A third wave of antibiotic resistance occurred with the development of methicillin-resistant strains of *S. aureus* outside of health care establishments, called Community-Associated MRSA (CA-MRSA). These linages are unrelated to HA-MRSA and arose from numerous genetically distinct lineages across the world. Some have spread internationally while others have remained restricted to certain geographic regions [36–38]. In the United States, USA300 is the most prevalent strain of CA-MRSA, but each continent appears to have its own predominant type [26].

2. *S. aureus* and Eye Infections

S. aureus is a leading cause of eye infections such as dacryocystitis, conjunctivitis, keratitis, cellulitis, corneal ulcers, blebitis, and endophthalmitis [39–41]. Contact lens wearers, especially those who wear extended-wear contact lenses or fail at proper lens hygiene, are at greater risk for the development of keratitis. *S. aureus* is one of the leading causes of bacterial keratitis, along with *Pseudomonas aeruginosa*. The number of eye infections appears to be rising due to increasing use of ocular surgical procedures, especially in elderly and diabetic populations. In the developed world, the rate of intraocular surgeries, for example for cataract removal and lens replacement, has been steadily increasing [42]. The leading infectious complication from cataract surgery is endophthalmitis, with an incidence of 0.01–0.3% [43,44]. Intravitreal injections are also being performed at greater frequencies for the treatment of neovascular eye disease, neurodegenerative diseases, and intraocular inflammation [45]. It was estimated that in 2014, 18 million intraocular injections of anti-vascular endothelial growth factor (VEGF) agents to treat age-related macular degeneration were performed [46]. The frequencies of injection-related complications has also increased [47], with endophthalmitis incidences after intravitreal injection reported to be 0.006–0.16% per injection and 0.7–1.3% over the course of treatment [48–51]. While these infection rates are statistically very low compared to diseases such as diabetes or cancer, because of the numbers of procedures, they number in the tens of thousands. In addition, eye infections can cause profound vision loss and disability, costing millions of healthcare dollars and reductions in quality of life for those affected.

S. aureus expresses an amalgam of toxins, enzymes, and secreted proteins whose virulence in humans and animal models has been well documented [13,52–55]. Studies on the contribution of these virulence factors to eye infections have focused on a few well-characterized toxins, such as α-toxin, β-toxin, γ-toxin, and PVL (Figure 2B). However, *S. aureus* has many more toxic factors in its armamentarium, some whose roles in eye infections have yet to be defined. Most studies of the activities of *S. aureus* toxins have focused on cell killing, but many recent studies have indicated that sublytic doses of some *S. aureus* toxins can have dramatic effects on target cells [56–59]. These reports have indicated that by altering the permeability of cell membranes, toxins can kill or manipulate the functions of the immune cells, and disrupt epithelial barriers to promote bacterial growth and spreading. By interacting with various cell surface proteins, toxins are able to target diverse cell types—mostly cells of the immune system which only express those receptors [56–60].

S. aureus elaborates many virulence factors which, depending upon the infection site in the eye, can manifest as serious but treatable infections such as conjunctivitis or dacryocystitis, or sight-threating infections such as corneal ulcers, endophthalmitis, or orbital cellulitis [39–41]. The eye is an especially vulnerable organ whose integrity, structure, and function are essential for proper sight. For the most part, our eyes resist the negative impact of infecting organisms through the concerted actions of blinking, tear film, and antimicrobial peptides and other enzymes which protect the ocular surface [55,61–63]. When *S. aureus* comes in contact with tissues of the eye and is able to circumvent these protections, infection results (Figure 2B). This review will discuss *S. aureus* and its toxins in the context of ocular infections and examine studies conducted in animal models to determine the roles and mechanisms of *S. aureus* toxins in eye infections. The strengths and limitations of these models and future steps for study will be discussed.

3. Membrane-Damaging Toxins

α-toxin is also known as α-hemolysin because it was originally named for its ability to cause β-hemolysis of red blood cells [11]. Generations of microbiology students are therefore condemned to having to remember that α-hemolysin/α-toxin causes β-hemolysis. α-toxin is a beta-barrel pore-forming toxin expressed by 95–100% of *S. aureus* isolates from various sites [53,64] and, in our small cohort of strains, 94% of ocular *S. aureus* isolates (Table 1). α-toxin has been widely studied for its role in the pathogenesis of central nervous system infections [65], endocarditis [66], endophthalmitis [52], keratitis [55,67,68], mastitis [69], pneumonia [70,71], sepsis [72,73], and skin and soft tissue infections [74,75].

Table 1. Evaluation of toxin genes in a random sampling of *Staphylococcus aureus* isolated from ocular infections at the Dean A. McGee Eye Institute (USA) from April 2011 to February 2018.

Gene	MSSA (7)	MRSA (9)	All (16)
hla	6 (86%)	9 (100%)	15 (94%)
hlb	4 (57%)	5 (56%)	8 (50%)
mecA	0	9	9

Numbers in parentheses in the top row indicate total number of isolates. Percentages in the table indicate percent positive of the column total by PCR, using primers described previously [64]. PCR for *mecA* was conducted to confirm clinical laboratory results. MSSA, methicillin-sensitive *Staphylococcus aureus*; MRSA, methicillin-resistant *Staphylococcus aureus*.

Monomers of α-toxin combine in the eukaryotic cell membrane to form a seven member (heptameric) pre-pore ring-like structure [76]. The mature pore functions as an ion channel, allowing passage of Ca^{2+}, K^+, ATP and other 1–4 kDa-sized molecules into and out of the cell [77,78]. α-toxin targets cells by binding to the receptor a disintegrin and metalloproteinase 10 (ADAM10), which is required for α-toxin-mediated cytotoxicity [73,76]. Binding of α-toxin to ADAM10 results in assembly of an α-toxin-ADAM10 complex in cholesterol/sphingolipid-rich caveolar rafts [76]. At high concentrations of α-toxin, cell death predominates. However, at sublytic concentrations, α-toxin binding of ADAM10 induces the activation of ADAM10 metalloprotease activity which cleaves E-cadherin adherens junctions and results in disruption of focal adhesions and disruption of tissue barriers [76]. Also, the α-toxin/ADAM10 complex stimulates the dephosphorylation of FAD, p130cas, paxillin and src, leading to the tissue disruption that is characteristic of *S. aureus* infections [5] and enhanced bacterial dissemination [60].

A characteristic feature of *S. aureus* infections of the cornea are epithelial defects (Figure 3). In an experimental rabbit model of *S. aureus* keratitis [67], as organisms replicate and form microcolonies, the corneal epithelial cells die in a zone around the colonies, exposing the underlying stroma, and forming a painful ulcer that is similar to that seen in human cases (Figure 3). The role of α-toxin as a major toxin responsible for corneal epithelial ulceration in keratitis was first demonstrated in an experimental rabbit model in which isogenic mutants of the *S. aureus* strain 8325-4 deficient in α-toxin were injected directly into the corneal stroma. At 15 hours post infection, eyes injected with α-toxin-deficient *S. aureus* had lower slit lamp scores, no epithelial erosions, and less inflammation as indicated by reduced myeloperoxidase activity, compared to the wild-type *S. aureus* parent strain [67]. Further work using this experimental model and an α-toxin-deficient isogenic mutant confirmed that the wild-type strain produced a more severe pathology than the α-toxin mutant [68]. Similar results were observed in the same rabbit model when infections were initiated by wild-type strain Newman and its α-toxin-deficient mutant [79], and in a comparison of *S. aureus* 8325-4 and its α-toxin-deficient mutant in young (6–7 weeks old) and aged mice (36–48 weeks old) using the corneal scarification method [80]. In the latter study, wild-type *S. aureus* produced a more severe pathology than the α-toxin deficient strain, and the pathology was more severe in the aged mice. Callegan et al. [52] used an intravitreal injection model of endophthalmitis in rabbits to demonstrate that infection with an α-toxin-deficient mutant of *S. aureus* 8325-4 resulted in significantly reduced retinal damage compared to the wild-type strain. Recently, Putra et al. [81] reported their comparison of *S. aureus* strain JE2 and its α-toxin-deficient mutant in an experimental mouse model of keratitis initiated after total corneal epithelial debridement. Corneal wound healing was significantly improved following infection with the α-toxin-deficient mutant strain. This mutant was also inefficient in infecting debrided mouse corneas, further confirming the important role of α-toxin in *S. aureus* keratitis. Although it is clear that an absence of α-toxin results in less pathology during *S. aureus* keratitis, it is not known whether α-toxin is acting toward cells or triggering the immune response by similar mechanisms in different species. Different mouse strains have different responses following experimental exposure to *S. aureus*, but whether α-toxin contributes to these differences has not been investigated.

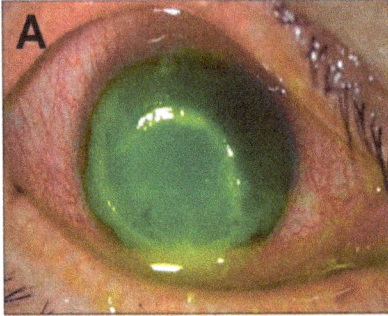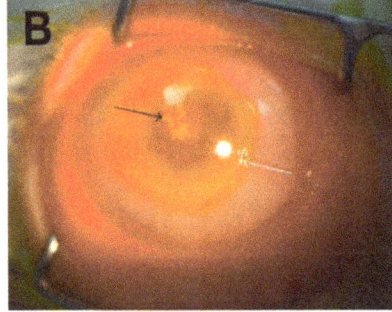

Figure 3. *S. aureus*-induced corneal ulcers in humans and rabbits. (**A**) A corneal ulcer with superficial stromal infiltration and anterior chamber hypopyon in a human eye infected with MRSA. Copyright © Lee et al., 2010 [82]. (**B**) *S. aureus* keratitis in a rabbit eye. Black arrow indicates inflammatory cell infiltration and staphylococcal microcolonies formed in the corneal stroma. White arrow indicates the edge of the epithelial erosion stained with fluorescein. Copyright © Marquart et al., 2011 [83]. These are open access articles distributed under the Creative Commons Attribution License, which permits unrestricted use, distribution, and reproduction in any medium, provided the original work is properly cited.

Topical application of purified α-toxin to the eyes of rabbits caused significant and dose-dependent inflammation of the conjunctiva and iris at six hours post application [68]. Injection of purified α-toxin into rabbit corneas also caused significant and dose-dependent inflammation of the cornea and iris [68]. These results were confirmed in the mouse keratitis model after corneal scarification. Again, corneal pathology was more severe in aged mice (36–38 weeks old) than in young mice (6–7 weeks old) [80]. However, when purified α-toxin was injected into the corneas of young (6–8 weeks old) and aged rabbits (about 30 months old), there was significantly greater disease in young rabbits than in aged ones [84]. Histological studies of rabbit eyes four hours after α-toxin injection showed epithelial cell death by necrosis and apoptosis, sloughing of viable cornea epithelial cells, severe corneal edema, and inflammatory cell migration from the tear film and from the limbal vessels into the cornea [85]. Intravitreal injection of purified α-toxin in a mouse model of endophthalmitis produced an inflammatory response, particularly via IL-1β, but only mild retinal damage and edema, and produced no significant decline in retinal function, as measured by a-or b-wave amplitudes via electroretinography [86].

The expression of α-toxin is tightly regulated. Control of toxin expression in vitro is influenced by the interplay of at least three global regulatory loci: the quorum-sensing systems *agr* (accessory gene regulator), *sarA* (staphylococcal accessory gene regulator), and *sae* (staphylococcal accessory protein effector). Loss of any one of these loci negatively impacts toxin production in vitro [87,88]. During in vivo growth, regulation appears to be more complicated. Booth et al. [89] reported that infection with *S. aureus* strain RN6390 with a defective *agr* resulted in a significant reduction in retinal pathology in a rabbit model of endophthalmitis. Infection with a double mutant of *agr* and *sar* resulted in an even greater reduction of pathology, but a deficiency in *sar* alone resulted in ocular pathology similar to that of the wild-type *S. aureus*. Using an implanted device model in guinea pigs, Goerke et al. [87] reported that mutation of *agr* or *sarA* in strains RN6390 and Newman did not impact the level of transcription of the gene for α-toxin, *hla*. However, loss of *sae* caused an almost total downregulation of α-toxin production in both strains. Xiong et al. [88] examined the α-toxin production of two *S. aureus* strains: RN6390, which lacks a functional *sigB* regulon, and SH1000, which has a repaired *sigB* regulon. *SigB* negatively influences the production of α-toxin in vitro [90]. Xiong et al. [88] reported that in vitro isogenic mutants in RN6390 of *agr*, *sarA*, and *agr/sarA* had significant reductions in α-toxin production. Mutation of *sae* completely abolished α-toxin production in both strains during in vitro growth. SH1000 produced less α-toxin overall than RN6390 but the results of the mutations were similar, except that mutating *sarA* did not reduce α-toxin production. During in vivo growth with RN6390, using an infective endocarditis model in rabbits, only mutation of *sae* resulted in a significant reduction in α-toxin production, compared to the wild-type strain. With SH1000 there was no significant reduction in toxin production with any of the mutant strains. Even with the *sae* mutation, both strains produced measurable amounts of αtoxin during in vivo growth. These data indicate that during in vivo growth other factors, in addition to *agr*, *sarA* and *sae*, are involved in α-toxin regulation. O'Callaghan et al. [68] injected *agr*-deficient mutants of strain 8325-4 into the corneas of rabbits and reported that the *agr* mutant was even less virulent than the α-toxin mutant. Girgis et al. [80] also reported that the *agr* mutant was less virulent than the α-toxin mutant in the mouse keratitis scarification model. Both authors surmised that these outcomes were likely due to deficiencies in α-toxin and all the other toxins under its control in the *agr*-deficient mutants.

The aforementioned studies confirm an important role for α-toxin in the pathogenesis of *S. aureus* keratitis and endophthalmitis. A rational strategy for improving the therapeutic outcome of keratitis would be to block the effect of α-toxin on the cornea. As nearly all ocular *S. aureus* isolates have been reported to possess the *hla* gene (Table 1 and [64]), this virulence factor is a viable target. Passive and active immunization of rabbits against α-toxin has been shown to be effective at reducing corneal disease in the rabbit model of keratitis [79]. Active immunization against α-toxin prevented the formation of corneal epithelial erosions [79]. Whether immunization strategies would be effective in limiting the effects of α-toxin in endophthalmitis or other ocular *S. aureus* infections remains an open question. Recently, nanoparticles composed of erythrocyte membranes surrounding a biologically inert poly-lactic-co-glycolic core have been developed to function as a type of decoy to neutralize pore-forming toxins [91]. These nanoparticles, termed nanosponges, neutralized α-toxin and reduced hemolytic activity in vitro, protected mice from developing staphylococcal α-toxin induced skin lesions, and decreased mortality after systemic injection of a lethal dose of α-toxin [91]. In a mouse model of *S. aureus* endophthalmitis, Coburn et al. [92] reported that intraocular injection of nanosponges alone following infection did not result in improved retinal function compared to untreated mice. However, injection of nanosponges in conjunction with the fourth-generation fluoroquinolone gatifloxacin resulted in decreased inflammation, less damage to the retinal architecture, and significantly improved retinal function compared to untreated or gatifloxacin only treated mice following intraocular infection with an MRSA ocular isolate [92]. These studies suggested that nanosponges might be a viable adjunctive therapy for intraocular infections caused by *S. aureus*. Together, these therapeutics studies imply that blocking α-toxin would improve the therapeutic outcome of ocular *S. aureus* infections.

4. Bi-Component Toxins

In *S. aureus*, bi-component leukotoxins are composed of a pair of proteins designated S (slow) and F (fast) based on their elution speeds [93]. Initial binding of the leukotoxin receptor is mediated by the S-component, followed secondarily by binding of the F-component, resulting in formation of a lytic pore-forming octamer (β-barrel) of alternating subunits that insert into the cell membrane, resulting in osmotic imbalance and cell lysis [56,94]. Human *S. aureus* isolates can encode up to five leukocidin toxins: Panton–Valentine leukocidin (PVL), γ-hemolysin AB (HlgAB), γ-hemolysin CB (HlgCB), LukED, and LukAB (also referred to as LukGH) [94]. Elucidation of the role of these toxins in human disease has been hampered by the species and cell specificities of some of the toxins [95].

Panton–Valentine leukocidin (PVL) is a prophage-encoded toxin present in about 5% of *S. aureus* isolates and most CA-MRSA strains [53]. PVL binds to complement component C5a anaphylatoxin chemotactic receptors to target neutrophils, monocytes, macrophages, natural killer cells, dendritic cells, and T lymphocytes [60,94]. C5aR1 is highly expressed on phagocytic cells, but C5aR2 has low expression on neutrophils [94–96]. The cytotoxicity of PVL can be blocked with C5aR1 antagonists [94,96]. PVL is active against human and rabbit neutrophils, but macaque, cow, and mouse neutrophils are resistant to its effects [93]. PVL stimulates inflammasome activation in monocytes and primary macrophages at sublethal doses [58]. *S. aureus* β-toxin, δ-toxin, γ-toxin, LukDE, and PSMα3 synergize with PVL to amplify IL-1β release to trigger inflammation [58]. Based on species specificity, the most appropriate animal models for the study of PVL are in rabbits and humanized mice with engrafted primary human haematopoietic cells [95].

Not deterred by an acknowledged lack of lytic activity by PVL against mouse neutrophils, Zaidi et al. [97] tested PVL-deficient isogenic mutants of *S. aureus* USA400 and USA300 (currently the predominant CA-MRSA strain in the United States) in an experimental scratch model of mouse keratitis. Inactivation of PVL in the LAC strain of USA300 resulted in reduced corneal opacity and reduced corneal bacterial counts compared to the wild-type strain. However, results with USA400 strains depended on the strain tested. Infection with the isogenic PVL-deficient mutant of MW2 and its wild-type parent strain resulted in similar degrees of corneal opacity and corneal bacterial counts. Infection with the isogenic PVL-deficient mutant of NRS 193 had reduced corneal opacity and reduced corneal bacterial counts compared to its wild-type parent. Infection with the isogenic PVL-deficient mutant of NRS 194 had reduced corneal opacity, but did not have reduced corneal bacterial counts, as compared to its wild-type parent strain. In this study, topical applications of neutralizing polyclonal antibodies against PVL resulted in significantly reduced corneal opacity and bacterial numbers in corneas infected with USA300 strains LAC and SF8300, but in corneas infected with USA400 strains MW2 and NRS 193, the bacterial loads were unaffected and there was no reduction in corneal opacity. Corneas infected with strain NRS 194 showed a reduction in corneal opacity, but bacterial counts were unaffected following the application of anti-PVL antibodies [97].

Siqueira et al. [98] demonstrated that intravitreal injections of six different combinations of the PVL and γ-toxin subunits (S and F) from *S. aureus* ATCC strain 49,775 caused acute inflammatory reactions involving the posterior and anterior chambers as well as the conjunctiva and eyelids. The clinical and histological signs of inflammation began by 4 h after injection and persisted for the 5 days of the experiment. The toxicity was dose-dependent. Certain combinations were more toxic than others. The most virulent combination was the PV-S subunit plus γ-toxin-F subunit, although it is not known if this combination occurs naturally. Liu et al. [99] injected PVL into rabbit eyes and identified retinal ganglion cells as the primary targets of the toxin, and these cells were the only ones identified to express C5a receptors. Binding of PVL triggered increased IL-6 expression in the retina and apoptosis of microglial cells. Liu et al. [100] demonstrated similar results using a rabbit retinal explant model. However, in addition to PVL binding to retinal ganglion cells, the toxin also co-localized to horizontal cells which did not express C5a receptor, and there was no significant increase in IL-6 as compared to the control explants. Peterson et al. [101] recently reported that the PVL gene was detected in only 14.7% of *S. aureus* keratitis isolates in their study, so it is unclear how clinically relevant this toxin is to ocular disease.

γ-toxin consists of two bicomponent pore-forming toxins: HlgAB and HlgCB, which are encoded in the core genome and are reported to be present in 99.8% of clinical isolates [53,102]. During in vitro culture, γ-toxin is highly upregulated in the presence of blood [103] and upon phagocytosis by neutrophils [104]. HlgAB targets the chemokine receptors CXCR1, CXCR2, and CCR2 [105], while HlgCB targets the compliment component C5a receptors C5aR1 and C5aR2 [97]. These receptors are highly expressed on phagocytic cells [105]. HlgCB is active against neutrophils of human, rabbit, macaque, and cow [105]. Mouse neutrophils are resistant to HlgCB. In mice only CCR2 is able to act as a receptor for HlgAB [105].

Supersac et al. [106] published a study on the role of γ-toxin using a rabbit model of endophthalmitis infected with the *S. aureus* strain Newman, which the authors believed did not produce α-toxin or β-toxin [107]. After intravitreal injection of either a γ-toxin mutant of Newman, the wild-type, or a γ-toxin complemented strain, all three strains produced a strong inflammatory response, with the only difference being that the γ-toxin mutant did not cause inflammation of the eyelid [106]. More recent studies have shown that while strain Newman has a truncated copy of β-toxin, it does have the α-toxin gene [108], is a weak producer of α-toxin in vitro [79,87], and is a strong producer of α-toxin in vivo [87]. Using the rabbit model of endophthalmitis, Callegan et al. [52] showed with an isogenic mutant of γ-toxin in strain 8325-4 that loss of this gene had no positive effect on retinal function. Dajcs et al. [79], using a rabbit model of keratitis and a γ-toxin mutant of the Newman strain, reported significantly lower slit lamp scores and reduced numbers of PMN in the corneal stroma as compared to the wild-type strain or the genetically rescued strain. Outcomes here appear to be model- and strain-dependent.

Phenol-soluble modulins (PSMs), of which δ-toxin is a member, are a family of small (20–44 amino acids) amphipathic secreted peptides which are only produced by the members of the genus *Staphylococcus* [109]. All *S. aureus* have the *hld* gene and two loci with genes coding for *psm* α and *psmβ* [110]. However, the widely used *S. aureus* strain 8325-4 is unique for not having the gene for *psmα* [111]. PSMs have multiple roles in *S. aureus* pathogenesis, such as facilitating biofilm dissemination, cytolytic activity, and proinflammatory activity [110,112]. PSMα induces the release of TLR2-activating lipoproteins from bacterial cells [113]. While micromolar concentrations of PSMs cause cytolysis, nanomolar concentrations cause inflammation [114]. PSMs are detected by the formyl peptide receptor 2 (FPR2) [114,115] on mammalian cells, resulting in the attraction and activation of neutrophils and inflammation [110]. The mechanism of cytolysis is not clear, but PSMs appear to have a detergent-like action on the cell membrane [112]. Cytolysis is likely nonspecific and receptor independent [116]. Many species of staphylococci express PSM, but *S. aureus* expresses highly cytotoxic forms, such as PSMα1–α4 [116]. Although PSMs are among the most abundant proteins secreted in staphylococcal culture filtrates [109], we are unaware of any studies using animal models to examine their role in eye infections.

5. Enzymes

β-toxin is a neutral sphingomyelinase, not a hemolytic toxin per se. β-toxin is able to hydrolyze the plasma membrane lipid sphingomyelin [117] and is responsible for α-hemolysis on blood agar plates. Reports on the frequency of the *hlb* gene in *S. aureus* clinical isolates vary. Although the *hlb* gene was reported to be present in 39–57% of clinical isolates [64,118] and, in our small cohort, 50% of ocular isolates (Table 1), van Wamel et al. [119] reported that in 90% of the isolates examined, the β-toxin gene had been inactivated by phage φSa3 insertion. However, Salgado-Pabon et al. [120] reported that during in vivo and in vitro growth, phage φSa3 can excise from a subpopulation of *S. aureus* and translocate to atypical sites in the chromosome, restoring β-toxin production. β-toxin is toxic to human monocytes, but is inactive at equal concentrations against human erythrocytes, fibroblasts, granulocytes, and lymphocytes [121].

β-toxin has not been shown epidemiologically to be associated with a specific disease or infection severity criteria [112]. However, O'Callaghan et al. [68] demonstrated in a rabbit model of keratitis that topical application of β-toxin caused significant conjunctival inflammation, and corneal injection of purified β-toxin resulted in a rapid edematous reaction in the sclera. Intrastromal injection of an isogenic β-toxin-deficient mutant of *S. aureus* strain 8325-4 caused less scleral edema than its wild-type parental strain, but slit lamp scores, epithelial erosions, and intrastromal ulcers were similar to that of keratitis cause by the wild-type strain [68]. Similar results were reported in an experimental rabbit model of endophthalmitis. *S. aureus* 8325-4 and the same β-toxin-deficient mutant were compared for their virulence following intravitreal injection, and the β-toxin-deficient mutant resulted in significantly reduced retinal damage compared to the wild-type strain [52]. This reduction in damage, however, was not as significant as the reduction of damage in eyes infected with the α-toxin-deficient mutant of the same wild-type parent strain. Together, these studies suggest that β-toxin may contribute to ocular changes in the cornea and within the eye, but its expression in the eye is not essential for complete virulence.

6. Discussion

The role of α-toxin in eye infections has been well established by studies comparing α-toxin-deficient isogenic mutants with their wild-type parental strains in models of keratitis and endophthalmitis in both rabbits and mice (Table 2). Ocular administration of purified α-toxin has also been tested. α-toxin's role as a major virulence factor in ocular *S. aureus* infections is clear cut. However, the studies of the role of the bi-component toxins in ocular infections have been less conclusive. Studies examining the role of PVL were conducted in the mouse [97], the cells of which have been reported to be resistant to PVL [105]. The studies by Supersac et al. [106] and Dajcs et al. [79] both used a γ-toxin-deficient isogenic mutant of the Newman strain in a rabbit model. While Supersac et al. [106] reported a strong inflammatory response following intravitreal injection of wild-type or γ-toxin-deficient Newman, Dajcs et al. [79] reported significantly less inflammation following instrastromal injection of the same γ-toxin-deficient Newman mutant. These two studies suggest a model-dependent role for the virulence of γ-toxin. Studies of the importance of the enzyme β-toxin in corneal and intraocular infection models have reported less virulence in β-toxin-deficient mutants [67]; however, the extent to which ocular virulence is muted in β-toxin-deficient or γ-toxin-deficient ocular infections is not nearly as significant as that of ocular infections caused by α-toxin-deficient *S. aureus*.

Except for α-toxin, our understanding of the role of *S. aureus* toxins in eye infections is limited. Limitations that should be accounted for in such studies are the species-specific sensitivities to several of the bi-component toxins, and the variability in toxin types and the amounts produced by the various isolates commonly used to study *S. aureus* infections. Many of the studies reviewed in this paper used *S. aureus* isolate 8325-4, which has a published genome, numerous mutants, and has been tested in many non-ocular infection models. *S. aureus* 8325-4 was isolated in 1960, before the development of CA-MRSA and HA-MRSA, has been phage cured, is known to lack phenol soluble modulin α3, and has shown significant variation in α-toxin production depending on which lab isolate is used [111]. The use of clinically relevant ocular isolates could provide insight into which genes are important in eye infections and give insight into the interplay of the various toxins produced by the pathogens. However, isolates should first be characterized and compared for relevance to the disease model. Most of the data reviewed in this paper was generated using animal models of infection (Table 2). Although obvious, the many physiological similarities and differences must be accounted for when interpreting and translating data to humans [122–124]. All animal infection models have limitations, but most can provide useful information. Animal models not only provide us with the control necessary to initiate various aspects of the infection process, but also facilitate study of the host response to infection, an essential player absent in cell or tissue culture models [125]. Understanding the limitations of these models is important in evaluating the usefulness of the data it produces. For example, rabbit keratitis models using intrastromal injections of bacteria are technically simple, highly controlled, and the output is highly reproducible [55]. However, this model does not replicate the normal manner in which humans develop bacterial keratitis and it bypasses important steps in the infection process. Mouse models of bacterial keratitis suffer from the limitation that only certain mouse strains reliably develop infections [80]. Some mouse strains clear the infections without treatment, unlike infections in humans. The debridement of the mouse corneal epithelium to prepare it for application of bacteria [81] is also not the most accurate model of how the human cornea is damaged prior to infection, unless there is a large wound from chemical or mechanical trauma. To date, the scratch model, similar to that which is used in studies of *Pseudomonas* keratitis [126], is as close an approximation of what may happen to a human cornea prior to *S. aureus* infection that has been attempted. Mouse and rabbit models of bacterial endophthalmitis initiated by intravitreal injections are an approximation of traumatic endophthalmitis, but they lack the traumatic aspect, which may include significant tissue damage and the presence of blood in the eye [125]. Injection or topical application of purified toxin (Table 3) is a strategy that may approximate the toxin concentrations needed for tissue damage and may be helpful in studying the mechanisms by which toxins affect these tissues, but these studies may lack physiological relevance because the amounts and timing of toxin production have not been quantified, bacteria do not produce a single bolus of toxin in infected tissue, and different strains produce different complements of toxin during infection.

Table 2. Studies analyzing the role of *S. aureus* toxins in ocular infections using staphylococcal strains specifically deficient in the toxin of interest.

Toxin Absent	*S. Aureus* Strain	Ocular Infection Model	Result	References
α-toxin	8325-4	Keratitis rabbit	↓ Slit lamp scores ↓ Myeloperoxidase activity ↓ Corneal erosions	[67]
α-toxin	8325-4	Keratitis rabbit	↓ Slit lamp scores ↓ Inflammation	[68]
β-toxin	8325-4	Keratitis rabbit	↓ Scleral edema	[68]
γ-toxin	Newman	Endophthalmitis rabbit	↓ Lid inflammation	[106]
α-toxin	Newman	Keratitis rabbit	↓ Slit lamp scores ↓ Inflammation	[81]
γ-toxin	Newman	Keratitis rabbit	↓ Slit lamp scores ↓ Corneal PMN	[81]
α-toxin	8325-4	Endophthalmitis rabbit	↓ Retinal damage	[52]
β-toxin	8325-4	Endophthalmitis rabbit	↓ Retinal damage	[52]
γ-toxin	8325-4	Endophthalmitis rabbit	No change	[52]
α-toxin	8325-4	Keratitis mouse	More severe in aged mice	[80]
PVL	Various USA 300 and 400 strains	Keratitis mouse	Enhanced virulence in a subset of MRSA strains	[97]
α-toxin	JE2	Keratitis mouse	↑ Corneal healing	[81]

Table 3. Summary of studies analyzing the role of *S. aureus* toxins on ocular tissue using purified toxins applied directly to the eye.

Toxin	Inoculation Method	Ocular Model	Result	References
α-toxin	Topical	Rabbit cornea	Inflammation of conjunctiva and iris	[68]
α-toxin	Injection	Rabbit cornea	Inflammation of cornea and iris, corneal epithelial defect	[68]
β-toxin	Topical	Rabbit cornea	Inflammation of conjunctiva	[68]
β-toxin	Injection	Rabbit cornea	Scleral edema	[68]
γ-toxin	Injection	Rabbit cornea	Acute inflammatory reactions	[98]
PVL	Injection	Rabbit cornea	Acute inflammatory reactions	[98]
α-toxin	Injection	Rabbit cornea	↑SLE score, edema, epithelia cell death	[85]
α-toxin	Topical	Mouse cornea	Corneal pathology more severe in aged mice	[80]
α-toxin	Injection	Rabbit cornea	Corneal pathology more severe in young rabbits	[81]
α-toxin	Injection	Mouse Vitreous	Mild retinal damage, no reduction in retinal function	[89]

A common and relatively straightforward method for examining the importance of a toxin to the pathogenesis of a disease is to create an isogenic mutant specifically deficient in that toxin and examine how the pathology of the disease is altered in the preferred model. While the absence of a single toxin may significantly alter pathogenicity if an organism produces few toxins, for *S. aureus* and many other pathogens, this strategy is rather like removing one part of a car to determine how much damage it does when it hits an object. This practice will provide us with valuable information about how important some parts of the car are, but it will vastly underestimate the importance of most of them. However, an informative start has been made into understanding the roles of *S. aureus* toxins in eye infections. The toxins of *S. aureus* appear to have robust cytotoxic actions on ocular tissues and perhaps subtoxic activities in targeting the both the innate and adaptive branches of the immune system to both modulate the immune response and to block the development of a protective response. A more complete understanding of these activities will be facilitated by developing and using models that are more physiologically relevant and approach the environments and mechanisms by which *S. aureus* infects the human eye.

Author Contributions: Conceptualization, R.A., M.H.M., and M.C.C.; Investigation, R.A.; Resources, M.C.C.; Writing—Original Draft Preparation, R.A. and M.C.C.; Writing—Review and Editing, R.A., F.C.M., P.S.C., and M.C.C.; Visualization, R.A., M.H.M., and M.C.C.; Project Administration, M.C.C.; Funding Acquisition, M.C.C.

Funding: Our research and some of the studies mentioned herein were supported by National Institutes of Health grants R01EY028810, R01EY024140, R01EY025947 and R21EY028066 (to MCC), NEI Vision Core Grant P30EY027125 (to MCC, OUHSC), a Presbyterian Health Foundation Equipment Grant (to Robert E. Anderson, OUHSC), and an unrestricted grant to the Dean A. McGee Eye Institute from Research to Prevent Blindness.

Conflicts of Interest: The authors declare no conflict of interest.

References

1. Ogston, A. Report upon micro-organisms in surgical diseases. *Br. Med. J.* **1881**, *1*, 369–375. [CrossRef] [PubMed]
2. Fetsch, A. *Staphylococcus Aureus*; Academic Press: Cambridge, MA, USA, 2018; p. 3.
3. Rosenbach, F.J. *Mikro-organismen bei den Wund-Infection-krankheiten des Menschen*; Wiesbaden, J.F. Bergmann: Plano, TX, USA, 1884; pp. 19–21.
4. Hooker, C. Diphtheria, Immunization and the Bundaberg Tragedy: A Study of Public Health in Australia. *Health Hist.* **2000**, *2*, 52–78. [CrossRef]
5. Berube, B.J.; Wardenburg, J.B. Staphylococcus aureus α-Toxin: Nearly a Century of Intrigue. *Toxins* **2013**, *5*, 1140–1166. [CrossRef] [PubMed]
6. Burnet, F.M. The Exotoxins of Staphylococcus Pyogenes Aureus. *J. Pathol. Bacteriol.* **1929**, *32*, 717–734. [CrossRef]

7.   Burnet, F.M. The Production of Staphylococcal Toxin. *J. Pathol. Bacteriol.* **1930**, *33*, 1–16. [CrossRef]

8.   Burky, E.L. Studies on Cultures and Broth Filtrates of Staphylococci. *J. Immunol.* **1933**, *25*, 419–437.

9.   Woolpert, O.C.; Dack, G.M. Relation of Gastro-Intestinal Poison to Other Toxic Substances Produced by Staphylococci. *J. Infect. Dis.* **1933**, *52*, 6–19. [CrossRef]

10.   Bigger, J.W. The Production of Staphylococcal Haemolysin With Observations on Its Mode of Action. *J. Pathol. Bacteriol.* **1933**, *36*, 87–114. [CrossRef]

11.   Glenny, A.T.; Stevens, M.F. Staphylococcus Toxins and Antitoxins. *J. Pathol. Bacteriol.* **1935**, *40*, 201–210. [CrossRef]

12.   Kumar, S.; Lindorfer, R.K. The Characterization of Staphylococcal Toxins. *J. Exp. Med.* **1962**, *30*, 1095–1106. [CrossRef]

13.   Bernheimer, A.W.; Schwartz, L.L. Isolation and Composition of Staphylococcal Alpha Toxin. *J. Gen. Microbiol.* **1963**, *30*, 455–468. [CrossRef] [PubMed]

14.   Lowy, F.D. Staphylococcus Aureus Infections. *N. Engl. J. Med.* **1998**, *339*, 520–532. [CrossRef] [PubMed]

15.   Sollid, J.U.E.; Furberg, A.S.; Hanssen, A.M.; Johannessen, M. Staphylococcus aureus: Determinates of human carriage. *Infect. Genet. Evolut.* **2014**, *21*, 531–541. [CrossRef] [PubMed]

16.   Wertheim, H.F.L.; Vos, M.C.; Ott, A.; van Belkum, A.; Voss, A.; Kluytmans, J.; van Keulen, P.; Vandenbroucke-Grauls, C.; Meester, M.; Verbrugh, H. Risk and outcome of nosocomial *Staphylococcus aureus* bacteremia in nasal carriers versus non-carriers. *Lancet* **2004**, *364*, 703–705. [CrossRef]

17.   Lebon, A.; Moll, H.A.; Tavakol, M.; van Wamel, W.J.; Jaddoe, V.; Hofman, A.; Verbrugh, H.A.; van Belkum, A. Correlation of Bacterial Colonization Status between Mother and Child: The Generation R. Study. *J. Clin. Microbiol.* **2010**, *48*, 960–962. [CrossRef] [PubMed]

18.   Gorwitz, R.J.; Kruszon-Moran, D.; McAllister, S.K.; McQuillan, G.; McDougal, L.K.; Fosheim, G.E.; Jensen, B.J.; Killgore, G.; Tenover, F.C.; Kuehnert, M.J. Changes in the Prevalence of Nasal Colonization with Staphylococcus aureus in the United States 2001–2004. *J. Infect. Dis.* **2008**, *197*, 1226–1234. [CrossRef] [PubMed]

19.   Uemura, E.; Kakinohana, S.; Higa, N.; Toma, C.; Nakasone, N. Comparative characterization of *Staphylococcus aureus* isolates from throats and noses of healthy volunteers. *Jpn. J. Infect. Dis.* **2004**, *57*, 21–24.

20.   Saxena, S.; Singh, K.; Talwar, V. Methicillin-resistant *Staphylococcus aureus* prevalence in community in the East Delhi area. *Jpn. J. Infect. Dis.* **2003**, *56*, 54–56.

21.   Omuse, G.; Kariuki, S.; Revathi, G. Unexpected absence of methicillin-resistant Staphylococcus aureus nasal carriage by healthcare workers in a tertiary hospital in Kenya. *J. Hosp. Infect.* **2012**, *80*, 71–73. [CrossRef]

22.   Hamdan-Partida, A.; Sainz-Espunes, T.; Bustos-Martinez, J. Characterization and Persistence of Staphylococcus aureus Strains Isolated from the Anterior Nares and Throats of Healthy Carriers in a Mexican Community. *J. Clin. Microbiol.* **2010**, *48*, 1701–1705. [CrossRef]

23.   Eriksen, N.; Espersen, F.; Rosdahl, V.T.; Jensen, K. Carriage of Staphylococcus aureus among 104 healthy persons during a 19-month period. *Epidemiol. Infect.* **1995**, *115*, 51–60. [CrossRef] [PubMed]

24.   Kluytmans, J.; van Belkum, A.; Verbruch, H. Nasal Carriage of Staphylococcus aureus: Epidemiology, Underlying Mechanisms, and Associated Risks. *Clin. Microbiol. Rev.* **1997**, *10*, 505–520. [CrossRef] [PubMed]

25.   Acton, D.S.; Plat-Sinnige, M.J.; van Wamel, W.; de Groot, N.; van Belkum, A. Intestinal carriage of Staphylococcus aureus: how does its frequency compare with that of nasal carriage and what is its impact? *Eur. Clin. Microbiol. Infect. Dis.* **2009**, *28*, 115–127. [CrossRef] [PubMed]

26.   Chambers, H.F.; DeLeo, F.R. Waves of Resistance: Staphylococcus aureus in the Antibiotic Era. *Nat. Rev. Microbiol.* **2009**, *7*, 629–641. [CrossRef] [PubMed]

27.   Gaynes, R. The Discovery of Penicillin–New Insights After More Than 75 years of Clinical Use. *Emerging Infect. Dis. J.* **2017**, *23*, 849–853. [CrossRef]

28.   Fleming, A. *Penicillin its Practical Applications*, 2nd ed.; Butterworth & Co.: London, UK, 1950; p. 15.

29.   Kirby, W.M.M. Extraction of a highly potent penicillin inactivator from penicillin resistant staphylococci. *Science* **1944**, *99*, 452–453. [CrossRef]

30.   Barber, M.; Rozwadowska-Dowzenko, M. Infection by Penicillin-Resistant Staphylococci. *Lancet* **1948**, *23*, 641–644. [CrossRef]

31.   Blair, J.E.; Carr, M. Distribution of Phage Groups of *Staphylococcus aureus* in the Years 1927 through 1947. *Science* **1960**, *132*, 1247–1248. [CrossRef] [PubMed]

32. Jevons, M.P.; Parker, M.T. The evolution of new hospital strains of Staphylococcus aureus. *J. Clin. Pathol.* **1964**, *17*, 243–250. [CrossRef]
33. Chambers, H.F. Methicillin-Resistant Staphylococci. *Clin. Microbiol. Rev.* **1988**, *1*, 173–186. [CrossRef]
34. Barber, M. Methicillin-resistant staphylococci. *J. Clin. Pathol.* **1961**, *14*, 385–393. [CrossRef] [PubMed]
35. Lee, A.S.; de Lencastre, H.; Garau, J.; Kluytmans, J.; Malhotra-Kumar, S.; Peschel, A.; Harbarth, S. Methicillin-resistant Staphylococcus aureus. *Nat. Rev.* **2018**, *4*, 1188–1196.
36. Vandensch, F.; Naimi, T.; Enright, M.C.; Lina, G.; Nimmo, G.R.; Heffernan, H.; Liassine, N.; Bes, M.; Greenland, T.; Reverdy, M.; et al. Community-Acquired Methicillin-Resistant Staphylococcus aureus Carrying Panton-Valentine Leukocidin Genes: Worldwide Emergence. *Emerg. Infect. Dis.* **2003**, *9*, 978–984. [CrossRef] [PubMed]
37. Mediavilla, J.R.; Chen, L.; Mathema, B.; Kreiswirth, B.N. Global epidemiology of community-associated methicillin resistant Staphylococcus aureus (CA-MRSA). *Curr. Opin. Microbiol.* **2012**, *15*, 588–595. [CrossRef] [PubMed]
38. Boswihi, S.S.; Udo, E.E. Methicillin-resistant Staphylococcus aureus: An update on the epidemiology, treatment options and infection control. *Curr. Med. Res. Pract.* **2018**, *8*, 18–24. [CrossRef]
39. Blomquist, P.H. Methicillin-Resistant Staphylococcus aureus Infections of the Eye and Orbit. Trans. *Am. Ophthalmol. Soc.* **2006**, *104*, 322–345.
40. Hsiao, C.; Chuang, C.; Tan, H.; Ma, D.; Lin, K.; Chang, C.; Huang, Y. Methicillin-Resistant Staphylococcus aureus Ocular Infection: A 10-Year Hospital-Based Study. *Ophthalmology* **2012**, *119*, 522–527. [CrossRef] [PubMed]
41. Chuang, C.; Hsiao, C.; Tan, H.; Ma, D.; Lin, K.; Chang, C.; Huang, Y. Staphylococcus aureus Ocular Infection: Methicillin-Resistance, Clinical Features, and Antibiotic Susceptibilities. *PLoS ONE* **2012**, *8*, e42437. [CrossRef]
42. Brian, G.; Taylor, H. Cataract blindness-challenges for the 21st century. *Bull. World Health Organ.* **2001**, *79*, 249–256.
43. Endophthalmitis Study Group. Prophylaxis of postoperative endophthalmitis following cataract surgery: Results of the ESCRS multicenter study and identification of risk factors. *J. Cataract Refract. Surg.* **2007**, *33*, 978–988. [CrossRef]
44. Taban, M.; Behrens, A.; Newcomb, R.L.; Nobe, M.Y.; Saedi, G.; Sweet, P.M.; McDonnell, P.J. Acute endophthalmitis following cataract surgery: A systemic review of the literature. *Arch. Ophthalmol.* **2005**, *123*, 613–620. [CrossRef] [PubMed]
45. Peyman, G.A.; Lad, E.M.; Moshfeghi, D.M. Intravitreal injection of therapeutic agents. *Retina* **2009**, *29*, 875–912. [CrossRef] [PubMed]
46. Rupenthal, I.D. Sector Overview Ocular Drug Delivery: Exciting Times Ahead. Available online: http://ondrugdelivery.com/publications/54/Sector_Overview.pdf (accessed on 1 June 2019).
47. Sadaka, A.; Durand, M.L.; Gilmore, M.S. Bacterial endophthalmitis in the age of outpatient intravitreal therapies and cataract surgeries: Host-microbe interactions in intraocular infection. *Prog. Retin. Eye Res.* **2012**, *31*, 316–331. [CrossRef] [PubMed]
48. VEGF Inhibition Study in Ocular Neovascularization (V.I.S.I.O.N.) Clinical Trial Group; D'Amico, D.J.; Masonson, H.N.; Patel, M.; Adamis, A.P.; Cunningham, E.T., Jr.; Guyer, D.R.; Katz, B. Pegaptanib sodium for neovascular age-related macular degeneration: Two-year safety results of the two prospective, multicenter, controlled clinical trials. *Ophthalmology* **2006**, *113*, 992–1001.e6.
49. Diago, T.; McCannel, C.A.; Bakri, S.J.; Pulido, J.S.; Edwards, A.O.; Pach, J.M. Infectious endophthalmitis after intravitreal injection of antiangiogenic agents. *Retina* **2009**, *29*, 601–605. [CrossRef] [PubMed]
50. Klein, K.S.; Walsh, M.K.; Hassan, T.S.; Halperin, L.S.; Castellarian, A.A.; Roth, D.; Driscoll, S.; Prenner, J.L. Endophthalmitis after anti-VEGF injections. *Ophthalmology* **2009**, *116*, 1225. [CrossRef]
51. Sampat, K.M.; Garg, S.J. Complications of intravitreal injections. *Curr. Opin. Ophthalmol.* **2010**, *21*, 178–183. [CrossRef] [PubMed]
52. Callegan, M.C.; Engelbert, M.; Parke, D.W.; Jett, B.D.; Gilmore, M.S. Bacterial Endophthalmitis: Epidemiology, Therapeutics, and Bacterium-Host Interactions. *Clin. Microbiol. Rev.* **2002**, *15*, 111–124. [CrossRef]
53. Grumann, D.; Nubel, U.; Borker, B.M. Staphylococcus aureus toxins–Their functions and genetics. *Infect. Genet. Evolut.* **2014**, *21*, 583–592. [CrossRef]

54. Marquart, M.E.; O'Callaghan, R.J. Infectious Keratitis: Secreted Bacterial Proteins That Mediate Corneal Damage. *J. Ophthalmol.* **2013**, *2013*, 369094. [CrossRef]
55. O'Callaghan, R.J. The Pathogenesis of Staphylococcus aureus Eye Infections. *Pathogens* **2018**, *7*, 9. [CrossRef] [PubMed]
56. Alonzo, F.; Torres, V.J. Bacterial Survival Amidst and Immune Onslaught: The Contribution of the Staphylococcus aureus Leukotoxins. *PLoS Pathog.* **2013**, *9*, e1003143. [CrossRef] [PubMed]
57. Inoshima, I.; Inoshima, N.; Wilke, G.A.; Powers, M.E.; Frank, K.M.; Wang, Y.; Wardenburg, J.B. Staphylococcus aureus pore-forming toxin subverts the activity of ADAM10 to cause lethal infection in mice. *Nat. Med.* **2011**, *17*, 1310–1314. [CrossRef] [PubMed]
58. Perret, M.; Badiou, C.; Lina, G.; Burbaud, S.; Benito, Y.; Bes, M.; Cottin, V.; Couzon, F.; Juruj, C.; Dauwalder, O.; et al. Cross-talk between Staphylococcus aureus leukocidins-intoxicated macrophages and lung epithelial cells trigger chemokine secretion in an inflammasome-dependent manner. *Cell. Microbiol.* **2012**, *14*, 1019–1036. [CrossRef] [PubMed]
59. Seilie, E.S.; Wardenburg, J.B. Staphylococcus aureus pore-forming toxins: The interface of pathogen and host complexity. *Semin. Cell Dev. Biol.* **2017**, *72*, 101–116. [CrossRef] [PubMed]
60. Peraro, M.D.; van der Goot, F. Pore-forming toxins: Ancient, but never really out of fashion. *Nat. Rev.* **2016**, *14*, 77–92. [CrossRef] [PubMed]
61. Ueta, M.; Kinoshita, S. Innate immunity of the ocular surface. *Brain Res. Bull.* **2010**, *81*, 219–228. [CrossRef] [PubMed]
62. Ueta, M.; Kinoshita, S. Ocular surface inflammation is regulated by innate immunity. *Prog. Retin. Eye Res.* **2012**, *31*, 551–575. [CrossRef] [PubMed]
63. Foulsham, W.; Coco, G.; Amouzegar, A.; Chauhan, S.K.; Dana, R. When Clarity Is Crucial: Regulating Ocular Surface Immunity. *Trends Immunol.* **2018**, *39*, 288–301. [CrossRef]
64. Booth, M.C.; Pence, L.M.; Mahasreshti, P.; Callegan, M.C.; Gilmore, M.S. Clonal Associations among Staphylococcus aureus Isolates from Various Sites of Infection. *Infect. Immun.* **2001**, *69*, 345–352. [CrossRef]
65. Kielian, T.; Cheung, A.; Hickey, W.F. Diminished Virulence of an Alpha-Toxin Mutant of Staphylococcus aureus in Experimental Brain Abscesses. *Infect. Immun.* **2001**, *69*, 6902–6911. [CrossRef] [PubMed]
66. Bayer, A.; Ramos, M.D.; Menzies, B.E.; Yeaman, M.R.; Shen, A.J.; Cheung, A.L. Hyperproduction of Alpha-Toxin by Staphylococcus aureus Results in Paradoxically Reduced Virulence in Experimental Endocarditis: A Host Defense Role for Platelet Microbicidal Proteins. *Infect. Immun.* **1997**, *65*, 4652–4660. [PubMed]
67. Callegan, M.C.; Engel, L.S.; Hill, J.M.; O'Callaghan, R.J. Corneal Virulence of Staphylococcus aureus: Roles of Alpha-Toxin and Protein A in Pathogenesis. *Infect. Immun.* **1994**, *62*, 2478–2482. [PubMed]
68. O'Callaghan, R.J.; Callegan, M.C.; Moreau, J.M.; Green, L.C.; Foster, T.J.; Hartford, O.M.; Engel, L.S.; Hill, J.M. Specific Roles of Alpha-toxin and Beta-Toxin during Staphylococcus aureus Corneal Infection. *Infect. Immun.* **1997**, *65*, 1571–1578. [PubMed]
69. Jonsson, P.; Lindberg, M.; Haraldsson, I.; Wadstrom, T. Virulence of Staphylococcus aureus in a Mouse Mastitis Model: Studies of Alpha Hemolysin, Coagulase, and Protein A as Possible Virulence Determinates with Protoplast Fusion and Gene Cloning. *Infect. Immun.* **1985**, *49*, 765–769. [PubMed]
70. Wardenburg, J.B.; Patel, R.J.; Schneewind, O. Surface Proteins and Exotoxins Are Required for the Pathogenesis of Staphylococcus aureus Pneumonia. *Infect. Immun.* **2007**, *75*, 1040–1044. [CrossRef] [PubMed]
71. Wardenburg, J.B.; Bae, T.; Otto, M.; Deleo, F.R.; Schneewind, O. Poring over pores: Alpha-hemolysin and Panton-Valentine leukocidin in Staphylococcus aureus pneumonia. *Nat. Med.* **2007**, *13*, 1405–1406. [CrossRef] [PubMed]
72. Menzies, B.E.; Kernodle, D.S. Passive Immunization with Antiserum to a Nontoxic Alpha-Toxin Mutant from Staphylococcus aureus Is Protective in a Murine Model. *Infect. Immun.* **1996**, *64*, 1839–1841. [PubMed]
73. Powers, M.E.; Kim, H.K.; Wang, Y.; Wardenburg, J.B. ADAM10 Mediates Vascular Injury Induced by Staphylococcus aureus α-Hemolysin. *J. Infect. Dis.* **2012**, *206*, 352–356. [CrossRef]
74. Kennedy, A.D.; Wardenburg, J.B.; Gardner, D.J.; Long, D.; Whitney, A.R.; Braughton, K.R.; Schneewind, O.; DeLeo, F.R. Targeting of Alpha-Hemolysin by Active or Passive Immunization Decreases Severity of USA300 Skin Infection in a Mouse Model. *J. Infect. Dis.* **2010**, *202*, 1050–1058. [CrossRef]

75. Patel, A.H.; Nowland, P.; Weavers, E.D.; Foster, T. Virulence of Protein A-Deficient and Alpha-Toxin-Deficient Mutants of Staphylococcus aureus Isolated by Allele Replacement. *Infect. Immun.* **1987**, *55*, 3103–3110. [PubMed]

76. Wilke, G.A.; Wardenburg, J.B. Role of a disintegrin and metalloprotease 10 in Staphylococcus aureus α-hemolysin-mediated cellular injury. *Proc. Natl. Acad. Sci. USA* **2010**, *107*, 13473–13478. [CrossRef] [PubMed]

77. Bhakdi, S.; Tranum-Jensen, J. Alpha-toxin of Staphylococcus aureus. *Microbiol. Rev.* **1991**, *55*, 733–751. [PubMed]

78. Parker, M.W.; Feil, S.C. Pore-forming protein toxins: From structure to function. *Prog. Biophys. Mol. Biol.* **2005**, *88*, 91–142. [CrossRef] [PubMed]

79. Dajcs, J.J.; Austin, M.S.; Sloop, G.D.; Moreau, J.M.; Hume, E.B.H.; Thompson, H.W.; McAleese, F.M.; Foster, T.J.; O'Callaghan, R.J. Corneal Pathogenesis of Staphylococcus aureus Strain Newman. *Investig. Ophthalmol. Vis. Sci.* **2002**, *43*, 1109–1115.

80. Girgis, D.O.; Sloop, G.D.; Reed, J.M.; O'Callaghan, R.J. Effects of Toxin Production in a Murine Model of Staphylococcus aureus Keratitis. *Investig. Ophthalmol. Vis. Sci.* **2005**, *46*, 2064–2070. [CrossRef] [PubMed]

81. Putra, I.; Rabiee, B.; Anwar, K.N.; Gidfar, S.; Shen, X.; Babalooee, M.; Ghassemi, M.; Afsharkhamseh, N.; Bakhsh, S.; Missiakas, D.; et al. Staphylococcus aureus Alpha-Hemolysin Impairs Corneal Epithelial Wound Healing and Promotes Intracellular Bacterial Invasion. *Exp. Eye Res.* **2019**, *181*, 263–270. [CrossRef]

82. Lee, K.; Lee, H.; Kim, M. Two Cases of Corneal Ulcer due to Methicillin-Resistant Staphylococcus aureus in High Risk Groups. *Korean J. Ophthalmol.* **2010**, *24*, 240–244. [CrossRef] [PubMed]

83. Marquart, M.E. Animal Models of Bacterial Keratitis. *J. Biomed. Biotechnol.* **2011**, *2011*, 1–12. [CrossRef]

84. O'Callaghan, R.J.; McCormick, C.C.; Caballero, A.R.; Marquart, M.E.; Gatlin, H.P.; Fratkin, J.D. Age-Related Differences in Rabbits during Experimental Staphylococcus aureus Keratitis. *Investig. Ophthalmol. Vis. Sci.* **2007**, *48*, 5125–5131. [CrossRef]

85. Moreau, J.M.; Sloop, G.D.; Engel, L.S.; Hill, J.M.; O'Callaghan, J. Histopathological studies of staphylococcal alpha-toxin: Effects on rabbit corneas. *Curr. Eye Res.* **1997**, *16*, 1221–1228. [CrossRef] [PubMed]

86. Kumar, A.; Kumar, A. Role of Staphylococcus aureus Virulence Factors in Inducing Inflammation and Vascular Permeability in a Mouse Model of Bacterial Endophthalmitis. *PLoS ONE* **2015**, *10*, e0128423. [CrossRef]

87. Goerke, C.; Fluckiger, U.; Steinhuber, A.; Zimmerli, W.; Wolz, C. Impact of the regulatory loci agr, sarA, and sae of Staphylococcus aureus on the induction of α-toxin during device-related infection resolved by direct quantitative transcript analysis. *Mol. Microb.* **2001**, *40*, 1439–1447. [CrossRef]

88. Xiong, Y.Q.; Willard, J.; Yeaman, M.R.; Cheung, A.L.; Bayer, A.S. Regulation of Staphylococus aureus α-Toxin Gene (hla) Expression by agr, sarA, and sae In Vitro and in Experimental Infective Endocarditis. *J. Infect. Dis.* **2006**, *194*, 1267–1275. [CrossRef] [PubMed]

89. Booth, M.C.; Cheung, A.L.; Hatter, K.L.; Jett, B.D.; Callegan, M.C.; Gilmore, M.S. Staphylococcal Accessory Regulator (sar) in Conjunction with agr Contributes to Staphylococcus aureus Virulence in Endophthalmitis. *Infect. Immun.* **1997**, *65*, 1550–1556. [PubMed]

90. Horsburgh, M.J.; Aish, J.L.; White, I.J.; Shaw, L.; Lithgow, J.K.; Foster, S.J. σB Modulates Virulence Determinant Expression and Stress Resistance: Characterization of a Functional rsbU Strain Derived from Staphylococcus aureus 8325-4. *J. Bacteriol.* **2002**, *184*, 5457–5467. [CrossRef] [PubMed]

91. Hu, C.M.; Fang, R.H.; Copp, J.; Luk, B.T.; Zhang, L. A biomimetic nanosponge that absorbs pore-forming toxins. *Nat. Nanotechnol.* **2013**, *8*, 336–340. [CrossRef]

92. Coburn, P.S.; Miller, F.C.; LaGrow, A.L.; Land, C.; Mursalin, H.; Livingston, E.; Amayem, O.; Chen, Y.; Gao, L.; Callegan, M.C. Disarming Pore-Forming Toxins with Biomimetic Nanosponges in Intraocular Infections. *mSphere* **2019**, *4*, e00262-19. [CrossRef]

93. Gravet, A.; Colin, D.A.; Keller, R.; Giradot, R.; Monteil, H.; Prevost, G. Characterization of a novel structural member, LukE-LukD, of the bi-component staphylococcal leucotoxins family. *FEBS Lett.* **1998**, *136*, 202–208. [CrossRef]

94. Spaan, A.N.; Schiepers, A.; de Haas, C.J.C.; van Hooijdonk, D.; Badiou, C.; Contamin, H.; Vandenesch, F.; Lina, G.; Gerard, N.P.; Gerard, C.; et al. Differential Interactions of the Staphylococcal Toxins Panton-Valentine Leukocidin and γ-Hemolysin CB with Human C5a Receptors. *J. Immunol.* **2015**, *195*, 1034–1043. [CrossRef]

95. Spaan, A.N.; van Strijp, J.A.G.; Torres, V.J. Leukocidins: Staphylococcal bi-component pore-forming toxins find their receptors. *Nat. Rev.* **2017**, *15*, 435–447. [CrossRef] [PubMed]

96. Spaan, A.N.; Henry, T.; van Rooijen, W.; Perret, M.; Badiou, C.; Aerts, P.C.; Kemmink, J.; de Haas, C.; van Kessel, K.; Vandenesch, F.; et al. The Staphyloccocal Toxin Panton-Valentine Leukocidin Targets Human C5a Receptors. *Cell Host Microbe* **2013**, *13*, 584–594. [CrossRef] [PubMed]

97. Zaidi, T.; Zaidi, T.; Yoong, P.; Pier, G.B. Staphylococcus aureus Corneal Infections: Effect of the Panton-Valentine Leukocidin (PVL) and Antibody to PVL on Virulence and Pathology. *Investig. Ophthalmol. Vis. Sci.* **2013**, *54*, 4430–4438. [CrossRef] [PubMed]

98. Siqueira, J.A.; Speeg-Schatz, C.; Freitas, F.I.S.; Sahel, J.; Monteil, H.; Prevost, G. Channel-forming leucotoxins from Staphylococcus aureus cause severe inflammatory reactions in a rabbit eye model. *J. Med. Microbiol.* **1997**, *46*, 486–494. [CrossRef] [PubMed]

99. Liu, X.L.; Heitz, P.; Roux, M.; Keller, D.; Bourcier, T.; Sauer, A.; Pervost, G.; Gaucher, D. Panton-Valentine Leukocidin Colocalizes with Retinal Ganglion and Amacrine Cells and Acivates Glial Reactions and Microglial Apoptosis. *Sci. Rep.* **2018**, *8*, 2953. [CrossRef] [PubMed]

100. Liu, X.L.; Roux, M.J.; Picaud, S.; Keller, D.; Sauer, A.; Heitz, P.; Pervost, G.; Gaucher, D. Panton-Valentine Leucocidin Proves Direct Neuronal Targeting and Its Early Neuronal and Glial Impacts a Rabbit Retinal Explant Model. *Toxins* **2018**, *10*, 455. [CrossRef] [PubMed]

101. Peterson, J.C.; Durkee, H.; Miller, D.; Maestre-Mesa, J.; Arboleda, A.; Aguilar, M.C.; Relhan, N.; Flynn, H.W.; Amescua, G.; Parel, J.; et al. Molecular epidemiology and resistance profiles among healthcare- and community-associated Staphylococcus aureus keratitis isolates. *Infect. Drug Resist.* **2019**, *12*, 831–843. [CrossRef] [PubMed]

102. von Eiff, C.; Friedrich, A.W.; Peters, G.; Becker, K. Prevalence of genes encoding for members of the staphylococcal leukotoxin family among clinical isolates of Staphylococcus aureus. *Diagn. Microbiol. Infect. Dis.* **2004**, *49*, 157–162. [CrossRef]

103. Malachowa, N.; Whitney, A.R.; Kobayashi, S.D.; Sturdevant, D.E.; Kennedy, A.D.; Braughton, K.R.; Shabb, D.W.; Diep, B.A.; Chambers, H.F.; Otto, M.; et al. Global Changes in Staphylococcus aureus Gene Expression in Human Blood. *PLoS ONE* **2011**, *6*, e18617. [CrossRef]

104. Voyich, J.M.; Braughton, K.R.; Sturdevant, D.E.; Whitney, A.R.; Said-Salim, B.; Porcella, S.F.; Long, R.D.; Dorward, D.W.; Gardener, D.J.; Kreiswirth, B.N.; et al. Insights into Mechanisms Used by *Staphylococcus aureus* to Avoid Destruction by Human Neutrophils. *J. Immunol.* **2005**, *175*, 3907–3919. [CrossRef]

105. Spaan, A.N.; Vrieling, M.; Wallet, P.; Badiou, C.; Reyes-Robles, T.; Ohneck, E.A.; Benito, Y.; de Haas, C.; Day, C.J.; Jennings, M.P.; et al. The staphylococcal toxins γ-haemolysin AB and CB differentially target phagocytes by employing specific chemokine receptors. *Nat. Commun.* **2014**, *5*, 5438. [CrossRef] [PubMed]

106. Supersac, G.; Piemont, Y.; Kubina, M.; Prevost, G.; Foster, T.J. Assessment of the role of γ-toxin in experimental endophthalmitis using a hlg-deficient mutant of Staphylococcus aureus. *Microb. Pathog.* **1998**, *24*, 241–251. [CrossRef] [PubMed]

107. Dassy, B.; Hogan, T.; Foster, T.J.; Fournier, J. Involvement of the accessory gene regulator (agr) in expression of the type 5 capsular polysaccharide by Staphylococcus aureus. *Microbiology* **1993**, *139*, 1301–1306. [CrossRef] [PubMed]

108. Baba, T.; Bae, T.; Schneewind, O.; Takeuchi, F.; Hiramatsu, K. Genome Sequence of Staphylococcus aureus Strain Newman and Comparative Analysis of Staphylococcal Genomes: Polymorphism and Evolution of Two Major Pathogenicity Islands. *J. Bacteriol.* **2008**, *190*, 300–310. [CrossRef] [PubMed]

109. Periasamy, S.; Chatterjee, S.S.; Cheung, G.Y.C.; Otto, M. Phenol-soluble modulins in staphylococci What are they originally for? *Commun. Integr. Biol.* **2012**, *5*, 275–276. [CrossRef] [PubMed]

110. Wang, R.; Braughton, K.R.; Kretschmer, D.; Bach, T.L.; Queck, S.Y.; Li, M.; Kennedy, A.D.; Dorward, D.W.; Kelbanoff, S.J.; Peschel, A.; et al. Identification of novel cytolytic peptides as key virulence determinates for community-associated MRSA. *Nat. Med.* **2007**, *13*, 1510–1514. [CrossRef]

111. Baek, K.T.; Frees, D.; Renzoni, A.; Barras, C.; Rodriguez, N.; Manzano, N.; Kelley, W.L. Genetic Variation in the Staphylococcus aureus 8325 Strain Lineage Revealed by Whole-Genome Sequencing. *PLoS ONE* **2013**, *8*, e77122. [CrossRef]

112. Vandenesch, F.; Lina, G.; Henry, T. Staphylococcus aureus hemolysins, bi-component leukocidins, and cytolytic peptides: A redundant arsenal of membrane-damaging virulence factors? *Front. Cell. Infect. Microbiol.* **2012**, *2*, 12. [CrossRef]

113. Hanzelmann, D.; Joo, H.; Franz-Wachel, M.; Hertlein, T.; Stevanovic, S.; Macek, B.; Wolz, C.; Gotz, F.; Otto, M.; Krestschmer, D.; et al. Toll-like receptor 2 activation depends on lipopeptide shedding by bacterial surfactants. *Nat. Commun.* **2016**, *7*, 12304. [CrossRef]

114. Kretschmer, D.; Gleske, A.; Rautenberg, M.; Wang, R.; Koberle, M.; Bohn, E.; Schoneberg, T.; Rabiet, M.; Boulay, F.; Klebanoff, S.J.; et al. Human Formyl Peptide Receptor 2 Senses Highly Pathogenic Staphylococcus aureus. *Cell Host Microbe* **2010**, *7*, 463–473. [CrossRef]

115. Weiss, E.; Hanzelmann, D.; Fehlhaber, B.; Klos, A.; von Leowenich, F.; Liese, J.; Peschel, A.; Kretschmer, D. Formyl-peptide receptor 2 governs leukocyte influx in local Staphylococcus aureus infections. *FASEB J.* **2018**, *32*, 26–36. [CrossRef] [PubMed]

116. Otto, M. Staphylococcus aureus toxins. *Curr. Opin. Microbiol.* **2014**, *17*, 32–37. [CrossRef] [PubMed]

117. Freer, J.H.; Arbuthnott, J.P. Toxins of Staphylococcus aureus. *Pharmacol. Ther.* **1983**, *19*, 55–106. [CrossRef]

118. Aarestrup, F.M.; Larsen, H.D.; Eriksen, N.H.R.; Elsberg, C.S.; Jensen, N.E. Frequency of α-and β-haemolysin in Staphylococcus aureus of bovine and human origin. *Apmis* **1999**, *107*, 425–430. [CrossRef] [PubMed]

119. van Wamel, W.; Rooijakkers, S.; Ruyken, M.; van Kessel, K.; van Strijp, J. The Innate Immune Modulators Staphylococcal Complement Inhibitor and Chemotaxis Inhibitory Protein of Staphylococcus aureus Are Located on β-Hemolysin-Converting Bacteriophages. *J. Bacteriol.* **2006**, *188*, 1310–1315. [CrossRef] [PubMed]

120. Salgado-Pabon, W.; Herrera, A.; Vu, B.G.; Stach, C.S.; Merriman, J.A.; Spaulding, A.R.; Schlievert, P.M. Staphylococcus aureus β-toxin Production is Common in Strains with the β-toxin Gene Inactivation by Bacteriophage. *J. Infect. Dis.* **2014**, *210*, 784–792. [CrossRef] [PubMed]

121. Walev, I.; Weller, U.; Strauch, S.; Foster, T.; Bhakdi, S. Selective Killing of Human Monocytes and Cytokine Release Provoked by Sphingomyelinase (Beta-Toxin) of Staphylococcus aureus. *Infect. Immun.* **1996**, *64*, 2974–2979.

122. Mestas, J.; Hughes, C.C.W. Of Mice and Not Men: Differences between Mouse and Human Immunology. *J. Immunol.* **2004**, *172*, 2731–2738. [CrossRef]

123. Seok, J.; Warren, H.S.; Cuenca, A.G.; Mindrinos, M.N.; Baker, H.V.; Xu, W.; Richards, D.R.; McDonald-Smith, G.P.; Gao, H.; Hennessy, L.; et al. Genomic responses in mouse models poorly mimic human inflammatory diseases. *Proc. Natl. Acad. Sci. USA* **2013**, *110*, 3507–3512. [CrossRef]

124. Warren, H.S.; Tompkins, R.G.; Modawer, L.L.; Seok, J.; Xu, W.; Mindrinos, M.N.; Maier, R.V.; Xiao, W.; Davis, R.W. Mice are not men. *Proc. Natl. Acad. Sci. USA* **2015**, *112*, E345. [CrossRef]

125. Astley, R.A.; Coburn, P.S.; Parkunan, S.M.; Callegan, M.C. Modeling intraocular bacterial infections. *Prog. Retina Eye Res.* **2016**, *54*, 30–48. [CrossRef] [PubMed]

126. Kwon, B.; Hazlett, L.D. Association of CD4+ T Cell-Dependent Keratitis with Genetic Susceptibility to Pseudomonas aeruginosa Ocular Infection. *J. Immunol.* **1997**, *159*, 6283–6290. [PubMed]

Review

Exploring the Role of *Staphylococcus Aureus* Toxins in Atopic Dermatitis

Fabio Seiti Yamada Yoshikawa, Josenilson Feitosa de Lima, Maria Notomi Sato, Yasmin Álefe Leuzzi Ramos, Valeria Aoki [†] and Raquel Leao Orfali [*,†]

Laboratory of Dermatology and Immunodeficiencies (LIM-56), Department of Dermatology, University of Sao Paulo Medical School, Sao Paulo-SP 01246-903, Brazil; faseiti@gmail.com (F.S.Y.Y.); jnilsonflima@yahoo.com.br (J.F.d.L.); marisato@usp.br (M.N.S.); yasmimleuzzi@gmail.com (Y.Á.L.R.); valeria.aoki@gmail.com (V.A.)
* Correspondence: raquelleao@hotmail.com; Tel.: +5511-2661-8036
† These authors shared the mentorship, critical revision and supervision of this review.

Received: 8 May 2019; Accepted: 30 May 2019; Published: 5 June 2019

Abstract: Atopic dermatitis (AD) is a chronic and inflammatory skin disease with intense pruritus and xerosis. AD pathogenesis is multifactorial, involving genetic, environmental, and immunological factors, including the participation of *Staphylococcus aureus*. This bacterium colonizes up to 30–100% of AD skin and its virulence factors are responsible for its pathogenicity and antimicrobial survival. This is a concise review of *S. aureus* superantigen-activated signaling pathways, highlighting their involvement in AD pathogenesis, with an emphasis on skin barrier disruption, innate and adaptive immunity dysfunction, and microbiome alterations. A better understanding of the combined mechanisms of AD pathogenesis may enhance the development of future targeted therapies for this complex disease.

Keywords: *Staphylococcus aureus*; enterotoxins; atopic dermatitis; innate immunity; adaptive immunity; microbiome

Key Contribution: A better understanding of the role/mechanisms of *S. aureus* in the pathogenesis of AD may contribute to the development of future targeted therapies for this complex disease.

Introduction

Atopic dermatitis (AD) is a prevalent, chronic, inflammatory, and immune-mediated skin disease [1]. Complex interactions among susceptibility genes encoding skin barrier molecules [2], inflammatory response elements, environmental factors, and infectious agents (especially *Staphylococcus aureus* and herpes virus) that lead to dysbiosis of the microbial community resident in AD skin [3,4], together with the altered immunologic status of the host, are crucial elements in the pathophysiology of AD [5,6]. Bacterial infection is a powerful trigger for AD flares and has become a matter of concern due to the widespread occurrence of antibiotic-resistant strains (methicillin-resistant *S. aureus*; MRSA) [7,8].

S. aureus is a Gram-positive bacterium present in 20–30% of healthy subjects. Rates of *S. aureus* carriage in AD skin reach 30–100% [9,10]. *S. aureus* produces many virulence factors which determine its pathogenicity and antimicrobial survival, including secreted toxins, enzymes, and cell-surface-associated antigens [11]. Together, these factors allow this bacterium to elude the host's natural defenses.

Single-lipid membranes, surrounded by a peptidoglycan and lipoteichoic acid layer anchored by diacylglycerol, are the components of *S. aureus*' cell wall [11]. Of note, *S. aureus* products include a myriad of components that play specific roles in the inflammatory/immune response, as follows: (a) Superantigens (e.g., staphylococcal enterotoxins (SE)A–U and toxic shock syndrome toxin (TSST)-1)—proteins with high mitogenic properties, leading to T- and B-cell expansions causing clonal deletion and massive cytokine production. (b) Cytotoxins (e.g., α-toxin and leukocidins)—trigger cytokine production, hemolysis, and leukocyte cell death within targeting specific cell surface receptors. (c) Enzymes (e.g., β-toxin)—trigger cytotoxicity resulting in cell death, inflammation, and tissue barrier disruptions. (d) Adhesins—cell wall receptors for epidermal and dermal laminin and fibronectin. (e) Other enzymes (e.g., proteases and/or nucleases)—mediate host protein degradations that can also act on self-proteins to degrade biofilms for bacterial dissemination. Altogether, these toxins and enzymes provide critical nutrients that are essential for bacterial growth and survival, targeting different aspects of the host's immune response and therefore contributing to *S. aureus* virulence (Figure 1A) [9,11,12].

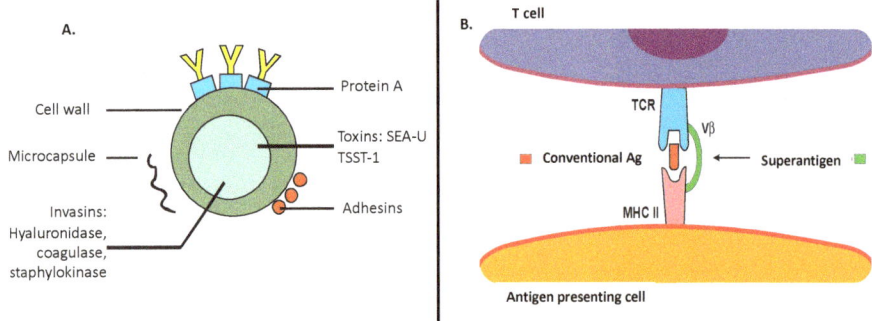

Figure 1. (**A**) *Staphylococcus aureus* main compounds that contribute to enhanced bacterial virulence factor. SE—staphylococcal enterotoxins; TSST-1—toxic shock syndrome toxin-1. (**B**) *S. aureus* as a superantigen: direct binding to the major histocompatibility complex (MHC) class II molecule, with specificity for the T-cell receptor (TCR)-Vβ chain.

S. aureus may act as a conventional antigen or as a superantigen to activate T cells. Unlike conventional antigens, superantigens do not need to be processed and can bind directly to the surface of the major histocompatibility complex (MHC) class II molecule with T-cell receptor (TCR) Vβ-chain specificity, generating a polyclonal T-cell stimulation (Figure 1B) [9,13,14]. Our group previously studied the effects of distinct antigens and mitogens (tetanic toxoid, *Candida albicans*, SEA, and phytohemagglutinin) in the proliferative response of peripheral blood mononuclear cells (PBMCs) from adults with AD and identified a decreased PBMC proliferation response to these stimuli, suggesting a compromised immune profile due to staphylococcal chronic skin colonization [13]. In this review, we focus on the role/mechanisms of *S. aureus* in the pathogenesis of AD.

1. Role of *S. Aureus* Toxins in the Disruption of the Skin Barrier and Innate Immunity

The skin is the first barrier to pathogens and, as a frontline defender, shows a remarkable immune function due to the presence of a diverse set of cells such as keratinocytes, dendritic cell (DCs), mast cells (MCs), and lymphocytes [15]. *S. aureus* has higher adhesion to corneocytes from AD patients than to those in non-AD individuals. This enhanced binding is mediated by the interaction between bacterial adhesins, such as protein A, and host surface receptors [16]. Moreover, staphylococcal toxins do have immune stimulatory abilities, and there is higher production of such toxins in strains isolated from AD individuals [17].

Most of the innate immune response relies on the identification of nonself molecules by pattern recognition receptors (PRRs). Toll-like receptors (TLRs) are one of the most critical classes of PRRs, particularly bacterial motifs. As membrane-associated receptors, they are expressed at the cell surface where they can survey the extracellular environment for pathogen clues, but are also found in endosomes, thus tracking intracellular stimuli [18]. TLR2 has been identified as a key receptor for

S. aureus recognition due to its ability to identify lipoproteins, which are abundantly expressed in the cell wall of Gram-positive bacteria [18]. Interestingly, keratinocytes from AD patients show a reduced response to TLR2 agonists, with lower production of IL-6, IL-8, CCL20, and matrix metalloproteinase-9 (MMP-9) [19]. Although the mechanisms involved in this diminished response are still unknown, TLR2 expression in these cells are equivalent in both AD and healthy subjects. The dysfunction in TLR response in AD patients can be associated with deficiencies in the signaling components. Some studies point out that alterations in the myeloid differentiation protein (MyD88) pathway is needed for the development of SEB-induced AD-like phenotype [20,21].

In parallel with TLRs, another remarkable class of PRRs is the nucleotide-binding oligomerization domain (NOD)-like receptor (NLR) family. NLRs mainly work by forming multicomponent platforms known as inflammasomes, which drive the activation of protease caspase-1; *S. aureus* components activate the NLRP3 inflammasome [22]. In AD, *S. aureus* α-hemolysin is capable of activating this complex in keratinocytes, and inflammasome activation is greatly compromised due to the reduced expression of its components, driven by the action of T helper (Th)2 cytokines [23]. Nonetheless, keratinocytes may circumvent this limitation by producing IL-1β and IL-18 in an inflammasome-independent fashion in response to other *S. aureus* toxins, such as phenol-soluble modulins (PSM), which directly cause cell lysis and cytokine extravasation [24].

Moreover, α-hemolysin contributes to disruption of the skin barrier by interacting with lipid sphingomyelin, generating pores in the cell membrane that culminate in keratinocyte lysis. AD patients show a higher susceptibility to α-hemolysin action once Th2 cytokines promote the downregulation of acid sphingomyelinase, which increases the availability of α-hemolysin targets [25]. In addition, AD patients show defective filaggrin expression, contributing to disease severity. Filaggrin is highly expressed in differentiated keratinocytes and promotes the secretion of acid sphingomyelinase and resistance to α-hemolysin activity [26,27].

The α-toxin can also compromise the keratinocyte layer by altering E-cadherin integrity. Curiously, this effect can be blocked by pharmacological activation of the G-protein-coupled estrogen receptor (GPER), which ultimately promotes repression of the host protein ADAM10, a receptor for α-hemolysin responsible for cleaving E-cadherin [28]. Moreover, *S. aureus* could compromise the keratinocyte biology at the transcriptional level. Brauweiler et al. (2017) showed that *S. aureus* lipoteichoic acid (LTA) profoundly affects the keratinocyte genetic program, repressing genes associated with cell differentiation by interfering with the protein p63, which is a master transcription regulator in skin [29].

Using a murine model of *S. aureus* epicutaneous exposure to induce an AD-like disease, Liu et al. (2017) uncovered the role of PSMα in promoting IL-36 secretion by keratinocytes, which directly activates T cells to produce IL-17 and drives skin inflammation [30]. Interestingly, Baldry et al. (2018), using the same approach, suggested that another PSM, such as δ-toxin, would be more relevant to induce inflammation once the use of a mutant strain for δ-toxin or the pharmacological inhibition of its production with solonamide could efficiently block AD development [31]. In support of the latter findings, Matsuo et al. (2018) showed that δ-toxin, when combined with ovalbumin, could promote an AD-like phenotype without any other bacterial component, inducing production of the Th2-recruiting chemokines CCL17 and CCL22 in keratinocytes [32].

In sharp contrast to the epicutaneous model, previous systems such as tape stripping, in combination with chemically induced AD-like lesions, did not indicate a remarkable contribution of *S. aureus* toxins in AD pathogenesis. It is possible that the effects of toxins such as SEA were mostly coadjuvant, with mild contributions to the cytokine pool and disease phenotype due to the harsh stress caused by this method [33]. However, evidence from human skin biopsies did show a high expression of SEA in AD tissues, which suggests an indisputable contribution of these toxins to immune activation [34].

In addition to keratinocytes, *S. aureus* toxins also interfere in the activity of immune cells, particularly those which reside in or are recruited to the skin. Saloga et al. (1996) investigated the immune response kinetics after a single intracutaneous dose of SEB and showed the activation of resident Langerhans cells, degranulation of MCs, and recruitment of eosinophils followed by the influx of mononuclear cells (monocytes/macrophages and lymphocytes) [35]. Thus, staphylococcal toxins alone may exert multiple actions on skin immunity.

A milestone work from Nakamura et al. (2013) demonstrated that the δ-toxin is one of the main activators of MCs, triggering their degranulation and reinforcing the causal link between *S. aureus* colonization and AD development [31]. δ-Toxin promotes MC degranulation by activating the receptor Mas-related G-protein-coupled receptor member X2 (MRGPRX2), promoting Ca^{2+} influx and release of granule content. As MRGPRX2 is also expressed in keratinocytes, this suggests that the contribution of δ-toxin to AD pathogenesis can be extended to other cell types [36]. Interestingly, SEB was initially proposed to inhibit IL-4 production in the MC line HMC-1 [37], but whether this effect applies to primary MCs is still unknown.

Other granulocytes can also infiltrate AD lesions. Basophils can release histamine in response to a diverse set of *S. aureus* toxins in an IgE-dependent manner [38,39], while eosinophils respond to SEB, protein A, and peptidoglycan through the receptor CD48 with secretion of the enzyme eosinophil peroxidase and the neutrophil chemoattractant IL-8 [40]. Staphylococcal toxins inhibit eosinophil apoptosis, therefore not only triggering but also perpetuating the allergic response by postponing cell death [41].

DCs belongs to a key immune population in the initiation and polarization of T-cell responses due to their ability to present antigens and to secrete polarizing factors according to different stimuli [42,43]. Concerning their role in AD pathogenesis, SEB-stimulated DCs develop a Th2-polarizing phenotype, which supports the atopic profile in these patients [44]. In turn, Th2 products may affect the innate response as described by Kasraie et al. (2016), who showed that IL-31, a T-cell cytokine involved in pruritus, activates monocytes and macrophages, enhancing the inflammatory response [45].

Curiously, the altered DC phenotype in AD patients is not limited to resident skin populations. Kapitány et al. (2017) observed that CD1c$^+$ blood pre-DCs, precursors of dermal DCs from AD patients, have an altered response (either constitutively or after SEB stimulation), with heightened production of Th2-associated chemokines [46]. These authors also noticed that these cells are more premature, which could be the result of the particular serum cytokine signature in those patients or due to an altered hemopoiesis with early release of immature cells. These results reinforce the notion that AD is far beyond a restricted dermatological disease but rather a condition of systemic immune dysfunction. Indeed, some works showed that epicutaneous SEB sensitization can enhance Th17-driven lung inflammation by inducing the production of the IL-17 polarizing cytokine IL-6 [47] and that α-toxin can help infection by viruses, such as herpes simplex virus 1, by facilitating viral entry into the host cell [48].

Intriguingly, while SEB induced apoptosis of the human monocytic cell line THP-1 [49], similar to what was described for T cells [50], TSST-1 showed the opposite effect, protecting primary monocytes from death [51]. The proposed mechanism of action for SEB-induced death relies on induction of TNF-α, which activates the extrinsic apoptotic pathway [49], while TSST-1 promotes GM-CSF (Granulocyte Macrophage Colony-Stimulating Factor) secretion, promoting cell survival [51]. The scenario that prevails in the host may depend on multiple factors, such as the host immune response or differences in the toxin profile secreted among different *S. aureus* strains.

SEB efficiently activates monocytes, inducing production of TNF-α in a TLR2- and TLR4-dependent way, in nonatopic individuals compared to AD patients, corroborating the anergic/exhausted profile described in AD [52]. Similarly, α-toxin was a strong inducer of CXCL10 in macrophages from healthy individuals but not in AD counterparts [53], which could contribute to a dampened Th1 response in atopic subjects. A possible explanation could rely on the reduced expression of immune receptors such as the lower expression of TLR2 in AD macrophages [54], which can compromise the triggering of an effective immune response, allowing bacterial overgrowth. Although α-toxin was shown to promote TLR2 expression in monocytes from non-AD subjects, thus enhancing their cytokine response [55], the effects on AD cells can have a divergent outcome, supporting the atopic response.

The host genetic background must be considered for a full appreciation of the immune mechanisms triggered by staphylococcal toxins. Krogman et al. (2017), using transgenic mice expressing different alleles of HLA-DR (Human Leukocyte Antigen – DR isotype) molecules, showed how the immune response changes in response to TSST-1, with some alleles favoring an increased inflammatory response and tissue damage [56]. Nieburh et al. (2008) identified a single-nucleotide polymorphism in the TLR2 gene from AD patients that increases the production of IL-6 and IL-12p70 by monocytes [57], potentially aggravating their inflammatory reaction. Those genetic differences may help to explain why AD patients show very intense *S. aureus* proliferation and respond with deleterious inflammation, even though the bacterium is a component of skin microbiota. Another variable that should be considered is the age of the host, since atopic children may exhibit different response kinetics to bacterial antigens than adults [52].

Newer evidence suggests that in addition to their inflammation-promoting roles, staphylococcal toxins may also counterbalance the host regulatory mechanisms. Though glucocorticoids (GC) are a common therapeutic strategy in AD management, utilizing the glucocorticoid receptor (GR) to dampen the inflammatory reaction, some patients become unresponsive to treatment. Huang et al. (2018) showed that SEB can block GC action by impeding the nuclear translocation of GR in keratinocytes, favoring the persistence of inflammation [58]. Also, staphylococcal enterotoxins can modulate the induction of myeloid-derived suppressor cells (MDSCs). While lower doses of toxins favor MDSC generation, higher doses (the likely scenario in AD patients who show *S. aureus* overgrowth) modulate MDSC response, thus circumventing inflammation control [59].

2. *S. Aureus*: Straight Relationship with Adaptive Immunity in AD

Decreased skin microbiota diversity associated with high susceptibility to *S. aureus* colonization [10,60,61] is described in the skin of AD patients. *S. aureus* is able to secrete more than 20 different toxins that play an important role in AD adaptive immune response. Around 60% of isolated *S. aureus* strains are capable of secreting exotoxins [62], acting on distinct T- and B-cell pathways. As superantigens, they can stimulate cytokine secretion and T-cell proliferation; as conventional allergens, they can induce production of staphylococcal-specific IgE exotoxin [63].

Classically, AD immune pathogenesis is described as a dysfunction of the Th1/Th2 balance [64]. Acute flares of AD are characterized by Th2 and Th22 cell infiltrates, whereas Th1 cells are detected in chronic lesions [65]. However, there are reports of novel subsets of human Th cells, such as Th17 and Th22 cells [66], that are present in the inflammatory response of AD [5,27,64,67]. AD has several immune subtype profiles, all having a common Th2/Th22 polarization, but also displays differential immune skewing, such as increased Th17 that has been identified in the skin of intrinsic, Asian, and early pediatric AD patients [68]. Th17-derived IL-17 is able to coordinate local tissue inflammation through upregulation of proinflammatory cytokines and chemokines, including IL-6, TNF-α, IL-1β, CXCL1, CCL2, CXCL2, CCL7, and CCL20 [67]. In cooperation with IL-17, IL-22 triggers antimicrobial peptide production and initiates an acute phase response [69,70]. In recent publications, there was expression of Th22 cells and IL-22-producing CD8$^+$ T cells present in acute and chronic AD skin lesions [65,71].

The major Th2 cytokine involved in AD acute flares is IL-4, which mediates enhanced expression of fibronectin and fibrinogen, working as adhesion molecules for *S. aureus* and therefore contributing to AD skin bacterial chronic colonization. AD patients with colonized skin by *S. aureus* have increased disease severity, elevated eosinophil blood counts, total IgE serum levels, CCL17 and periostin (Th2 biomarkers) plasma levels, and suppressed activity of Treg cells [72] when compared with those without *S. aureus* colonization [73], corroborating the chronic immune activation described in AD.

Th2 cells can also incite B cells, which leads to IgE and IL-5 production, stimulating maturation and survival of eosinophils in some types of AD [65]. Furthermore, purified eosinophils from AD patients stimulated with TLR2/6 agonist and SEB showed decreased levels of TIMP-1, TIMP-2, and CCL5, revealing a potential breakdown in the remodeling process mediated by eosinophils [74].

B-cell activation can also be modulated by *S. aureus* exotoxins (working as allergens) and stimulating exotoxin-specific IgE antibody production. SEA- or SEB-specific IgE levels are significantly elevated in the plasma of children with AD and correlate with disease severity when compared with non-AD children [75]. Enhanced production of IgE antibodies against *S. aureus* antigens is also described in adults with AD and is associated with asthma severity [76]. B cells are also a relevant source of cytokine secretion by a T-cell-independent pathway. Parcina et al. (2013) demonstrated that protein A, an important virulence factor of *S. aureus*, induced increased B-cell proliferation and regulatory B cells (Breg) secreting IL-10 via the plasmacytoid dendritic cells (pDC) pathway [77], suggesting a tolerogenic immune profile in AD.

Pruritus is one of the major symptoms in AD patients and it is associated with impaired quality of life, as demonstrated in studies evaluating the patients' perspectives of the disease using PO-SCORAD [78,79]. IL-31 is a Th2 cytokine associated with AD severity and pruritus [80,81]. Staphylococcal toxins, such as α-toxin and SEB, are capable of inducing a potent secretion of IL-31 by CD4$^+$ T cells in AD patients [45,82].

In Figure 2, we have summarized the noteworthy findings in AD pathogenesis related to chronic staphylococcal enterotoxin activation described in previous studies by our group.

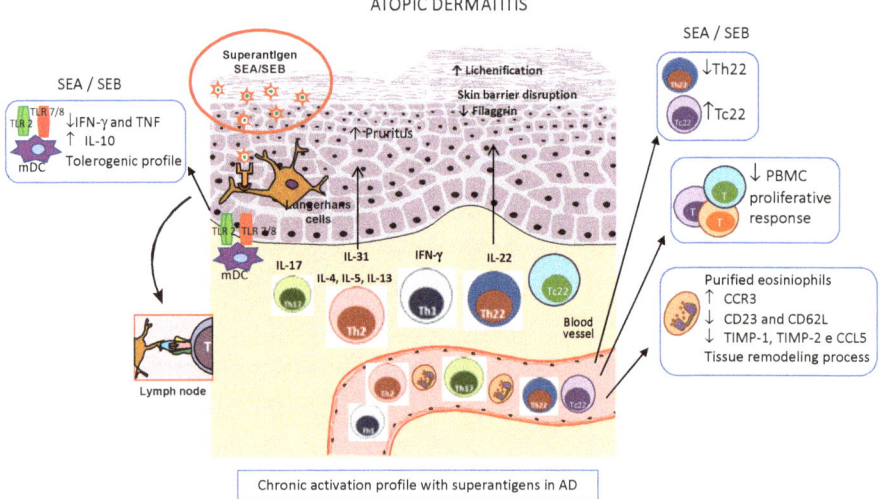

Figure 2. Superantigen-activated dendritic cells stimulate T helper (Th)2 cells to produce IL-4, IL-5, IL-13, and IL-31, leading to skin barrier disruption, decreased antimicrobial peptide production, impaired keratinocyte differentiation, and pruritus. In chronic atopic dermatitis (AD), there is an enrollment of Th1, Th22, and Th17 subsets that leads to epidermal thickening and abnormal keratinocyte proliferation (lichenification). Effects of staphylococcal enterotoxins in AD: 1. Dysfunctional CD4$^+$ IL-22-secreting T cells and upregulated Tc22 cells. 2. Reduced peripheral blood mononuclear cell (PBMC) proliferative response corroborating an exhausted immune profile. 3. Increased frequency of CCR3$^+$ and decreased expression of CD23 and CD62L receptors, and TIMP-1, TIMP-2, and CCL5 in purified eosinophils of AD patients even in a nonstimulated condition, indicating a potential breakdown in the tissue remodeling process in AD mediated by eosinophils. 4. Enhanced frequency of IL-10 under TLR4 and decreased frequency of IFN-γ and TNF under TLR2 and 7/8 stimulation in classic mDC (myeloid dendritic cells), indicating a tolerogenic profile in AD. All these findings together corroborate the chronic activated profile related to superantigens in AD pathogenesis.

3. Microbiome and AD

An assembly of microbiota resides our human skin and cohabits in an established balance [10]. The molecular approach to the human microbiome reveals high diversity of skin microbiota within and between distinct topographical regions [10,61,83,84].

In patients with AD, the skin microbiota is altered by endogenous factors, such as skin barrier protein mutations (filaggrin, among others) or exogenous stimuli, such as soaps, topical corticosteroids, and antibiotics, leading to a modified/noneffective response of the host to allergens, pathogens, and tissue damage [85].

The microbiome can exert both beneficial and harmful influences in AD skin once it interacts with the local immune system [86]. *S. aureus* is directly correlated with increased expression of IL-4, IL-13, IL-22, IL-31, TSLP, and other cytokines and decreased expression of cathelicidin, evidencing the impact of skin dysbiosis on disease exacerbation [87]. An example of a beneficial relationship between bacteria and the skin is that the commensal bacterium *Staphylococcus epidermidis* modulates TLR3-dependent inflammation by initiating a TLR2-mediated cross-talk mechanism to suppress inflammation, indicating that more microbiome diversity seems to be more beneficial once a diverse ecosystem is more resistant [88], suggesting that the timing of exposure to commensal bacteria may affect the development of tolerance.

When analyzing lesional skin and *S. aureus* colonization in AD subjects, there is evidence of loss of microbial diversity during acute flares in patients with AD, in contrast with microbiome diversity restoration after successful anti-inflammatory treatment [61]. Kennedy et al. showed relevant findings such as the absence of *S. aureus* colonization of children prior to AD onset and colonization of antecubital fossa with commensal staphylococci at month two in infants, which is associated with decreased incidence of AD at the age of one year [89].

Furthermore, the characterization of the microbiome in children and adults with AD still requires more studies. Shi et al. [90] recently published data on microbiome studies performed in both pediatric and adult patients with AD. These authors detected significant differences in the microbiome profile of AD individuals, with two defined patterns according to age (young children and adults–teenagers). Microbiome diversity was increased in nonlesional skin of young patients when compared with adults. *Staphylococcus* was abundant in nonlesional and lesional AD skin in both age groups, suggesting susceptibility to pathogen colonization [90].

Recent studies indicate that certain strains of commensal flora composed of coagulase-negative staphylococci (CoNS) compete with *S. aureus* on the skin, increasing antimicrobial peptide production [10,91]. Additional studies in AD individuals are needed to confirm whether colonization with commensal bacteria will modulate the uncontrolled immune response with clinical improvement of the cutaneous lesions in AD patients.

4. Perspectives Targeting *S. Aureus* in AD

Increasing evidence corroborates that *S. aureus* has a pivotal role in AD pathogenesis, correlating with disease flares and severity. However, many questions remain unresolved, especially regarding the best strategy against *S. aureus* in AD skin.

Well-established AD treatment options, such as narrowband UVB phototherapy, systemic cyclosporine, topical corticosteroids, and calcineurin inhibitors, indicate that therapeutic success is a reflection of the immunomodulatory mechanism of action suppressing staphylococcal-enterotoxin-activated T cells, improving the skin barrier function, and consequently decreasing *S. aureus* colonization [92]. Recent assays have evaluated the effect of dupilumab (anti-IL-4Rα) on the host–microbe interface in atopic dermatitis (available online in clinicaltrials.gov, identifier: NCT03389893) [92]. Likewise, other monoclonal antibodies are under investigation, for example, nemolizumab (CIM331), a humanized antibody against the IL-31 receptor [93] (clinicaltrials.gov, identifier: NCT01986933). Furthermore, it is important for future research to address the interactions between the gut and skin microbiomes, including *S. aureus* species, and therapies targeting the immune system via microbioma modulation

A recent AD clinical trial utilizing topical application of commensal *Roseomonas mucosa* demonstrated significant decreases in disease severity scores, topical steroid requirements, and *S. aureus* colonization, with no adverse events or treatment complications [94].

We also have to consider that vaccines against *S. aureus* represent a possible novel approach to manipulating the AD skin microbiome. Next-generation anti-*S. aureus* vaccines will require an association between targeting specific effector T-cell subsets combined with inducing specific neutralizing antitoxin antibodies [10,92,95].

One remarkable point for future therapeutic target strategies for AD should address the interaction between host and pathogen, focusing on the pathogenic role of staphylococcal enterotoxins in modulating cytokine release by activated effector T cells and affecting population-level responses to pathogens.

Author Contributions: Conceptualization, Writing—Review and Editing, Supervision: R.L.O., M.N.S. and V.A.; Writing—Original Draft Preparation: F.S.Y.Y. and J.F.L.; Review of literature: Y.Á.L.R.; All authors contributed to discussing and reviewing the manuscript.

Funding: This research was funded by Fundação de Amparo à Pesquisa do Estado de São Paulo (FAPESP), grants number 2016/24161-1 and 2018/23211-0.

Acknowledgments: We are thankful to the Laboratory of Dermatology and Immunodeficiencies (LIM-56), Department of Dermatology, University of Sao Paulo Medical School.

Conflicts of Interest: The authors declare no conflict of interests.

References

1. Leung, D.Y.M. New insights into atopic dermatitis: Role of skin barrier and immune dysregulation. *Allergol. Int.* **2013**, *62*, 151–161. [CrossRef] [PubMed]
2. Elias, P.M.; Schmuth, M. Abnormal skin barrier in the etiopathogenesis of atopic dermatitis. *Curr. Opin. Allergy Clin. Immunol.* **2009**, *9*, 437–446. [CrossRef] [PubMed]
3. Salava, A.; Lauerma, A. Role of the skin microbiome in atopic dermatitis. *Clin. Trans. Allergy* **2014**, *4*, 33. [CrossRef] [PubMed]
4. Nakatsuji, T.; Chen, T.H.; Narala, S.; Chun, K.A.; Two, A.M.; Yun, T.; Shafiq, F.; Kotol, P.F.; Bouslimani, A.; Melnik, A.V.; et al. Antimicrobials from human skin commensal bacteria protect against staphylococcus aureus and are deficient in atopic dermatitis. *Sci. Transl. Med.* **2017**, *9*. [CrossRef] [PubMed]
5. Hayashida, S.; Uchi, H.; Moroi, Y.; Furue, M. Decrease in circulating th17 cells correlates with increased levels of ccl17, ige and eosinophils in atopic dermatitis. *J. Dermatol. Sci.* **2011**, *61*, 180–186. [CrossRef] [PubMed]
6. De Benedetto, A.; Agnihothri, R.; McGirt, L.Y.; Bankova, L.G.; Beck, L.A. Atopic dermatitis: A disease caused by innate immune defects? *J. Investig. Dermatol.* **2009**, *129*, 14–30. [CrossRef]
7. Saeed, K.; Marsh, P.; Ahmad, N. Cryptic resistance in staphylococcus aureus: A risk for the treatment of skin infection? *Curr. Opin. Infectious Dis.* **2014**, *27*, 130–136. [CrossRef]
8. Turner, N.A.; Sharma-Kuinkel, B.K.; Maskarinec, S.A.; Eichenberger, E.M.; Shah, P.P.; Carugati, M.; Holland, T.L.; Fowler, V.G., Jr. Methicillin-resistant staphylococcus aureus: An overview of basic and clinical research. *Nat. Rev. Microbiol.* **2019**, *17*, 203–218. [CrossRef]
9. Krakauer, T. Update on staphylococcal superantigen-induced signaling pathways and therapeutic interventions. *Toxins* **2013**, *5*, 1629–1654. [CrossRef]
10. Paller, A.S.; Kong, H.H.; Seed, P.; Naik, S.; Scharschmidt, T.C.; Gallo, R.L.; Luger, T.; Irvine, A.D. The microbiome in patients with atopic dermatitis. *J. Allergy Clin. Immunol.* **2019**, *143*, 26–35. [CrossRef]
11. Oliveira, D.; Borges, A.; Simoes, M. Staphylococcus aureus toxins and their molecular activity in infectious diseases. *Toxins* **2018**, *10*, 252. [CrossRef] [PubMed]
12. Tam, K.; Torres, V.J. Staphylococcus aureus secreted toxins and extracellular enzymes. *Microbiol. Spectr.* **2019**, *7*, 2. [CrossRef] [PubMed]

13. Orfali, R.L.; Sato, M.N.; Takaoka, R.; Azor, M.H.; Rivitti, E.A.; Hanifin, J.M.; Aoki, V. Atopic dermatitis in adults: Evaluation of peripheral blood mononuclear cells proliferation response to staphylococcus aureus enterotoxins a and b and analysis of interleukin-18 secretion. *Exp. Dermatology* **2009**, *18*, 628–633. [CrossRef] [PubMed]

14. Tuffs, S.W.; Haeryfar, S.M.M.; McCormick, J.K. Manipulation of innate and adaptive immunity by staphylococcal superantigens. *Pathogens* **2018**, *7*, 53. [CrossRef] [PubMed]

15. Matejuk, A. Skin immunity. *Arch. Immunol. Ther. Exp. (Warsz)* **2018**, *66*, 45–54. [CrossRef]

16. Cole, G.W.; Silverberg, N.L. The adherence of staphylococcus aureus to human corneocytes. *Arch. Dermatol.* **1986**, *122*, 166–169. [CrossRef] [PubMed]

17. Ezepchuk, Y.V.; Leung, D.Y.; Middleton, M.H.; Bina, P.; Reiser, R.; Norris, D.A. Staphylococcal toxins and protein a differentially induce cytotoxicity and release of tumor necrosis factor-alpha from human keratinocytes. *J. Investig. Dermatol.* **1996**, *107*, 603–609. [CrossRef]

18. Kumar, H.; Kawai, T.; Akira, S. Pathogen recognition by the innate immune system. *Int. Rev. Immunol.* **2011**, *30*, 16–34. [CrossRef]

19. Niebuhr, M.; Heratizadeh, A.; Wichmann, K.; Satzger, I.; Werfel, T. Intrinsic alterations of pro-inflammatory mediators in unstimulated and tlr-2 stimulated keratinocytes from atopic dermatitis patients. *Exp. Dermatol.* **2011**, *20*, 468–472.

20. Kissner, T.L.; Ruthel, G.; Cisney, E.D.; Ulrich, R.G.; Fernandez, S.; Saikh, K.U. Myd88-dependent pro-inflammatory cytokine response contributes to lethal toxicity of staphylococcal enterotoxin b in mice. *Innate Immun.* **2011**, *17*, 451–462. [CrossRef]

21. Fassbender, S.; Opitz, F.V.; Johnen, S.; Forster, I.; Weighardt, H. Myd88 contributes to staphylococcal enterotoxin b-triggered atopic dermatitis-like skin inflammation in mice. *J. Investig. Dermatol.* **2017**, *137*, 1802–1804. [PubMed]

22. Munoz-Planillo, R.; Franchi, L.; Miller, L.S.; Nunez, G. A critical role for hemolysins and bacterial lipoproteins in staphylococcus aureus-induced activation of the nlrp3 inflammasome. *J. Immunol.* **2009**, *183*, 3942–3948. [PubMed]

23. Niebuhr, M.; Baumert, K.; Heratizadeh, A.; Satzger, I.; Werfel, T. Impaired nlrp3 inflammasome expression and function in atopic dermatitis due to th2 milieu. *Allergy* **2014**, *69*, 1058–1067. [PubMed]

24. Syed, A.K.; Reed, T.J.; Clark, K.L.; Boles, B.R.; Kahlenberg, J.M. Staphlyococcus aureus phenol-soluble modulins stimulate the release of proinflammatory cytokines from keratinocytes and are required for induction of skin inflammation. *Infect. Immun.* **2015**, *83*, 3428–3437. [PubMed]

25. Brauweiler, A.M.; Goleva, E.; Leung, D.Y.M. Th2 cytokines increase staphylococcus aureus alpha toxin-induced keratinocyte death through the signal transducer and activator of transcription 6 (stat6). *J. Investig. Dermatol.* **2014**, *134*, 2114–2121. [CrossRef]

26. Brauweiler, A.M.; Bin, L.; Kim, B.E.; Oyoshi, M.K.; Geha, R.S.; Goleva, E.; Leung, D.Y. Filaggrin-dependent secretion of sphingomyelinase protects against staphylococcal alpha-toxin-induced keratinocyte death. *J. Allergy Clin. Immunol.* **2013**, *131*, 421–427. [PubMed]

27. Batista, D.I.S.; Perez, L.; Orfali, R.L.; Zaniboni, M.C.; Samorano, L.P.; Pereira, N.V.; Sotto, M.N.; Ishizaki, A.S.; Oliveira, L.M.S.; Sato, M.N.; et al. Profile of skin barrier proteins (filaggrin, claudins 1 and 4) and th1/th2/th17 cytokines in adults with atopic dermatitis. *J. Eur. Acad. Dermatol. Venereol.* **2015**, *29*, 1091–1095. [CrossRef]

28. Triplett, K.D.; Pokhrel, S.; Castleman, M.J.; Daly, S.M.; Elmore, B.O.; Joyner, J.A.; Sharma, G.; Herbert, G.; Campen, M.J.; Hathaway, H.J.; et al. Gper activation protects against epithelial barrier disruption by staphylococcus aureus alpha-toxin. *Sci. Rep.* **2019**, *9*, 1343.

29. Brauweiler, A.M.; Hall, C.F.; Goleva, E.; Leung, D.Y.M. Staphylococcus aureus lipoteichoic acid inhibits keratinocyte differentiation through a p63-mediated pathway. *J. Investig. Dermatol.* **2017**, *137*, 2030–2033.

30. Liu, H.; Archer, N.K.; Dillen, C.A.; Wang, Y.; Ashbaugh, A.G.; Ortines, R.V.; Kao, T.; Lee, S.K.; Cai, S.S.; Miller, R.J.; et al. Staphylococcus aureus epicutaneous exposure drives skin inflammation via il-36-mediated t cell responses. *Cell Host Microbe* **2017**, *22*, 653–666 e655.

31. Baldry, M.; Nakamura, Y.; Nakagawa, S.; Frees, D.; Matsue, H.; Nunez, G.; Ingmer, H. Application of an agr-specific antivirulence compound as therapy for staphylococcus aureus-induced inflammatory skin disease. *J. Infect. Dis.* **2018**, *218*, 1009–1013.

32. Matsuo, K.; Nagakubo, D.; Komori, Y.; Fujisato, S.; Takeda, N.; Kitamatsu, M.; Nishiwaki, K.; Quan, Y.S.; Kamiyama, F.; Oiso, N.; et al. Ccr4 is critically involved in skin allergic inflammation of balb/c mice. *J. Investig. Dermatol.* **2018**, *138*, 1764–1773. [PubMed]

33. Kim, B.S.; Choi, J.K.; Jung, H.J.; Park, K.H.; Jang, Y.H.; Lee, W.J.; Lee, S.J.; Kim, S.H.; Kang, H.Y.; Kim, J.M.; et al. Effects of topical application of a recombinant staphylococcal enterotoxin a on dncb and dust mite extract-induced atopic dermatitis-like lesions in a murine model. *Eur. J. Dermatol.* **2014**, *24*, 186–193.

34. Lee, H.W.; Kim, S.M.; Kim, J.M.; Oh, B.M.; Kim, J.Y.; Jung, H.J.; Lim, H.J.; Kim, B.S.; Lee, W.J.; Lee, S.J.; et al. Potential immunoinflammatory role of staphylococcal enterotoxin a in atopic dermatitis: Immunohistopathological analysis and in vitro assay. *Ann. Dermatol.* **2013**, *25*, 173–180. [PubMed]

35. Saloga, J.; Leung, D.Y.; Reardon, C.; Giorno, R.C.; Born, W.; Gelfand, E.W. Cutaneous exposure to the superantigen staphylococcal enterotoxin b elicits a t-cell-dependent inflammatory response. *J. Investig. Dermatol.* **1996**, *106*, 982–988. [PubMed]

36. Azimi, E.; Reddy, V.B.; Lerner, E.A. Brief communication: Mrgprx2, atopic dermatitis and red man syndrome. *Itch (Philadelphia, Pa.)* **2017**, *2*, e5.

37. Ackermann, L.; Pelkonen, J.; Harvima, I.T. Staphylococcal enterotoxin b inhibits the production of interleukin-4 in a human mast-cell line hmc-1. *Immunology* **1998**, *94*, 247–252.

38. Jorgensen, J.; Bach-Mortensen, N.; Koch, C.; Fomsgaard, A.; Baek, L.; Jarlov, J.O.; Espersen, F.; Jensen, C.B.; Skov, P.S.; Norn, S. Bacteria and endotoxin induce release of basophil histamine in patients with atopic dermatitis. In vitro experiments with s. Aureus, teichoic acid, e. Coli and e. Coli lps. *Allergy* **1987**, *42*, 395–397.

39. Wehner, J.; Neuber, K. Staphylococcus aureus enterotoxins induce histamine and leukotriene release in patients with atopic eczema. *Br. J. Dermatol.* **2001**, *145*, 302–305.

40. Minai-Fleminger, Y.; Gangwar, R.S.; Migalovich-Sheikhet, H.; Seaf, M.; Leibovici, V.; Hollander, N.; Feld, M.; Moses, A.E.; Homey, B.; Levi-Schaffer, F. The cd48 receptor mediates staphylococcus aureus human and murine eosinophil activation. *Clin. Exp. Allergy* **2014**, *44*, 1335–1346.

41. Wedi, B.; Wieczorek, D.; Stunkel, T.; Breuer, K.; Kapp, A. Staphylococcal exotoxins exert proinflammatory effects through inhibition of eosinophil apoptosis, increased surface antigen expression (cd11b, cd45, cd54, and cd69), and enhanced cytokine-activated oxidative burst, thereby triggering allergic inflammatory reactions. *J. Allergy Clin. Immunol.* **2002**, *109*, 477–484. [PubMed]

42. Mildner, A.; Jung, S. Development and function of dendritic cell subsets. *Immunity* **2014**, *40*, 642–656. [CrossRef] [PubMed]

43. dos Santos, V.G.; Orfali, R.L.; Titz, T.D.; Duarte, A.J.D.; Sato, M.N.; Aoki, V. Evidence of regulatory myeloid dendritic cells and circulating inflammatory epidermal dendritic cells-like modulated by toll-like receptors 2 and 7/8 in adults with atopic dermatitis. *Int. J. Dermatol.* **2017**, *56*, 630–635. [CrossRef] [PubMed]

44. Mandron, M.; Aries, M.F.; Brehm, R.D.; Tranter, H.S.; Acharya, K.R.; Charveron, M.; Davrinche, C. Human dendritic cells conditioned with staphylococcus aureus enterotoxin b promote th2 cell polarization. *J. Allergy Clin. Immunol.* **2006**, *117*, 1141–1147. [CrossRef] [PubMed]

45. Kasraie, S.; Niebuhr, M.; Werfel, T. Interleukin (il)-31 induces pro-inflammatory cytokines in human monocytes and macrophages following stimulation with staphylococcal exotoxins. *Allergy* **2010**, *65*, 712–721. [CrossRef] [PubMed]

46. Kapitany, A.; Beke, G.; Nagy, G.; Doan-Xuan, Q.M.; Bacso, Z.; Gaspar, K.; Boros, G.; Dajnoki, Z.; Biro, T.; Rajnavolgyi, E.; et al. Cd1c+ blood dendritic cells in atopic dermatitis are premature and can produce disease-specific chemokines. *Acta Derm. -Venereol.* **2017**, *97*, 325–331. [CrossRef] [PubMed]

47. Yu, J.; Oh, M.H.; Park, J.U.; Myers, A.C.; Dong, C.; Zhu, Z.; Zheng, T. Epicutaneous exposure to staphylococcal superantigen enterotoxin b enhances allergic lung inflammation via an il-17a dependent mechanism. *PLoS ONE* **2012**, *7*, e39032.

48. Bin, L.; Kim, B.E.; Brauweiler, A.; Goleva, E.; Streib, J.; Ji, Y.; Schlievert, P.M.; Leung, D.Y. Staphylococcus aureus alpha-toxin modulates skin host response to viral infection. *J. Allergy Clin. Immunol.* **2012**, *130*, 683–691 e682. [CrossRef]

49. Zhang, X.; Shang, W.; Yuan, J.; Hu, Z.; Peng, H.; Zhu, J.; Hu, Q.; Yang, Y.; Liu, H.; Jiang, B.; et al. Positive feedback cycle of tnfalpha promotes staphylococcal enterotoxin b-induced thp-1 cell apoptosis. *Front. Cell. Infection Microbiol.* **2016**, *6*, 109.

50. Kedzierska, A.; Kaszuba-Zwoinska, J.; Slodowska-Hajduk, Z.; Kapinska-Mrowiecka, M.; Czubak, M.; Thor, P.; Wojcik, K.; Pryjma, J. Seb-induced t cell apoptosis in atopic patients–correlation to clinical status and skin colonization by staphylococcus aureus. *Arch. Immunol. Ther. Exp.* **2005**, *53*, 63–70.

51. Bratton, D.L.; May, K.R.; Kailey, J.M.; Doherty, D.E.; Leung, D.Y. Staphylococcal toxic shock syndrome toxin-1 inhibits monocyte apoptosis. *J. Allergy Clin. Immunol.* **1999**, *103*, 895–900. [CrossRef]

52. Mandron, M.; Aries, M.F.; Boralevi, F.; Martin, H.; Charveron, M.; Taieb, A.; Davrinche, C. Age-related differences in sensitivity of peripheral blood monocytes to lipopolysaccharide and staphylococcus aureus toxin b in atopic dermatitis. *J. Investig. Dermatol.* **2008**, *128*, 882–889. [CrossRef] [PubMed]

53. Kasraie, S.; Niebuhr, M.; Kopfnagel, V.; Dittrich-Breiholz, O.; Kracht, M.; Werfel, T. Macrophages from patients with atopic dermatitis show a reduced cxcl10 expression in response to staphylococcal alpha-toxin. *Allergy* **2012**, *67*, 41–49. [CrossRef] [PubMed]

54. Niebuhr, M.; Lutat, C.; Sigel, S.; Werfel, T. Impaired tlr-2 expression and tlr-2-mediated cytokine secretion in macrophages from patients with atopic dermatitis. *Allergy* **2009**, *64*, 1580–1587. [CrossRef] [PubMed]

55. Niebuhr, M.; Schorling, K.; Heratizadeh, A.; Werfel, T. Staphylococcal alpha-toxin induces a functional upregulation of tlr-2 on human peripheral blood monocytes. *Exp. Dermatol.* **2015**, *24*, 381–383. [CrossRef] [PubMed]

56. Krogman, A.; Tilahun, A.; David, C.S.; Chowdhary, V.R.; Alexander, M.P.; Rajagopalan, G. Hla-dr polymorphisms influence in vivo responses to staphylococcal toxic shock syndrome toxin-1 in a transgenic mouse model. *HLA* **2017**, *89*, 20–28. [CrossRef] [PubMed]

57. Niebuhr, M.; Langnickel, J.; Draing, C.; Renz, H.; Kapp, A.; Werfel, T. Dysregulation of toll-like receptor-2 (tlr-2)-induced effects in monocytes from patients with atopic dermatitis: Impact of the tlr-2 r753q polymorphism. *Allergy* **2008**, *63*, 728–734. [CrossRef]

58. Huang, K.; Ran, L.; Wang, W.; Zhou, R.; Cai, X.; Li, R.; Li, Y.; Zhou, C.; He, W.; Wang, R. Glucocorticoid insensitivity by staphylococcal enterotoxin b in keratinocytes of allergic dermatitis is associated with impaired nuclear translocation of the glucocorticoid receptor alpha. *J. Dermatological Sci.* **2018**, *92*, 272–280. [CrossRef]

59. Stoll, H.; Ost, M.; Singh, A.; Mehling, R.; Neri, D.; Schafer, I.; Velic, A.; Macek, B.; Kretschmer, D.; Weidenmaier, C.; et al. Staphylococcal enterotoxins dose-dependently modulate the generation of myeloid-derived suppressor cells. *Front. Cell. Infection Microbiol.* **2018**, *8*, 321. [CrossRef]

60. Wichmann, K.; Uter, W.; Weiss, J.; Breuer, K.; Heratizadeh, A.; Mai, U.; Werfel, T. Isolation of alpha-toxin-producing staphylococcus aureus from the skin of highly sensitized adult patients with severe atopic dermatitis. *Br. J. Dermatol.* **2009**, *161*, 300–305. [CrossRef]

61. Kong, H.H.; Oh, J.; Deming, C.; Conlan, S.; Grice, E.A.; Beatson, M.A.; Nomicos, E.; Polley, E.C.; Komarow, H.D.; Program, N.C.S.; et al. Temporal shifts in the skin microbiome associated with disease flares and treatment in children with atopic dermatitis. *Genome Res.* **2012**, *22*, 850–859. [CrossRef] [PubMed]

62. Roesner, L.M.; Werfel, T.; Heratizadeh, A. The adaptive immune system in atopic dermatitis and implications on therapy. *Expert. Rev. Clin. Immunol.* **2016**, *12*, 787–796. [CrossRef] [PubMed]

63. Breuer, K.; Wittmann, M.; Kempe, K.; Kapp, A.; Mai, U.; Dittrich-Breiholz, O.; Kracht, M.; Mrabet-Dahbi, S.; Werfel, T. Alpha-toxin is produced by skin colonizing staphylococcus aureus and induces a t helper type 1 response in atopic dermatitis. *Clin. Exp. Allergy* **2005**, *35*, 1088–1095. [CrossRef] [PubMed]

64. Eyerich, K.; Novak, N. Immunology of atopic eczema: Overcoming the th1/th2 paradigm. *Allergy* **2013**, *68*, 974–982. [CrossRef] [PubMed]

65. Biedermann, T.; Skabytska, Y.; Kaesler, S.; Volz, T. Regulation of t cell immunity in atopic dermatitis by microbes: The yin and yang of cutaneous inflammation. *Front. Immunol.* **2015**, *6*, 353. [CrossRef]

66. Auriemma, M.; Vianale, G.; Amerio, P.; Reale, M. Cytokines and t cells in atopic dermatitis. *Eur. Cytokine Netw.* **2013**, *24*, 37–44.

67. Koga, C.; Kabashima, K.; Shiraishi, N.; Kobayashi, M.; Tokura, Y. Possible pathogenic role of th17 cells for atopic dermatitis. *J. Investig. Dermatol.* **2008**, *128*, 2625–2630. [CrossRef]

68. D'Auria, E.; Banderali, G.; Barberi, S.; Gualandri, L.; Pietra, B.; Riva, E.; Cerri, A. Atopic dermatitis: Recent insight on pathogenesis and novel therapeutic target. *Asian-Pac. J. Allergy Immunol.* **2016**, *34*, 98–108.

69. Liang, S.C.; Tan, X.Y.; Luxenberg, D.P.; Karim, R.; Dunussi-Joannopoulos, K.; Collins, M.; Fouser, L.A. Interleukin (il)-22 and il-17 are coexpressed by th17 cells and cooperatively enhance expression of antimicrobial peptides. *J. Exp. Med.* **2006**, *203*, 2271–2279. [CrossRef]

70. Duhen, T.; Geiger, R.; Jarrossay, D.; Lanzavecchia, A.; Sallusto, F. Production of interleukin 22 but not interleukin 17 by a subset of human skin-homing memory t cells. *Nat. Immunol.* **2009**, *10*, 857–863. [CrossRef]
71. Orfali, R.L.; da Silva Oliveira, L.M.; de Lima, J.F.; de Carvalho, G.C.; Ramos, Y.A.L.; Pereira, N.Z.; Pereira, N.V.; Zaniboni, M.C.; Sotto, M.N.; da Silva Duarte, A.J.; et al. Staphylococcus aureus enterotoxins modulate il-22-secreting cells in adults with atopic dermatitis. *Sci. Rep.* **2018**, *8*, 6665. [CrossRef] [PubMed]
72. Laborel-Preneron, E.; Bianchi, P.; Boralevi, F.; Lehours, P.; Fraysse, F.; Morice-Picard, F.; Sugai, M.; Sato'o, Y.; Badiou, C.; Lina, G.; et al. Effects of the staphylococcus aureus and staphylococcus epidermidis secretomes isolated from the skin microbiota of atopic children on cd4+ t cell activation. *PLoS ONE* **2015**, *10*, e0141067.
73. Simpson, E.L.; Villarreal, M.; Jepson, B.; Rafaels, N.; David, G.; Hanifin, J.; Taylor, P.; Boguniewicz, M.; Yoshida, T.; De Benedetto, A.; et al. Patients with atopic dermatitis colonized with staphylococcus aureus have a distinct phenotype and endotype. *J. Investig. Dermatol.* **2018**, *138*, 2224–2233. [CrossRef] [PubMed]
74. Titz, T.D.; Orfali, R.L.; de Lollo, C.; dos Santos, V.G.; Duarte, A.J.D.; Sato, M.N.; Aoki, V. Impaired cd23 and cd62l expression and tissue inhibitors of metalloproteinases secretion by eosinophils in adults with atopic dermatitis. *J. Eur. Acad. Dermatol. Venereol.* **2016**, *30*, 2072–2076. [CrossRef] [PubMed]
75. Lin, Y.T.; Yang, Y.H.; Hwang, Y.W.; Tsai, M.J.; Tsao, P.N.; Chiang, B.L.; Shau, W.Y.; Wang, L.F. Comparison of serum specific ige antibodies to staphylococcal enterotoxins between atopic children with and without atopic dermatitis. *Allergy* **2000**, *55*, 641–646. [CrossRef] [PubMed]
76. Tanaka, A.; Suzuki, S.; Ohta, S.; Manabe, R.; Furukawa, H.; Kuwahara, N.; Fukuda, Y.; Kimura, T.; Jinno, M.; Hirai, K.; et al. Association between specific ige to staphylococcus aureus enterotoxins a and b and asthma control. *Ann. Allergy Asthma Immunol.* **2015**, *115*, 191. [CrossRef] [PubMed]
77. Parcina, M.; Miranda-Garcia, M.A.; Durlanik, S.; Ziegler, S.; Over, B.; Georg, P.; Foermer, S.; Ammann, S.; Hilmi, D.; Weber, K.J.; et al. Pathogen-triggered activation of plasmacytoid dendritic cells induces il-10-producing b cells in response to staphylococcus aureus. *J. Immunol.* **2013**, *190*, 1591–1602. [CrossRef]
78. Boccardi, D.; D'Auria, E.; Turati, F.; DI, M.V.; Sortino, S.; Riva, E.; Cerri, A. Disease severity and quality of life in children with atopic dermatitis: Po-scorad in clinical practice. *Minerva Pediatrica* **2017**, *69*, 373–380.
79. Coutanceau, C.; Stalder, J.F. Analysis of correlations between patient-oriented scorad (po-scorad) and other assessment scores of atopic dermatitis severity and quality of life. *Dermatology* **2014**, *229*, 248–255. [CrossRef]
80. Raap, U.; Wichmann, K.; Bruder, M.; Stander, S.; Wedi, B.; Kapp, A.; Werfel, T. Correlation of il-31 serum levels with severity of atopic dermatitis. *J. Allergy Clin. Immunol.* **2008**, *122*, 421–423. [CrossRef]
81. Sonkoly, E.; Muller, A.; Lauerma, A.I.; Pivarcsi, A.; Soto, H.; Kemeny, L.; Alenius, H.; Dieu-Nosjean, M.C.; Meller, S.; Rieker, J.; et al. Il-31: A new link between t cells and pruritus in atopic skin inflammation. *J. Allergy Clin. Immunol.* **2006**, *117*, 411–417. [CrossRef] [PubMed]
82. Niebuhr, M.; Werfel, T. Innate immunity, allergy and atopic dermatitis. *Curr. Opin. Allergy Clin. Immunol.* **2010**, *10*, 463–468. [CrossRef] [PubMed]
83. Grice, E.A.; Kong, H.H.; Conlan, S.; Deming, C.B.; Davis, J.; Young, A.C.; Program, N.C.S.; Bouffard, G.G.; Blakesley, R.W.; Murray, P.R.; et al. Topographical and temporal diversity of the human skin microbiome. *Science* **2009**, *324*, 1190–1192. [CrossRef] [PubMed]
84. Huttenhower, C.; Gevers, D.; Knight, R.; Abubucker, S.; Badger, J.H.; Chinwalla, A.T.; Creasy, H.H.; Earl, A.M.; FitzGerald, M.G.; Fulton, R.S.; et al. Structure, function and diversity of the healthy human microbiome. *Nature* **2012**, *486*, 207–214.
85. Kuo, I.H.; Yoshida, T.; De Benedetto, A.; Beck, L.A. The cutaneous innate immune response in patients with atopic dermatitis. *J. Allergy Clin. Immunol.* **2013**, *131*, 266–278. [CrossRef] [PubMed]
86. Werfel, T.; Allam, J.P.; Biedermann, T.; Eyerich, K.; Gilles, S.; Guttman-Yassky, E.; Hoetzenecker, W.; Knol, E.; Simon, H.U.; Wollenberg, A.; et al. Cellular and molecular immunologic mechanisms in patients with atopic dermatitis. *J. Allergy Clin. Immunol.* **2016**, *138*, 336–349. [CrossRef] [PubMed]
87. Nakatsuji, T.; Chen, T.H.; Two, A.M.; Chun, K.A.; Narala, S.; Geha, R.S.; Hata, T.R.; Gallo, R.L. Staphylococcus aureus exploits epidermal barrier defects in atopic dermatitis to trigger cytokine expression. *J. Investig. Dermatol.* **2016**, *136*, 2192–2200. [CrossRef]
88. Schommer, N.N.; Gallo, R.L. Structure and function of the human skin microbiome. *Trends Microbiol.* **2013**, *21*, 660–668. [CrossRef]

89. Kennedy, E.A.; Connolly, J.; Hourihane, J.O.; Fallon, P.G.; McLean, W.H.; Murray, D.; Jo, J.H.; Segre, J.A.; Kong, H.H.; Irvine, A.D. Skin microbiome before development of atopic dermatitis: Early colonization with commensal staphylococci at 2 months is associated with a lower risk of atopic dermatitis at 1 year. *J. Allergy Clin. Immunol.* **2017**, *139*, 166–172. [CrossRef]

90. Shi, B.; Bangayan, N.J.; Curd, E.; Taylor, P.A.; Gallo, R.L.; Leung, D.Y.M.; Li, H. The skin microbiome is different in pediatric versus adult atopic dermatitis. *J. Allergy Clin. Immunol.* **2016**, *138*, 1233–1236.

91. Parlet, C.P.; Brown, M.M.; Horswill, A.R. Commensal staphylococci influence staphylococcus aureus skin colonization and disease. *Trends Microbiol.* **2019**. [CrossRef] [PubMed]

92. Clowry, J.; Irvine, A.D.; McLoughlin, R.M. Next-generation anti-staphylococcus aureus vaccines: A potential new therapeutic option for atopic dermatitis? *J. Allergy Clin. Immunol.* **2019**, *143*, 78–81. [CrossRef] [PubMed]

93. Ruzicka, T.; Hanifin, J.M.; Furue, M.; Pulka, G.; Mlynarczyk, I.; Wollenberg, A.; Galus, R.; Etoh, T.; Mihara, R.; Yoshida, H.; et al. Anti-interleukin-31 receptor a antibody for atopic dermatitis. *New Engl. J. Med.* **2017**, *376*, 826–835. [CrossRef] [PubMed]

94. Myles, I.A.; Earland, N.J.; Anderson, E.D.; Moore, I.N.; Kieh, M.D.; Williams, K.W.; Saleem, A.; Fontecilla, N.M.; Welch, P.A.; Darnell, D.A.; et al. First-in-human topical microbiome transplantation with roseomonas mucosa for atopic dermatitis. *JCI Insight* **2018**, *3*, 3. [CrossRef] [PubMed]

95. Blicharz, L.; Rudnicka, L.; Samochocki, Z. Staphylococcus aureus: An underestimated factor in the pathogenesis of atopic dermatitis? *Adv. Dermatol. Allergol.* **2019**, *36*, 11–17. [CrossRef] [PubMed]

MDPI

St. Alban-Anlage 66

4052 Basel

Switzerland

Tel. +41 61 683 77 34

Fax +41 61 302 89 18

www.mdpi.com

Toxins Editorial Office

E-mail: toxins@mdpi.com

www.mdpi.com/journal/toxins

Ingram Content Group UK Ltd.
Milton Keynes UK
UKHW050637090523
421393UK00005B/30

9 783039 214259